Nanocomposite, Ceramic, and Thin Film Scintillators

Nanocomposite, Ceramic, and Thin Film Scintillators

edited by

Martin Nikl

PAN STANFORD PUBLISHING

Published by

Pan Stanford Publishing Pte. Ltd.
Penthouse Level, Suntec Tower 3
8 Temasek Boulevard
Singapore 038988

Email: editorial@panstanford.com
Web: www.panstanford.com

British Library Cataloguing-in-Publication Data
A catalogue record for this book is available from the British Library.

Nanocomposite, Ceramic, and Thin Film Scintillators

ISBN 978-981-4745-22-2 (Hardcover)
ISBN 978-981-4745-23-9 (eBook)

Printed in the USA

Contents

Preface

Few years ago, after my invited presentation at a conference, I was contacted by Pan Stanford Publishing with an invitation to write a book on the subject of my lecture that dealt with scintillator materials. This was a very pleasant surprise, of course, but it took me a while to digest and accept it and to shape the idea into a proposal suitable for consideration for the publisher. Instead of a monograph, which demands a lot of time, I decided to invite some of my collaborators and few other well-known teams in the field of scintillators to prepare the chapters focused on less common but attractive scintillator technologies and to review related research, open problems, and application potential. Bulk single crystals are the most often reviewed scintillator materials on which few books have already been published. Therefore, we decided to orient the content of this book toward nanocomposite, glass-ceramic, thin film, and few other preparation technologies that sometimes offer unexpected technological possibilities, open new ways of scintillator material exploration, and, of course, bring many questions to deal with. To capture the interest of students and non-specialists also, we have carefully prepared an introductory chapter as a quick guide to the field of scintillation materials.

I very much appreciate the careful and patient collaboration from the publisher and want to thank Stanford Chong, Sarabjeet Garcha, and Shivani Sharma. I sincerely thank all my colleagues who kindly accepted the invitation to contribute to the book; they have really made a great job. My deep thanks also go to a number of my collaborators in the home institute, in my country, and around the globe with whom I learned a lot; together we have obtained many exciting results and had a lot of fun during the joint studies. Finally, I want to thank to my wife Natasha for her everlasting support and patience as life with a scientist is definitely not that easy.

Martin Nikl

Fall 2016

Chapter 1

Introduction to Scintillators

Martin Nikl and Valentin V. Laguta
Institute of Physics, Czech Academy of Sciences,
Cukrovarnicka 10, 16253 Prague, Czech Republic
nikl@fzu.cz

1.1 Introduction

In the first chapter, we introduce the concept and function of a scintillator material and describe the X- (or gamma)-to-ultraviolet/visible photons' transformation. We briefly mention the history of scintillator research and development. We survey the most important scintillation characteristics and parameters, which are used to quantitatively measure the materials' performance. We overview the experimental techniques from the family of the optical and magnetic spectroscopies used for the study of both basic physical phenomena and practical characteristics in scintillators. Last but not the least, various material morphologies and related preparation techniques are mentioned, with emphasis on those that are in focus of this book.

Owing to the high practical importance and relatively long history of scintillators, a large amount of information was published

Nanocomposite, Ceramic, and Thin Film Scintillators
Edited by Martin Nikl

ISBN 978-981-4745-22-2 (Hardcover), 978-981-4745-23-9 (eBook)
www.panstanford.com

on the topic. Let us mention an excellent survey of luminescent materials written by Blasse and Grabmaier [1], the overview of the methodology used in the radiation measurement authored by Knoll [2], the monograph devoted to scintillator materials by Rodnyi [3] and Lecoq et al. [4], and the useful handbook on phosphor written by Shionoya and Yen [5]. Numerous featured review papers or book chapters on phosphor and scintillator materials and their applications exist in the scientific literature as well, for instance Refs. 6–13.

Scintillator material works as a spectral and energy transformer: it converts a high-energy photon from X- or gamma-ray range into a bunch of ultraviolet-visible (UV/vis) ones, i.e., to a flash of light. Eventually the accelerated charged particles (electrons, protons, or heavier ions) or even neutrons can be detected through their energy deposit in the interaction with the scintillator host, which is again converted into a light flash. In practice, the scintillator detector consists of two parts, (i) the scintillator material itself and (ii) the photodetector, which converts the mentioned UV/vis photons into an electrical signal I(t) [12] (Fig. 1.1).

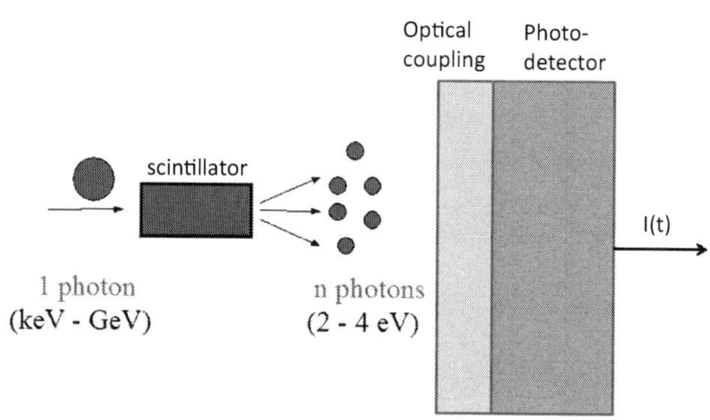

Figure 1.1 Principle of scintillator material and set-up of scintillation detector.

Dielectric or semiconductor wide-bandgap materials are employed for such a task. A phenomenological description of the scintillation mechanism and definition of efficiency criteria were already developed in the 1970s [14] and further refined later [15]. The scintillation mechanism can be divided into three consecutive sub-processes: *conversion, transport,* and *luminescence* (Fig. 1.2).

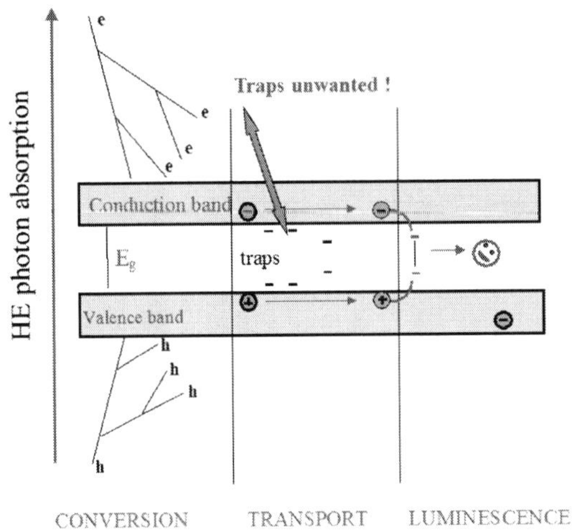

Figure 1.2 Sketch of scintillator mechanism.

Depending on the entering photon/particle energy the initial multistep interaction in the scintillator lattice occurs dominantly through (i) the photoelectric effect (below approx. 100 keV), (ii) Compton scattering effect within 200–8000 keV, and (iii) pair production above the latter limit. The hot electrons and deep holes that arise are gradually thermalized in the conduction and valence band edges, respectively. All of the conversion process lasts typically less than 1 ps; see Refs. 16 and 17 for a more detailed description of the process. In the transport process the separated electrons and holes can relax into exciton states or migrate individually through the host material, where they can be repeatedly trapped or even nonradiatively recombined at trapping levels arising because of lattice defects. Considerable delay in the charge carrier delivery to luminescent centers can be introduced owing to such trapping processes. This stage is the least predictable as point defects, flaws, surfaces, and interfaces can introduce energy levels into the forbidden gap and strongly modify/degrade scintillation performance. These phenomena are strongly dependent upon manufacturing technology [18]. During the final stage, the consecutive trapping of the electron and hole occurs at the luminescent center where they radiatively recombine. Nonradiative energy transfer toward the emission center via excitons is also possible.

It is not very surprising that the research on scintillation materials starts at the moment of X-ray discovery in November 1895 by W. C. Roentgen [19]. In the X-ray registration, a simple photographic film was found rather inefficient and that is why the search for materials able to convert this new invisible radiation into visible light started immediately in order to efficiently use sensitive photographic film-based detectors. $CaWO_4$ powder was employed for this purpose just a few months later, in early 1896, and together with ZnS-based powders introduced later, these powder phosphor materials came to be widely used for the detection of X-rays in such combined phosphor–film detectors [20].

It is interesting to note that in 1897 the cathode ray tube was invented, in which the energy of an accelerated electron beam is converted into visible light (this phenomenon is commonly called cathodoluminescence) by a phosphor material. The mechanism of this conversion is quite similar [21] to that of the X-ray conversion. Thus, at the end of the 19th century, two energetic (photon and particle) radiations were revealed together with γ-rays generated by natural radioactive elements. The latter phenomenon was discovered by H. Becquerel in uranium in early 1896 [22] and by M. C. Sklodowska and P. Curie in polonium and radium [23, 24]. Enormous application potential of all the above-named kinds of radiation thus stimulated the development of phosphor and scintillator materials necessary for their practical exploitation.

The history of bulk single-crystal scintillators begins in the late 1940s with the introduction of NaI:Tl and CsI:Tl [25, 26]. Since that time a number of material systems have been reported; see Ref. 27 for an overview. NaI:Tl and CsI:Tl, the first oxide-based $CdWO_4$ scintillator [28], and $Bi_4Ge_3O_{12}$ (BGO) [29] became widespread scintillators and are often used as "standard samples" to evaluate new materials under study. In the last two decades there has been considerable activity in this field triggered mainly by the need of high-energy physics and advanced imaging applications in science, medicine, and industry. The absolute majority of new single-crystal scintillators reported in this period is based on Ce^{3+}-doped and Pr^{3+}-doped materials, because of the fast decay time (typically 10–60 ns) and high quantum efficiency of the 5d-4f radiative transitions of these centers at room temperature (RT) and in some cases even at much higher ones [30]. For a survey of these materials, see Refs. 11–13, 31, and 32.

Despite the identical underlying physics, scientific communities working on phosphors and scintillators have been partially separated, mainly because of the different demands of related applications and different preparation technologies employed [6–9]. In general, materials are called phosphors when applied in the photon integrating (steady-state) detection mode, while scintillators are employed in the (X- or γ-ray) photon counting regime. At present, the separation between phosphor (powders) and scintillator (bulk) materials is somewhat suppressed as some materials are used in both detection modes, in powder, bulk, or other forms, depending on the application [12].

In the case of scintillators, X- or γ-ray photon counting consists of accumulating the generated light arriving soon after the initial conversion stage is accomplished (Fig. 1.1), since the scintillator works as a high-energy photon counter. Significantly delayed light such as that due to retrapping processes mentioned above cannot be technically exploited in the counting mode. The most important characteristics of scintillation materials are as follows:

(i) Scintillation efficiency
(ii) Light yield (LY)
(iii) Linearity of light response with the incident X(γ)-ray photon energy—energy resolution and nonproportionality
(iv) X(γ)-ray stopping power
(v) Scintillation response in time
(vi) Spectral matching between the scintillator and photodetector
(vii) Chemical and mechanical stability
(viii) Radiation resistance
(ix) Price

The overall scintillation efficiency of X(γ)-ray-to-light conversion is determined by both intrinsic and extrinsic material characteristics. The number of UV/visible photons, N_{ph}, produced in the scintillation conversion per energy E of incoming X(γ)-ray photon can be expressed as [14–16]:

$$N_{ph} = \frac{E}{\beta E_g} \times SQ \qquad (1.1)$$

where E_g represents the forbidden gap of the material, S and Q are quantum efficiencies of the transport and luminescence stages respectively, and β is a phenomenological parameter which is typically found to be between 2 and 3 for most materials. The relative efficiency can then be obtained as

$$\eta = \frac{E_{\text{vis}} N_{\text{ph}}}{E} \tag{1.2}$$

where E_{vis} is the energy of generated UV/vis photons. The most efficient material among the phosphors and scintillators today is ZnS:Ag with $\eta \approx 0.2$ and even more efficient materials could still be found within those with the narrower bandgap below 3 eV.

The LY of a scintillator is always a value inferior to that given by Eq. 1.1 since it represents only a fraction of generated visible photons, namely those arriving to the photodetector within a certain time gate defined by the detection electronics (so-called shaping time) after the high-energy photon absorption. The values of shaping time are usually set between 100 ns and 10 μs.

The linearity of light response with incident X(γ)-ray photon energy in scintillators is given by the dependence of LY on the incident photon energy. The value for LY is energy dependent: it is partly due to abrupt changes of the attenuation coefficient around the K and L edges of the elements constituting the compound, but it also follows from the non-equal conversion efficiency of the photoelectric effect and Compton scattering, which become progressively more important with the increasing energy of incoming X(γ)-rays. As a result, the energy resolution of a scintillating material (ability to resolve two photons of similar energy) is degraded with respect to the intrinsic limits based purely on statistical grounds [33]. In addition, especially at lower energies below a few tens of keV, the LY value systematically decreases because of increasing attenuation coefficient of the scintillator host: nanoregions with high density of elementary excitations arise, the (nonradiative) interaction of which introduces nonlinear energy losses [34]. Nonproportionality of scintillating materials has been systematically studied in recent years [35] and comprehensive models and explanations have appeared recently [36].

The X(γ)-ray stopping power (attenuation coefficient) of a material of a given thickness depends on its density, ρ, and its effective atomic number, Z_{eff} (for the calculation of Z_{eff}, see e.g. [7]). Considering only the interaction associated with the photoelectric effect, the stopping power is proportional to ρZ_{eff}^{3-4}, while in Compton effect and pair production it is proportional to ρZ_{eff} and ρZ_{eff}^2, respectively.

The kinetics of the light response of a scintillator is governed by the characteristics of the transport and luminescence stages in Fig. 1.2, as they are far slower than the initial conversion. The decay rate of the luminescence center itself is defined by its transition dipole moment from the excited-to-ground state and can be further enhanced by additional nonradiative quenching or energy transfer processes from the excited state. Such quenching or energy transfer, however, results in the decrease of parameter Q in Eq. 1.1 and the overall scintillation efficiency is consequently decreased. In the most simple case the emission intensity $I(t)$ is exponentially decaying as

$$I(t) \sim e^{-t/\tau} \tag{1.3}$$

where τ is the decay time. While decay times of the parity and/or spin forbidden transitions for most of the rare earth ions are typically on the order of several tens of μs up to ms, in the case of allowed 5d–4f transitions of Ce^{3+}, Pr^{3+}, and Nd^{3+} the values are typically within 5–60 ns. The allowed charge transfer transition to Yb^{3+} has been also systematically studied to provide another fast emission center in scintillators with somewhat longer radiative lifetime of the order of 100 ns [37, 38]. However, the Yb^{3+} charge transfer luminescence is heavily quenched in most of halide and oxide matrices at RT so that scintillation efficiency of such materials is very low. The 5d–4f transition of Eu^{2+} has gained renewed interest in mid-density scintillators, where ultrahigh light yield and excellent energy resolution can be achieved, e.g., in SrI_2:Eu [39]. Though the Eu^{2+} radiative decay time in various hosts is of the order of a few hundred ns up to about 2 μs, a favorable combination of ultrahigh light yield and energy resolution makes these materials prospective, e.g., in the field of homeland security measures [40].The highest radiative decay rates are offered by Wannier excitons in direct-gap semiconductors, where subnanosecond values of radiative lifetimes have been reported for ZnO, CuX, $CsPbX_3$ (X=Cl,Br), PbI_2, and HgI_2 [41]. The emission rate is enabled by the coherent nature of the exciton state spread over a (large) number of elementary cells [42].

The low binding energy of these excitonic states, however, results in partial ionization and consequently diminishing intensity of the prompt part of exciton-related emission at RT. In bulk materials, further efficiency decrease follows from the small Stokes shift, which leads to noticeable reabsorption losses [43].

Owing to the retrapping processes during the transport stage sketched in Fig. 1.2, the light emission response of a material under high energy excitation is often further complicated by slow non-exponential components. These processes are currently quantified by the afterglow, which is defined as a residual light intensity at a certain time (from a few milliseconds to minutes) after excitation cut-off [18]. In superfast imaging applications, afterglow phenomena even in the microsecond scale become important [44]. The delayed radiative recombination is responsible for slow components over a broad time scale in scintillator response (Fig. 1.3). Consequently, lowered LY values and deteriorated signal-to-background ratio is often observed [45]. Because of the repetitive nature of the measurement, the delayed recombination is reflected also by the elevated signal level before the rising edge of the decay curve [45, 46], which is illustrated in Fig. 1.3. In practical evaluation for both phosphors and scintillators, either simple 1/e or 1/10 decay times are sometimes provided, defined as the time when the light intensity decreases to 1/e or 1/10 of its initial value after an excitation flash, or so-called mean decay time can be calculated [9].

Spectral matching between the scintillator emission band and the photodetector spectral sensitivity dependence is an obvious requirement. Classical criteria were defined a time ago as near-UV-blue emission is optimum for a photomultiplier detector, while for a photodiode the green-red spectral region was considered the best. In recent years there has been an enormous development in semiconductor photodetectors and the latest generation of the back-illuminated "charge-coupled device" (CCD) photodetector shows enhanced sensitivity even down to 200 nm.

Chemical stability mainly concerns the hygroscopicity of materials, which in some cases inhibits operation in the open air, and hermetic sealing is necessary (NaI:Tl, CsI:Na, LaX_3:Ce, X=Cl,Br, SrI_2:Eu). Mechanical stability includes, e.g., the brittleness of a material, existence of cleavage planes in crystals, and similar obstacles which make the cutting and polishing procedures troublesome.

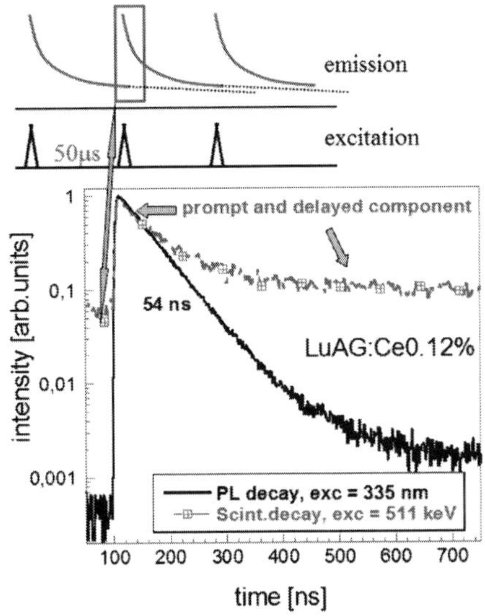

Figure 1.3 Normalized photoluminescence and spectrally unresolved scintillation decay of Ce-doped $Lu_3Al_5O_{12}$ single crystal. Upper part: Sketch of the measurement method. True background level is displayed in photoluminescence decay. Background enhancement in the repetitive decay measurement is due to long-lasting scintillation components from previous excitations. Reprinted from Ref. 12 with permission, © IOP Publishing. All rights reserved.

Radiation resistance of materials is a matter of concern mainly in the bulk scintillation materials [7]. Scintillation performance changes and instabilities arise due to the induced absorption resulting from material irradiation and creation of color centers [9]. Specifically, in the case of the induced absorption overlap with the emission spectrum, the re-absorption losses occur, less scintillation light arrives to the photodetector and the overall efficiency/LY value is degraded. An example of this behavior is given in Fig. 1.4. While this parameter has been mainly considered in the research of scintillators for high-energy physics, it should be noted that it can be also important in several medical imaging techniques [11] as well as in the case of industrial flaw detection or accelerator beam diagnostics applications.

The price of a scintillator material obviously depends on the volume manufactured, cost of raw materials, and other parameters (expensive technological components as crucibles, hot zone construction, energy consumption, etc.). While in case of large-volume production of $PbWO_4$ single crystals for calorimetric detectors at the Large Hadron Collider in CERN the price for 1 cm^3 of final scintillation elements was of the order of few US dollars, it can easily be 1–2 orders higher or even more in case of expensive manufacturing technologies of orthosilicate, rare earth halide, or isotope-enriched fluoride materials.

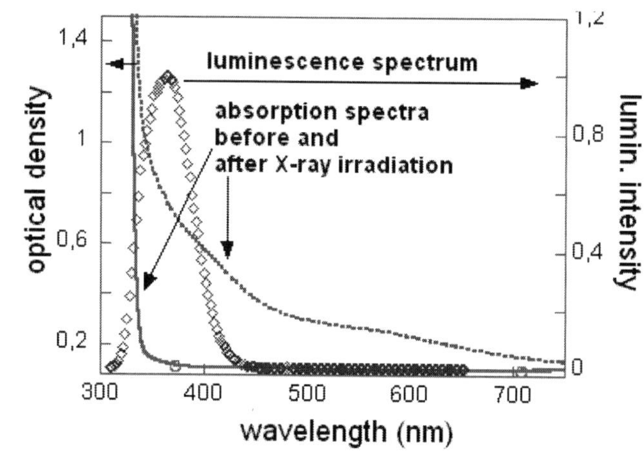

Figure 1.4 Absorption spectra of 2 cm thick $YAlO_3$:Ce single crystal before and after γ-ray irradiation (^{60}Co, 500 Gy dose) at RT. Overlap of the irradiation-induced absorption with luminescence spectrum of the material is demonstrated . Reprinted from Ref. 12 with permission, © IOP Publishing. All rights reserved.

Given the above-mentioned variety of characteristics to be studied and general complexity of solid state materials in which the atomistic processes of energy transfer and storage must be followed in the case of scintillators, a rather large number of experimental techniques must be employed for their study in a correlated manner. To understand the relation between material defects and the occurrence of traps in the material forbidden gap, their involvement in the energy capture and storage and interconnection with the production technology, one has to employ a set of specific characterization methods closely coordinated with the manufacturing technology. Correlated experiments using time-resolved luminescence spectroscopy, wavelength-resolved thermally stimulated luminescence, and electron paramagnetic resonance in an extended temperature interval (10–300 K at least) are very powerful tools for the above purpose. Furthermore, special "one-purpose" setups are used to measure practically important scintillation characteristics, especially light yield or scintillation decay; see also Ref. 2. The former is measured with the help of so-called pulse-height spectra using a multichannel analyzer and fast photomultiplier or a hybrid photomultiplier in the current regime [47], while the latter is obtained by the time-correlated single-photon counting method after the excitation by ^{22}Na isotope (511 keV photons) [48]. Nowadays, the scintillation decay is also routinely measured using the detection part based on a fast digital oscilloscope and fast photomultiplier in the current regime.

Time-resolved UV–visible emission spectroscopy enables detailed understanding of the luminescence center itself, its intracenter transitions, and possible nonradiative quenching or excited-state ionization phenomena [8, 32]. At the same time, its interconnection with the host lattice environment via energy transfer processes can be monitored. Selective pulse excitation in the VUV–UV–visible spectral region completed with X-ray or single-photon γ-ray excitations is used for this purpose.

Thermally stimulated luminescence (TSL) or thermoluminescence allows the study of point defects giving rise to localized energy levels (traps) in a material forbidden gap which are responsible for delayed recombination phenomena in scintillation mechanism. A very simplified scheme is proposed in Fig. 1.5 [49] for a systematic description. During irradiation (typically by X-rays) at a given temperature, free carriers are trapped at electron or hole centers. A subsequent linear heating cycle from irradiation temperature to a pre-selected upper limit allows the carriers in shallow-enough traps to be released by increasing thermal energy kT and recombine radiatively with carriers of the opposite charge firmly localized at recombination centers. Consequently, monitoring the emitted light as a function of sample temperature, so-called glow curve, a peak at a characteristic temperature appears the position of which is defined by the trap depth and its interaction with the surrounding. While different trapping processes can be separated through their related glow curve peaks, the measurement of the TSL emission spectrum can advice on the kind of recombination center.

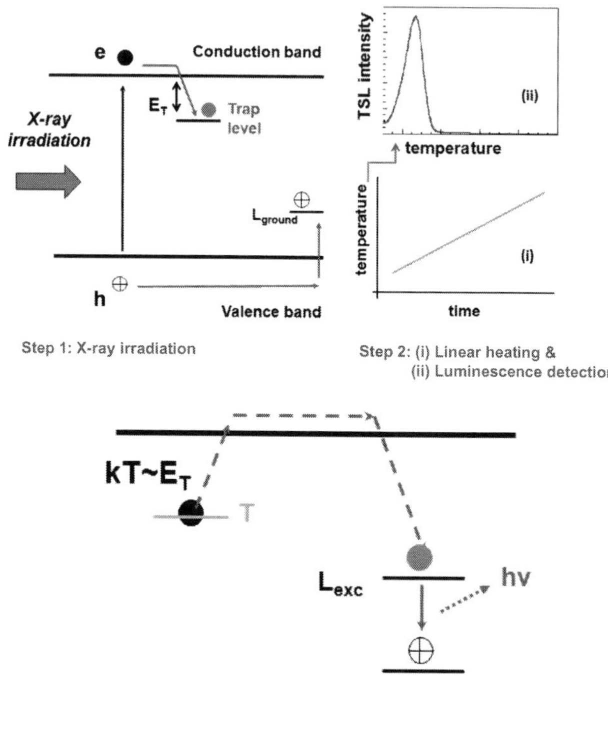

Figure 1.5 Sketch of the TSL process; see also Ref. 13. In this example, a trap for an electron is employed while the hole is localized directly at recombination center. However, the opposite arrangement is also possible.

In the simplest first-order kinetics model [49] the intensity $I^{TSL}(T)$ of emitted photons is

$$I^{TSL}(T) = sn_0 \exp(-E_T/kT)x$$
$$\exp\left[-\frac{s}{\beta}\int_{T_0}^{T}\exp(-E_T/kT'dT')\right] \tag{1.4}$$

Here n_0 (cm^{-3}) is the number of traps filled by irradiation, s (s^{-1}) is the frequency factor, β is the heating rate, and E_T (eV) is the trap depth. In the same simplified model, if instead of being heated the sample is held at a constant temperature after irradiation, the mean time spent by carriers in the trap before recombination is

$$\tau = 1/s \exp(E_T/kT) \tag{1.5}$$

This "detrapping" time determines the delayed recombination timing characteristics at a given temperature T. A detailed analysis of Eqs. 1.4 and 1.5 or of more general ones by numerical methods [49] can provide the trap parameters (E_T, s). The detrapping time τ at RT can thus be evaluated, allowing a direct comparison with slow components in scintillation decay or LY measurements. Spatial correlation between traps and recombination centers has also been evidenced in quite many materials [50], which allows also for the subgap (tunneling-driven) recombination processes and can further influence the scintillation response [51].

The nature and local structure of the defects, traps, and recombination centers mentioned above can be studied in a detailed manner using electron paramagnetic resonance (EPR) and related methods, such as electron nuclear double resonance (ENDOR) and optically detected magnetic resonance (ODMR) [52, 53]. All of these methods belong to so-called local probe methods and exploit the interaction of the magnetic moment of the unpaired electron(s) and nuclei in a defect with each other and with external magnetic fields.

In a magnetic resonance experiment, the position of the resonance lines reflects both local symmetry and a degree of lattice distortion around paramagnetic particle. In addition, the electron–nuclear interaction can give direct evidence for the nuclear (thus also the chemical) species coupled to the paramagnetic particle. This can be illustrated in Fig. 1.6 with an example related to the Pb^+ center in a $PbWO_4$ crystal [54] where an electron is localized at the Pb^+ p_z orbital around an oxygen vacancy. With application of an external magnetic field B_0, the lowest energy level of Pb^+ splits nearly linearly with the field strength. The microwave energy is absorbed by the sample when the frequency of the microwave irradiation is in resonance with the energy splitting. For lead ions with non-zero nuclear spin there is an additional splitting of the energy levels (hyperfine splitting) due to the interaction of electron and nuclear magnetic moments, and it gives rise to small-intensity satellites around the main resonance. Detection of hyperfine satellites in many cases allows unambiguous determination of the local structure of defects.

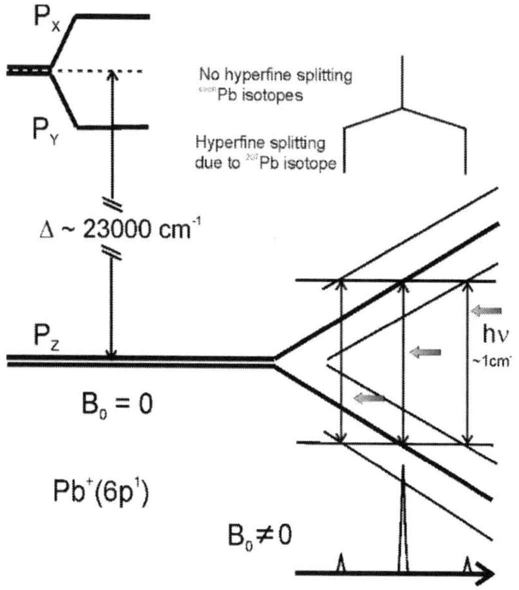

Figure 1.6 Sketch of an EPR measurement principle; see also Ref. 13.

In order to obtain the maximum amount of information from the EPR experiment, it should be performed on a single crystal. However, a great deal of information can still be obtained from experiment on powdered solids or on bulk ceramics, glass ceramics, or even composite systems. The essential feature of these experiments is that we are now dealing not with selectively oriented single crystal but with a large number of randomly oriented crystallites, which all contribute to the spectrum. This essentially reduces the spectrum resolution of a powder or ceramics. Spectral resolution enhancement for such spatially disordered systems can be achieved by using high-frequency microwaves, e.g., W-band (95 GHz) and even mm-wave 263 GHz. Further improvement of the information available from EPR experiment is provided by application of advanced pulsed EPR techniques, such as spin echo envelope modulation (ESEEM) spectroscopy and hyperfine sublevel correlation (HYSCORE) spectroscopy [55, 56]. Both techniques are useful tools for resolving small electron–nuclear couplings of distant nuclei or nuclei with weak magnetic moments that cannot be resolved in the ordinary EPR spectra. These techniques thus help to clarify the structure of a paramagnetic defect.

In favorable cases, one can obtain the following information from EPR experiments:

(i) symmetry of a defect,
(ii) concentration of defects,
(iii) identification of nuclei (atoms) constituting a defect,
(iv) charge state of paramagnetic ions involved,
(v) change of defect charge state under external influences (irradiation, temperature, etc.), and
(vi) dynamic properties of a defect, e.g., decay or growth of the concentration of defects after the irradiation is switched on (off). This last item allows the direct determination of the energy of the local electronic level of a defect created by irradiation and thus can be correlated with TSL characteristics.

Further refinement of the electronic structure of absorption or luminescence centers can be obtained using the technique of optically detected EPR via the magnetic circular dichroism (MCD) [53]. This technique provides a direct correlation between EPR and MCD by the "tagged MCD" method, when one keeps the magnetic field constant on one of the resonance positions and changes the wavelength at which the optical absorption or luminescence is detected. Therefore, ODMR offers the most compelling interrelation between atomic-scale defects and the related optical absorption or luminescence bands.

Recently, because of the essential improvement of the sensitivity and resolution, the solid state nuclear magnetic resonance (NMR) spectroscopy is successfully applied to the investigation of nonparamagnetic lattice defects. In particular, NMR provides a unique approach to quantify the site occupancy in crystalline and even amorphous materials [57]. The NMR signal in a properly designed experiment is directly proportional to the concentration of atoms in a given structural environment. This was demonstrated, for instance, in the study of antisite positions of lattice ions and oxygen vacancies in $YAlO_3$ crystals and incorporation of Sc ions in a $Lu_3Al_5O_{12}$ host [58, 59], and their distribution in the octahedral and dodecahedral sites [60].

Despite the best figure-of-merit of the single crystal scintillators, not nearly all intrinsically efficient compounds can be grown in the form of bulk single crystals with sufficiently big dimensions and manufacturing price affordable for practical applications. Too high melting temperatures, the presence of phase transitions between the melting point and room temperature or stoichiometry problems resulting from different volatility of binary constituents from the high temperature melt of complex compounds, are just a few examples of troubles which can prevent single-crystal preparation. Optical ceramics have been under development as an alternative to single-crystal materials to provide bulk optical elements in the case where single crystals cannot form or when ceramic materials show superior properties. They are transparent or translucent materials constituted by tight aggregating crystallite micrograins, each randomly oriented with respect to their neighbors. Their technology has greatly developed mainly within the last two decades owing to application of these materials in the field of solid state lasers [61], and issues dealing with the light scattering and material porosity have been successfully solved [62]. However, the application demands are higher in the case of scintillator with respect to laser optical ceramics because even the point defects and structural irregularities at the atomic scale can seriously limit the material performance due to the introduction of trapping levels in the material forbidden gap.

Development of optical ceramics for the scintillator applications was triggered by the needs of computed tomography (CT) medical imaging and started in the 1980s [63, 64]. The review of the results mainly achieved in the manufacturing and characterization of $(Y,Gd)_2O_3$:Eu, Gd_2O_2S:Pr,Ce,F, and $Gd_3Ga_5O_{12}$:Cr,Ce ceramics was reported in Ref. 65. It is worth mentioning that scintillation response of all these ceramics is slow, in the μs-to-ms time scale due to the forbidden character of radiative transitions of the emission centers used (Eu^{3+}, Pr^{3+}, Cr^{3+}, Tb^{3+}). More recently, fast optical ceramics based on the Ce- and Pr-doped YAG and LuAG have been reported in the literature [66–69]. In the last decade, the R&D in the field of such fast-scintillation ceramics became a hot topic in the hunt for new scintillation materials, and in addition to their classical manufacturing technologies, new ones such as spart plasma sintering (SPS) have appeared recently [70].

Another family of bulk scintillators based on so-called glass ceramics have been studied recently with the aim to develop a cheaper technological way to obtain bulk transparent scintillation elements [71, 72]. It is achieved by the directed crystallization of the scintillation nanophase inside a glass host using a targeted annealing procedure at the prepared parent glass element.

New X-ray-based imaging applications which require micrometer or submicrometer 2D resolution induced the development of thin-film scintillators with thickness in tens of micrometers [73]. Liquid-phase epitaxy, i.e., growth from a flux, is usually used to obtain high-quality single-crystal films [74]. Limitation is given by the need to use an isostructural high-quality substrate (e.g., $Y_3Al_5O_{12}$) and contamination of film by the flux constituents can occur [75].

Modern medical therapies such as photodynamical therapy have put the demand for nanopowder scintillators [76]. Functionalized nanopowder can be directed by blood flow, e.g., to tumor tissues and under excitation by X-rays the singlet oxygen is produced from functionalized surface of the grains which kills the cells around. Lanthanide-doped inorganic nanopowders are also considered in coming biomedical applications as luminescent nanoprobes [77].

Finally, a new strategy has been adopted and so-called nanocomposite materials have become a topic also in the scintillator field [78], again with the aim to prepare bulk transparent materials where the scintillation characteristics will be defined by a nanophase dispersed in a suitable host. In principle, these novel materials can include, e.g., organic-inorganic mixtures [79] and offer much higher flexibility in material composition and characteristics compared to glass ceramics mentioned above.

As already mentioned, mainly the single crystal and up to some extent optical ceramic scintillators were reviewed in recent literature, in the books [3, 4, 13, 80] and review articles [7–12]. In this book, in the chapters that follow, we are aiming to introduce and describe in more detail the research and development in thin-film scintillators, glass ceramics, and nanocomposite and optical ceramics prepared by spark plasma sintering. Furthermore, an example of an in-depth study of a powder phosphor material is included as well. Both technology description and various characterization aspects are provided together with application hints.

Acknowledgment

The support of Czech Science Foundation project no. P204/12/0805 is gratefully acknowledged.

References

1. Blasse, G., Grabmaier, B. C. (1994) *Luminescent Materials*. Springer: Berlin.
2. Knoll, G. F. (2000) *Radiation Detection and Measurement*. Wiley: New York.
3. Rodnyi, P. A. (1997) *Physical Processes in Inorganic Scintillators*. CRC Press: Boca Raton, US.
4. Lecoq, P., Annenkov, A., Gektin, A., Korzhik, M., Pedrini, C. (2006) *Inorganic Scintillators for Detector Systems*. Springer: Berlin.
5. Shionoya, S., Yen, W. M. (eds.) (1998) *Phosphor Handbook*. CRC Press: Boca Raton, US.
6. Brixner, L. H. (1987) New x-ray phosphors. *Mater Chem Phys* **16**, pp. 253–281.
7. Ishii, M., Kobayashi, M. (1992) Single crystals for radiation detectors. *Prog Cryst Growth Charact Mater* **23**, pp. 245–311.
8. Blasse, G. (1994) Scintillator materials. *Chem Mater* **6**, pp. 1465–1475.
9. Nikl, M. (2000) Wide band gap scintillation materials: Progress in the technology and material understanding. *Phys Status Solidi A* **178**, pp. 595–620.
10. Wojtowicz, A. J. (2002) Rare-earth-activated wide bandgap materials for scintillators. *Nucl Instrum Methods Phys Res A* **486**, pp. 201–207.
11. van Eijk, C. W. E(2002) Inorganic scintillators in medical imaging. *Phys Med Biol* **47**, pp. R85–R106.
12. Nikl, M. (2006) Scintillation detectors for x-rays. *Meas Sci Technol* **17**, pp. R37–R54.
13. Nikl, M., Vedda, A., Laguta, V. V. (2010) Single-crystal scintillation materials. In *Springer Handbook of Crystal Growth*, ed. Dhanaraj, G., Byrappa, K., Prasad, V., Dudley, M. Springer: Heidelberg, pp. 1663–1700.
14. Robbins, D. J. (1980) On predicting the maximum efficiency of phosphor systems excited by ionizing-radiation. *J Electrochem Soc.* **127**, pp. 2694–2702.

15. Lempicki, A., Wojtowicz, A. J., Berman, E. (1993) Fundamental limits of scintillator performance. *Nucl Instrum Methods Phys Res A* **333**, pp. 304–311.

16. Rodnyi, P. A., Dorenbos, P., van Eijk, C. W. E. (1995) Energy loss in inorganic scintillators. *Phys Status Solidi B* **187**, pp. 15–29.

17. Vasil'ev, A. N. (1996) Polarization approximation for electron cascade in insulators after high-energy excitation. *Nucl Instrum Methods Phys Res B* **107**, pp. 165–171.

18. Shepherd, J. A., Gruner, S. M., Tate, M. W., Tecotzky, M. (1997) Study of afterglow in x-ray phosphors for use on fast-framing charge-coupled device detectors. *Opt Eng* **36**, pp. 3212–3222.

19. Röntgen, W. C. (1896) On a new kind of rays. *Science* **3**, pp. 227–231

20. Fuchs, A. W. (1960) Radiography of 1896. *Image* **9**, pp. 4–17.

21. Ozawa, L., Itoh, M. (2003) Cathode ray tube phosphors. *Chem Rev* **103**, pp. 3835–3855.

22. Becquerel, H. (1896) Sur les radiations invisibles emises par les corps phosphorescents. *C R Acad Sci* **122**, pp. 501–503.

23. Curie, M. (1898) Rayons emises par les composes de l'uranium et du thorium. *C R Acad Sci* **126**, pp. 1101–1103.

24. Curie, P., Curie, M. (1898) Sur une substance nouvelle radio-active, continue dans la pechblende. *C R Acad Sci* **127**, pp. 175–178.

25. Hofstadter, R. (1949) The detection of gamma-rays with thallium-activated sodium iodide crystals. *Phys Rev.* **75**, pp. 796–810.

26. Van Sciver, W., Hofstadter, R. (1951) Scintillations in thallium-activated CaI_2 and CsI. *Phys Rev* **84**, pp. 1062–1063.

27. Weber, M. J. (2002) Inorganic scintillators: today and tomorrow. *J Lumin* **100**, pp. 35–45.

28. Gilette, R. H. (1950) *Rev Sci Instrum* **21**, pp. 294–301.

29. Weber, M. J., Monchamp, R. R. (1973) Luminescence of $Bi_4Ge_3O_{12}$: spectral and decay properties. *J Appl Phys* **44**, pp. 5495–5499.

30. W. Bachmann, C. Ronda, A. Meijerink, (2009) Temperature quenching of yellow Ce3+ luminescence in YAG:Ce. *Chem Mater* **21**, pp. 2077–2084.

31. van Eijk, C. W. E. (2001) Inorganic-scintillator development. *Nucl Instrum Methods Phys Res A* **460**, pp. 1–14.

32. Pejchal, J., Nikl, M., Mihóková, E., Mareš, J. A., Yoshikawa, A., Ogino, H., Schillemat, K. M., Krasnikov, A., Vedda, A., Nejezchleb, K., Múčka, V.,

(2009) Pr^{3+}-doped complex oxide single crystal scintillators. *J Phys D Appl Phys* **42**, 055117.

33. Dorenbos, P., de Haas, J. P. M., van Eijk, C. W. E, (1995) Non-proportionality in the scintillation response and the energy resolution obtainable with scintillation crystals. *IEEE Trans Nucl Sci* **2**, pp. 2190–2202.

34. Bizarri, G., Moses, W. W., Singh, J., Vasil'ev, A. N., Williams, R. T. (2009) An analytical model of nonproportional scintillator light yield in terms of recombination rates. *J Appl Phys* **105**, 044507.

35. Moses, W. W., Payne, S. A., Choong, W.-S., Hull, G., Reutter, B. W. (2008) Scintillator non-proportionality: present understanding and future challenges. *IEEE Trans Nucl Sci* **55**, pp. 1049–1053.

36. Wang, Z. , Williams, R. T., Grim, J. Q., Gao, F., Kerisit, S. (2013) Kinetic Monte Carlo simulations of excitation density dependent scintillation in CsI and CsI(Tl). *Phys Status Solidi B* **250**, pp. 1532–1540.

37. van Pieterson, L., Heeroma, M., de Heer, E., Meijerink, A. (2000) Charge transfer luminescence of Yb^{3+}. *J Lumin* **91**, pp. 177–193.

38. Nikl, M., Yoshikawa, A., Fukuda, T. (2004) Charge transfer luminescence in Yb^{3+}-containing compounds. *Opt Mater* **26**, pp. 545–549.

39. Cherepy, N. J., Hull, G., Drobshoff, A. D., Payne, S. A., van Loef, E., Wilson, C. M., Shah, K. S., Roy, U. N., Burger, A., Boatner, L. A., Choong, W.-S., Moses, W. W. (2008) Strontium and barium iodide high light yield scintillators. *Appl Phys Lett* **92**, 083508.

40. Milbrath, B. D., Peurrung, A. J., Bliss, M., Weber, W. J. (2008) Radiation detector materials: an overview. *J Mater Res* **23**, pp. 2561–2581.

41. Derenzo, S. E., Weber, M. J., Klintenberg, M. K. (2002) Temperature dependence of the fast, near-band-edge scintillation from CuI, HgI_2, PbI_2, ZnO:Ga and CdS:In. *Nucl Instrum Methods Phys Res A* **486**, pp. 214–219.

42. Wilkinson, J., Ucer, K. B., Williams, R. T. (2004) Picosecond excitonic luminescence in ZnO and other wide-gap semiconductors. *Radiat Meas* **38**, pp. 501–505.

43. Ehrentraut, D., Sato, H., Kagamitani, Y., Yoshikawa, A., Fukuda, T., Pejchal, J., Polak, K., Nikl, M., Odaka, H., Hatanaka, K., Fukumura, H. (2006) Fabrication and luminescence properties of single-crystalline, homoepitaxial zinc oxide films doped with tri- and tetravalent cations prepared by liquid phase epitaxy. *J Mater Chem* **16**, pp. 3369–3374.

44. Miller, S., Singh, B., Cool, S., Entine, G., Campbell, L., Bishel, R., Rushing, R., Nagarkar, V. V. (2011) Ultrahigh-speed X-ray imaging of hypervelocity projectiles. *Nucl Instr.Meth Phys Res A* **648** pp. S293–S296.

45. Nikl, M. (2005) Energy transfer phenomena in the luminescence of wide band-gap scintillator. *Phys Status Solidi A* **202**, pp. 201–206.

46. Nikl, M., Nitsch, K., Polak, K., Mihokova, E., Dafinei, I., Auffray, E., Lecoq, P., Reiche, P., Uecker, R. (1995) Slow components in the photoluminescence and scintillation decays of $PbWO_4$ single crystals. *Phys Status Solidi B* **195**, pp. 311–323.

47. Mares, J., D'Ambrosio, C. (2007) Hybrid photomultipliers: their properties and application in scintillation studies. *Opt Mater.* **30**, pp. 22–25.

48. Bollinger, L. M., Thomas, G. E. (1961) Measurement of the time dependence of scintillation intensity by a delayed coincidence method. *Rev Sci Instrum* **32**, pp. 1044–1050.

49. Chen, R., McKeever, S. W. S. (1997) *Theory of Thermoluminescence and Related Phenomena.* World Scientific: Singapore.

50. Clabau, F., Rocquefelte, X., Le Mercier, T., Deniard, P., Jobic, S., Whangbo, M.-H, (2006) Formulation of phosphorescence mechanisms in inorganic solids based on a new model of defect conglomeration. *Chem Mater* **18**, pp. 3212–3220.

51. Mihokova, E., Nikl, M. (2015) Luminescent materials: probing the excited state of emission centers by spectroscopic methods. *Meas Sci Technol* **26**, 012001.

52. Weil, J. A,; Bolton, J. R., Wertz, J. E. (1994) *Electron Paramagnetic Resonance: Elementary Theory and Practical Applications.* Wiley: New York.

53. Spaeth, J.-M., Niklas, J. R., Bartram, R. H. (1992) *Structural Analysis of Point Defects in Solids: An Introduction to Multiple Magnetic Resonance Spectroscopy.* Springer: Berlin.

54. Laguta, V. V., Martini, M., Vedda, A., Nikl, M., Mihokova, E., Bohacek, P., Rosa, J., Hofstatter, A., Meyer, B. K., Usuki, Y. (2001) Photoinduced Pb^+ center in $PbWO_4$: electron spin resonance and thermally stimulated luminescence study. *Phys Rev B* **64**, 165102.

55. Dikanov, S. A., Tsvetkov, Y. D. (1992) *Electron Spin Envelope Modulation (ESEEM) Spectroscopy.* CRC Press: Boca Raton, US.

56. Schweiger, A. and Jeschke, G. (2001) *Principles of Pulse Electron Paramagnetic Resonance.* Oxford University Press: New York.

57. Mackenzie, K. J. D., Smith, M. E. (2002) *Multinuclear Solid-State NMR of Inorganic Materials*. Pergamon Materials Series, vol. 6. Elsevier Science: Amsterdam.

58. Laguta, V. V., Nikl, M., Vedda, A., Mihokova, E., Rosa, J., Blazek, K. (2009) Hole and electron traps in the $YAlO_3$ single crystal scintillator. *Phys Rev B* **80**, 045114.

59. Babin, V., Gorbenko, V., Kondakova, I., Kärner, T., Laguta, V.V., Nikl, M., Zazubovich, S., Zorenko, Yu. (2011) Time-resolved spectroscopy of exciton states in single crystals and single crystalline films of $YAlO_3$ and $YAlO_3$:Ce. *J Phys D: Appl Phys* **44**, 315402.

60. Zorenko, Y., Gorbenko, V., Voznyak, T., Savchyn, V., Nizhankovskiy, S., Dan'ko, A., Puzikov, V., Laguta, V., Mares, J. A., Nikl, M., Nejezchleb, K., Batentschuk, M., Winnacker, A. (2012) Luminescent and scintillation properties of $Lu_3Al_5O_{12}$:Sc single crystal and single crystalline films. *Opt Mater* **34**, pp. 2080–2085.

61. Ikesue, A., Aung, Y. N. (2008) Ceramic laser materials. *Nat Photonics* **2**, pp. 721–727.

62. Krell, A., Klimke, J., T. Hutzler, (2009) Transparent compact ceramics: inherent physical issues. *Opt Mater* **31** pp. 1144–1150.

63. Ito, Y., Yamada, H., Yoshida, M., Fujii, H., Toda, G., Takeuchi, H., Tsukuda, Y. (1988) Hot isostatic pressed Gd_2O_2S:Pr, Ce, F translucent scintillator ceramics for x-ray computed tomography detectors. *Jpn J Appl Phys*, **27**, pp. L1371–L1373.

64. Grabmaier, B. C., Rossner, W., Leppert, J. (1992) Ceramic scintillators for x-ray computed-tomography. *Phys Status Solidi A* **130**, pp. K183–K187.

65. Greskovich, C., Duclos, S. (1007) Ceramic scintillators. *Annu Rev Mater Sci* **27**, pp. 69–88.

66. Zych, E., Brecher, C., Wojtowicz, A. J., Lingertat, H. (1997) Luminescence properties of Ce-activated YAG optical ceramic scintillator materials. *J Lumin* **75**, pp. 193–203.

67. Nikl, M., Mares, J. A., Solovieva, N., Li, H., Liu, X., Huang, L., Fontana, I., Fasoli, M., Vedda, A., D'Ambrosio, C. (2007) Scintillation characteristics of $Lu_3Al_5O_{12}$:Ce optical ceramics. *J Appl Phys* **101**, 033515.

68. Shi, Y., Nikl, M., Feng, X. Mares, J. A., Shen, Y., Beitlerova, A., Kucerkova, R., Pan, Y., Liu, Q. (2011) Microstructure, optical and scintillation characteristics of Pr^{3+} doped $Lu_3Al_5O_{12}$ optical ceramics. *J Appl Phys* **109,** 013522.

69. Yanagida, T., Fujimoto, Y., Kamada, K., Totsuka, D., Yagi, H., Yanagitani, T., Futami, Y., Yanagida, S., Kurosawa, S., Yokota, Y., Yoshikawa, A, Nikl,

M. (2012) Scintillation properties of transparent ceramic Pr:LuAG for different Pr concentration. *IEEE Trans Nucl Sci* **59**, pp. 2146–2151.

70. An, L., Ito, A., Goto, T. (2012) Effect of sintering temperature on the transparency and mechanical properties of lutetium aluminum garnet fabricated by spark plasma sintering. *J Eur Ceram Soc* **32**, pp. 3097–3102.

71. Johnson, J. A., Schweizer, S., Henke, B., Chen, G., Woodford, J., Newman, P. J., MacFarlane, D. R. (2006) Eu-activated fluorochlorozirconate glass-ceramic scintillators. *J Appl Phys* **100**, 034701.

72. Lee, G., Savage, N., Wagner, B., Zhang, Y., Jacobs, B., Menkara, H., Summers, C., Kang, Z. (2014) Synthesis and luminescence properties of transparent nanocrystalline GdF_3:Tb glass-ceramic scintillator. *J Lumin* **147** pp. 363–366.

73. Martin, T., Koch, A. (2006) Recent development in X-ray imaging with micrometer spatial resolution, *J Synchrotron Radiation* **13,** pp.180–194.

74. Kucera, M., Nitsch, K, Nikl, M, Hanus, M, Danis, S. (2010) Growth and characterization of YAG and LuAG epitaxial films for scintillation applications. *J Cryst Growth* **312**, pp. 1538–1545.

75. Babin, V., Gorbenko, V., Krasnikov, A., Makhov, A., Nikl, M., Zazubovich, S., Zorenko, Yu. (2010) Influence of lead-related centers on luminescence of Ce^{3+} and Pr^{3+} centers in single crystalline films of aluminium perovskites and garnets. *Radiat Meas* **45** pp. 415–418.

76. Juzenas, P., Chen, W., Sun, Y.-P., Coelho, M. A. N., Generalov, R., Generalova, N., Christensen, I. L. (2008) *Adv Drug Delivery Rev* **60**, 1600–1614.

77. Tu, D., Zheng, W., Liu, Y., Zhu, H., Chen, X. (2014) Luminescent biodetection based on lanthanide-doped inorganic nanoprobes. *Coordination Chem Rev* **273–274**, pp. 13–29.

78. McKigney, E. A., Del Sesto, R. E., Jacobsohn, L. G., Santi, P. A., Muenchausen, R. E., Ott, K. C., McCleskey, T. M., Bennett, B. L., Smith, J. F., Cooke, D. W. (2007) Nanocomposite scintillators for radiation detection and nuclear spectroscopy. *Nucl Instr Meth Phys Res A* **579** pp. 15–18.

79. Kishimoto, S., Shibuya, K., Nishikido, F., Koshimizu, M., Haruki, R., Yoda, Y. (2008) Subnanosecond time-resolved x-ray measurements using an organic-inorganic perovskite scintillator. *Appl Phys Lett* **93**, 261901.

80. Nikl, M., Yanagida, T., Yagi, H., Yanagitani, T., Mihokova, E., Yoshikawa, A. (2013) Optical ceramics for fast scintillator materials. In *Recent Advances in Ceramic Materials Research* , ed. Roa Rovira, J. J., Rubi, M. S. Nova Science Publishers, pp. 127–176.

Chapter 2

Nanocomposite Scintillators

Markus P. Hehlen,[a] Nickolaus A. Smith,[a] Michael W. Blair,[a] Andy Li,[b] Sy Stange,[a] Robert D. Gilbertson,[a] Edward A. McKigney,[a] and Ross E. Muenchausen[a]

[a]*Los Alamos National Laboratory, P.O. Box 1663, Los Alamos, NM 87545, USA*
[b]*Applied Research Associates, Inc., 801 North Quincy St. #700, Arlington, VA 22203, USA*
hehlen@lanl.gov

2.1 Introduction

The detection and characterization of ionizing radiation (α, β, γ, X-rays, neutrons) is critical to many scientific disciplines, including biology [17], archeology [55], material science [79], and astrophysics [18, 26], and it is at the heart of a plethora of commercial devices deployed in medicine [81], health physics [37], and oil exploration [91], just to name a few. Radiation detection also plays an increasingly important role in the discovery of illicit nuclear materials and associated activities, a capability that is critical to the areas of homeland security, nuclear nonproliferation, treaty verification, and threat reduction [22]. Detection of various forms

Nanocomposite, Ceramic, and Thin Film Scintillators
Edited by Martin Nikl

ISBN 978-981-4745-22-2 (Hardcover), 978-981-4745-23-9 (eBook)
www.panstanford.com

of ionizing radiation is accomplished by measuring the charges and the subsequent excitations and defects that are produced in a material when ionizing radiation interacts with it. There are methods that detect the total amount of radiation received during a defined period of time (i.e., radiation dosimeters) and methods that detect individual interactions in real time (i.e., radiation counters) [43]. The interaction of the radiation with matter can occur in gases (e.g., Geiger–Mueller counters) where the induced charges undergo avalanche multiplication in a strong electric field (Townsend discharge) to produce a measurable electric current pulse. A similar effect can be used in semiconductors (e.g., single-crystalline germanium), where the produced charge carriers can be separated in an electric field and detected as a current pulse. In some solids and liquids, the ionizing radiation can produce charges that subsequently create electronic excitations of luminescent centers (i.e., activators such as metal ions or organic dyes), resulting in the emission of light in the 2–5 eV photon energy range. This class of materials is referred to as *scintillators*, with the most common types being inorganic crystals and organic scintillators.

Inorganic solid scintillators are attractive for many radiation detection applications because their high Z offers good stopping power for the ionizing radiation, they produce an optical signal with short lifetime that can be detected with sensitive photodetectors and fast electronics, and they can operate at room temperature [43, 85]. Traditional solid scintillators encompass the class of inorganic ionic crystals, dating back to the discovery of Tl^{+}-doped NaI by Hofstadter in 1948 [29, 30]. The currently most efficient scintillators are rare earth halide single crystals such as $LaBr_3$ doped with Ce^{3+} as activator ($LaBr_3$:Ce^{3+}) [9] or SrI_2 doped with Eu^{2+} as activator (SrI_2:Eu^{2+}) [2]. One of the key scintillator properties is the photon yield, which describes the number of scintillation photons that are emitted per unit of ionizing energy absorbed by the material (typically expressed in units of optical photons emitted per MeV of ionizing radiation absorbed). A high photon yield not only gives the detector a high sensitivity for ionizing radiation but also enables the measurement of the energy of the incident radiation. In first approximation, the number of electron–hole pairs, $n_{e\text{-}h}$, generated as a result of a γ-ray absorption is proportional to the energy deposited in the material. The number of scintillation photons, n_p, subsequently emitted by the material is then given by

$$n_{\rm p} = \eta_{\rm c}\,\eta_{\rm i}\,n_{\rm e\text{-}h} = \eta_{\rm c}\eta_{\rm i}\frac{E_\gamma}{\beta E_{\rm g}} \tag{2.1}$$

(as shown in Fig. 2.1). In Eq. 2.1, E_γ is the energy of the absorbed γ-ray, E_g is the bandgap energy of the material, and β is a constant that is found to be close to 2.5 for a wide range of materials [21]. Here, η_c is the probability of capturing the electron–hole pair at the activator ion, and η_i is the intrinsic quantum efficiency of the activator. Some high-performance scintillators have $\eta_c\eta_i$ approaching 0.9 [25]. To first approximation, η_c, η_i, β, and E_g are independent of E_γ (proportional scintillator response), and therefore $n_p \propto E_\gamma$. That is, the pulse height received from the scintillator/photodetector assembly is directly proportional to the γ-ray energy that was deposited in the material. The spectrum of the incident radiation can then be measured by acquiring a histogram of pulse heights received from the scintillator/photodetector assembly over a period of time. If the energy resolution of the spectrum is sufficiently good, specific radioactive isotopes can be identified based on their known characteristic γ-ray emission energies. Such spectroscopic-grade radiation detectors are highly desired because they not only allow detection of the presence of a radioactive source but also provide valuable information on the types of radioisotopes contained in the source. This information is critical for applications in which specific radioisotopes need to be detected in the presence of radiation from natural background and benign radioisotopes.

Over the past few decades, the above considerations have driven the development of inorganic scintillators towards the growth of large single crystals of materials with high photon yield and good proportionality. This has proven to be challenging for many of the desired materials. A significant fraction of the cost of large single-crystal inorganics is due to the substantial infrastructure and energy required for growing, post-processing, and packaging large and often hygroscopic crystals. For example, achievable crystal volumes can be intrinsically limited in noncubic systems due to the anisotropy in the thermal conductivity and the coefficient of thermal expansion, α, along different crystal axes, such as in $LaCl_3$ where $\alpha_a/\alpha_c = 2.18$ [32]. This may cause cracking of the crystal during cooling and reduce the overall yield of the growth process. Furthermore, growing crystals from salt melts is energy intensive and intrinsically expensive. Many of the desired heavy metal halide crystals are also highly hygroscopic and require special precautions during growth, postprocessing, and packaging [86], further adding to manufacturing cost. As a result, high-performance scintillators continue to be an expensive system component with crystal volumes limited to typically 100–1,000 cm^3 [69]. These cost and size constraints have inspired research into alternative scintillator materials that could offer the good energy resolution of single-crystal scintillators at significantly lower cost.

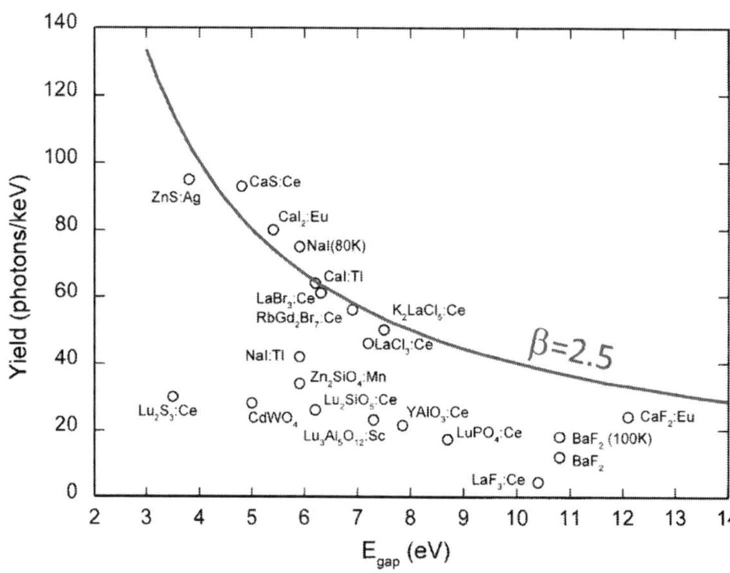

Figure 2.1 Scintillator photon yield as a function of bandgap energy for different scintillator materials. The solid line shows the ideal performance of scintillators with β = 2.5 (see Eq. 2.1). Reprinted from Ref. 21, Copyright 2002, with permission from Elsevier.

One particularly active avenue of such research pursues nanocomposite scintillators. Nanocomposites are a class of heterogeneous materials consisting of particles with nanometer-scale dimensions embedded in a solid matrix (see Fig. 2.2). A subclass of nanocomposites are optical nanocomposites that are designed such that ultraviolet (UV), visible, or near-infrared light can propagate through them with minimal optical scattering. Optical nanocomposites thus appear translucent or even transparent. Nanocomposite scintillators are optical nanocomposites that contain either organic or inorganic scintillators on the nanometer scale.

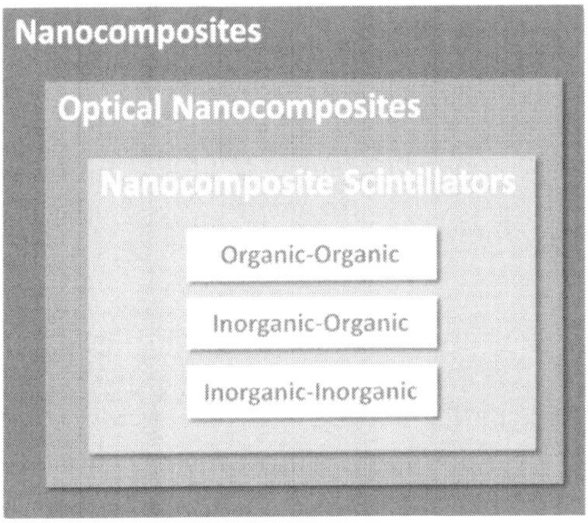

Figure 2.2 Definition of nanocomposite scintillators.

Various topics relating to nanocomposite scintillators will be examined in the following sections. In Section 2.2, the basic physical properties of scintillators will be reviewed, with focus on those that are particularly important for nanocomposite scintillators such as optical scattering (Section 2.2.1), energy deposition and energy resolution (Section 2.2.2), and energy transport (Section 2.2.3). In Section 2.3, the experimental procedures used to characterize the optical and scintillation properties of nanocomposites will be reviewed. The three classes of nanocomposite scintillators will be discussed in Section 2.4, including organic scintillating nanoparticles in an organic matrix (Section 2.4.1), inorganic scintillating nanoparticles in an organic matrix (Section 2.4.2), and inorganic scintillating nanoparticles in an inorganic matrix (Section 2.4.3). This review will conclude with a summary and an outlook in Section 2.5.

2.2 Physical Properties of Nanocomposite Scintillators

2.2.1 Optical Properties

2.2.1.1 Material requirements for radiation detectors

Many of the desired optical properties of nanocomposite scintillators are driven by the radiation detector device requirements. It is therefore helpful to briefly review the available materials, the basic structure of a radiation detector, and the processes that occur after an interaction with ionizing radiation.

The research and development of nanocomposite scintillators to date has focused primarily on materials derived from established bulk scintillators. As for bulk scintillators, high photon yield, short excited-state lifetime, good proportionality of the scintillation response, and an emission wavelength that matches the spectral responsivity of the photodetector are also desired attributes of nanocomposite scintillators. The list of potentially useful scintillators is ever growing [19, 74]. It encompasses a wide range of oxide and halide crystals that are in most cases doped with a metal ion (e.g., Ce^{3+}, Eu^{2+}, Bi^{3+}, Tl^{+}, or Pr^{3+}) that acts as the luminescent center (activator). In some cases, the crystals are undoped (self-activated), producing scintillation light via self-trapped excitons, excitons bound to isoelectronic hole traps, charge-transfer emissions, core–valence transitions, or donor–acceptor recombination. A current and extensive list of scintillator materials is being maintained by S. E. Derenzo et al. of Lawrence Berkeley National Laboratory [19]. Table 2.1 lists properties of some of the common commercially available γ-ray scintillators. While the choice of a scintillator material is generally driven by the requirements of a detection application, a variety of factors related to material growth, handling, physical properties, and cost have to be considered as well. In traditional inorganic bulk scintillators, the development efforts typically focus on the growth of large single crystals of the desired materials. In contrast, nanocomposite scintillators involve small particles of a scintillating material, leading to fundamentally different fabrication methods and challenges and thus potentially different material choices. Section 2.4 presents a detailed discussion of the various nanocomposite materials that have been considered as scintillators to date.

Table 2.1 Properties of selected commercially available γ-ray scintillators, including activator excited-state lifetime (τ), density (ρ), emission peak wavelength (λ_{em}), refractive index at λ_{em} (n), energy resolution (R), and photon yield (Y) [19, 58, 74]. Note that the energy resolution given for BC-408 plastic scintillator is somewhat misleading as there is no full energy peak. However, it serves as an estimate of the effectiveness of plastic scintillators [77].

Scintillator	τ [ns]	ρ [g/cm^3]	n	λ_{em} [nm]	R (at 662 keV) [%]	Y [photons/MeV]
$LaBr_3:Ce^{3+}$	30	5.1	2.31	380	2.9	75,000
NaI:Tl	230	3.67	1.85	415	7.1	45,000
LSO	40	7.4	1.82	420	7.9	33,000
YSO	42	4.44	~1.8	420	9.0	24,000
YAP	24	5.55	1.95	365	4.5	21,600
YAG	70	4.57	1.85	550	14.0	16,700
BGO	300	7.13	2.15	480	7.7	8,200
BC-408	10	1.032	1.58	425	~14	8,000
$LaF_3:Ce^{3+}$	30	5.9	1.63	330	16	4,500

Achieving a high photon yield in a scintillator is important because it is a key factor determining the energy resolution that can be achieved by the radiation detector. The energy resolution of a scintillation detector is defined as

$$R = \frac{\Delta E}{E_\gamma} \tag{2.2}$$

where ΔE is the full width at half maximum (FWHM) of the peak that is observed in the histogram of pulse heights when a scintillator is exposed to a monochromatic γ-ray source of energy E_γ [43]. R is commonly reported as a percentage at a particular energy (see Table 2.1). The shape of a single γ-ray emission line in the energy spectrum is typically described by a Gaussian line profile, and the

FWHM is related to the Gaussian linewidth parameter σ by FWHM = 2.35σ. As shown in Eq. (2.1), the number of photons created by a γ-ray interaction is given by the energy-transport efficiency ($1/\beta$), the energy-transfer efficiency to the activator (η_c), and the activator intrinsic quantum efficiency (η_i). Measuring the energy E_γ of an absorbed γ-ray via the number of scintillation photons it created, therefore, represents a discrete statistical process. As a result, the accuracy with which E_γ can be measured is fundamentally limited by statistical fluctuations in the process of converting one E_γ quantum into n_p optical photons. Increasing the light yield thus improves the energy resolution. Here it is assumed that counting statistics can be used to estimate the variance in the optical photon yield and that the scintillation photons are uncorrelated. The variance can be estimated by assuming that the production of scintillation photons is a Poisson process [43]. This yields $R = 2.35/\sqrt{N}$, where N is the number of counts obtained from the scintillator/photodetector device assembly. In practice, deviation from Poisson statistics is sometimes observed, and the measured energy resolution can be as much as 3–4 times better than expected. This is a result of the statistical nature of the thermalization process that follows the absorption of the γ-ray and during which a portion of the energy is converted to measureable scintillation photons while another portion is converted to thermal energy. This process is captured by the Fano factor, F, which is ≈ 1 for scintillators and around 0.15 for germanium, silicon, and gases [43]. Including the Fano factor, the energy resolution due to statistical fluctuations is given by

$$R = 2.35\sqrt{\frac{F}{N}} \tag{2.3}$$

There are a number of other factors that can degrade this fundamental energy resolution. They include thermal noise in the detector electronics, drift effects, incomplete absorption of the incident γ-ray by the material, and deviations of the scintillator response from the exact proportionality shown in Eq. 2.1 (i.e., variation of the photon yield with incident γ-ray energy [40]).

Figure 2.3 shows the primary components of a typical radiation detector that uses a scintillator: one surface of a scintillating material is optically coupled to a photodetector (e.g., photomultiplier tube), and the other surfaces are coated with a reflective material. An incident high-energy photon such as a γ-ray can deposit all or part of its energy within the scintillator. This γ-ray absorption is followed by rapid thermalization and transport of the energy in the bulk of the material $(1/\beta)$ and subsequent transfer of the energy to the activator ions (η_c). The excited activators then emit scintillation light isotropically inside the material (η_i).

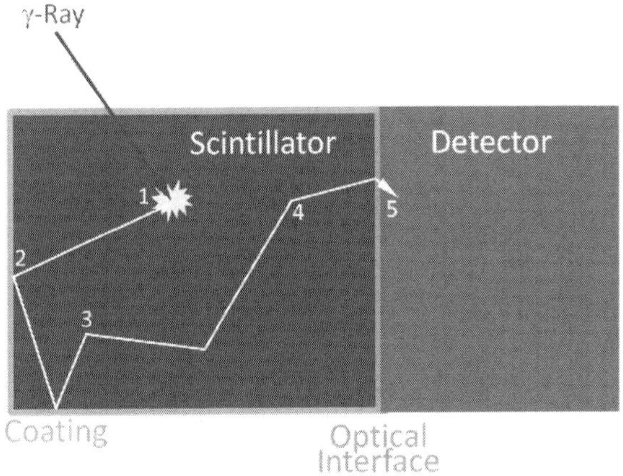

Figure 2.3 Schematic illustration of a scintillation-based radiation detector consisting of a coated scintillator material in optical contact with a photodetector. The energy deposited by a γ-ray causes emission of scintillation light (1) which can undergo reflection at the coated surfaces (2), reabsorption-reemission by the scintillator (3), and bulk scattering (4) before crossing the optical interface (5) and reaching the photodetector where it induces a photocurrent.

Several processes have to be considered as the scintillation light propagates through the material. The light can be reabsorbed by the activators themselves (self-absorption) because there typically exists a non-zero spectral overlap of the activator absorption spectrum, $\alpha(\lambda)$, and the activator luminescence spectrum, $g(\lambda)$. Reabsorption is followed either by nonradiative relaxation (i.e., loss of the electronic excitation of the activator to vibrational modes in the material) or by isotropic reemission of a photon, a competition that is determined by the quantum yield η_i. The scintillation light can also undergo reflection at the coated surfaces, and the associated losses are characterized by the coating single-surface absorption loss $\alpha_c(\lambda)$. It can also undergo scattering in the bulk of the scintillator at locations that have a discontinuity in the refractive index. Finally, the scintillation light has to exit the uncoated scintillator surface, traverse the scintillator/photodetector interface where losses can occur, and reach the photodetector where it is converted into electrical charge with efficiency $\eta_{\text{p-e}}$. Downstream electronics (not shown in Fig. 2.3) then detects the rate, and often the magnitude, of the charge pulses from the photodetector in order to determine the presence and the energy of the incident ionizing radiation. The efficiency of each of these steps contributes to the overall device performance. The more electrical charges (N) that are created by the photodetector per unit absorbed γ-ray energy the better is the energy resolution that can be achieved with the detector device [see Eq. 2.3].

In typical scintillators, the time it takes for the deposited γ-ray energy to thermalize and transfer to the activator, for the activator excited state to decay, and the emitted light to be detected by the photodetector is on the order of 20–100 ns. Such scintillation detectors ideally operate in a regime where the time between two γ-ray absorption events is long compared to this intrinsic time constant for a single event. Therefore, the scintillation events are generally well separated from each other in time (absence of pulse "pile-up"), allowing the acquisition of a γ-ray energy spectrum via the analysis of the individual detector pulses. Each scintillation event is characterized by a unique series of reabsorption, reemission, reflection, and scattering processes that give the associated scintillation light a characteristic path length L_i, which is defined as the distance between the original point of emission of the scintillation light and the final point of detection on the photodetector. The distribution of L_i and the associated average path length $\bar{L}$ is particularly important in the context of nanocomposite scintillators for two reasons. First, it determines the reabsorption loss in the scintillator (scales with $\bar{L}$) as well as the absorption loss at the coated surfaces (scales with the number of photon–surface interactions). Second, it influences the temporal shape of the charge pulse that is created by the photodetector. For a given scintillator size, a short average path length and a narrow path-length distribution are generally desired in order to achieve the best device performance. This requirement favors materials with good optical transparency for (i.e., minimal optical scattering of) the scintillation light.

2.2.1.2 Scattering in optical nanocomposites

Nanocomposite scintillators consist of scintillating nanoparticles with refractive index n_p embedded in a matrix with refractive index n_m. Note that the refractive index depends on both wavelength λ (dispersion) and temperature T (thermo-optic effect). In general, there exists a refractive index mismatch between the particles and the matrix, i.e., $n_\mathrm{p}(\lambda, T) \neq n_\mathrm{m}(\lambda, T)$. This optical heterogeneity on the nanometer scale causes light to be scattered as it propagates through the nanocomposite. Scattering thus increases the average path length $\overline{L}$ and broadens the path-length distribution. As a result, reabsorption and coating losses both increase and tend to degrade the overall detector performance. Furthermore, γ-ray energy deposited in the matrix usually does not generate any scintillation light. This may become problematic if a γ-ray deposits energy in both the non-scintillating matrix and the scintillating nanoparticles, an effect that degrades energy resolution. Therefore, one of the primary objectives of the design and fabrication of nanocomposite scintillators is the minimization of light scattering in composites with high nanoparticle loading.

Scattering of light by small particles is a well understood phenomenon and has been the topic of theoretical and experimental studies for well over a century [12]. However, quantitative descriptions of practical materials that typically consist of a collection of very many arbitrarily shaped particles are exceedingly difficult to obtain. Two important assumptions are therefore often made in order to render the problem more tractable. The first is to assume the particles to be homogeneous spheres of arbitrary radius and refractive index. The interaction of an electromagnetic wave with a homogeneous sphere allows one to expand the electromagnetic wave in terms of spherical harmonics from which the fields internal to the sphere and scattered by the sphere can be calculated. This approach is generally known as Mie's theory [73]. While many practical problems deal with nonspherical particles, it should be noted that Mie's theory is often a valuable first-order approximation even for cases involving nonspherical particles [12]. In a collection of particles, each particle is excited by the external field plus the field scattered by all the other particles, a situation that is very challenging to treat analytically. The second approximation of single scattering therefore assumes the particles to be independent of each other [12]. In that case, the total scattered field is simply the sum of the fields scattered by the individual particles, each exposed to the external field only. This is a good approximation if the concentration of particles is very low. In the case of nanocomposite scintillators however, where the nanoparticles may amount to up to half of the volume of the composite (see Section 2.2.2), this approximation may be inadequate, and multiple scattering must likely be considered by a quantitative treatment.

Mie's theory in the single-scattering approximation provides a rigorous albeit numerically laborious framework to describe light scattering in dilute optical nanocomposites. The treatment can be further and greatly simplified by assuming that the particle size is small compared to the wavelength. This regime is known as Rayleigh scattering. More precisely, the Rayleigh regime applies if $m \cdot x < 1$, where $x = 2\pi r/\lambda$ [12]. Here, r is the radius of the sphere, λ_0 is the wavelength of the light in vacuum, $\lambda = \lambda_0/n_m$, and $m = n_p/n_m$. The $m \cdot x < 1$ criterion states that the particle must be small enough such that it sees a uniform electric field at any point in time. In this case, the irradiance scattered by a single particle is

$$I_s = \frac{8\pi^4 r^6}{\lambda^4 D^2}\left(\frac{m^2-1}{m^2+2}\right)^2 (1+\cos^2\theta)I_i \tag{2.4}$$

where I_i is the irradiance of the unpolarized incident light, D is the distance between the scattering sphere and the observation point, and θ is the scattering angle (see Fig. 2.4) [12]. For example, one can assume Rayleigh scattering to be an adequate assumption for a fused silica sphere (n_p = 1.4584) with a 10 nm radius embedded in a polystyrene matrix (n_m = 1.5915) and illuminated by light with λ_0 = 589.29 nm (sodium D lines) because $m \cdot x \approx 0.16$ (neglecting any losses). Integrating Eq. 2.4 over 4π yields the single-particle Rayleigh scattering cross section

$$\sigma_s = \pi r^2 Q_{sca} = \frac{2\pi^5 d^6}{3\lambda^4}\left(\frac{m^2-1}{m^2+2}\right)^2 \tag{2.5}$$

where Q_{sca} is the scattering efficiency and d is the sphere diameter [12].

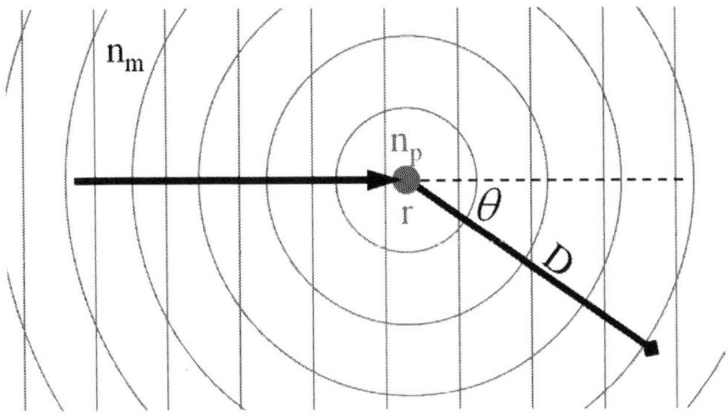

Figure 2.4 Schematic illustration of Rayleigh scattering of a plane wave propagating in a medium with refractive index $n_{\rm m}$ by a small spherical particle of radius r and refractive index $n_{\rm p}$. The distance D and angle θ between the particle and a point of observation are indicated.

Note that there would be no loss of scintillation light in a nanocomposite if the activator internal quantum yield $\eta_{\rm i}$ = 1 (i.e., each re-absorption event led to re-emission) and the coating single-surface absorption loss $\alpha_{\rm c}(\lambda)$ = 0. In this ideal case, an increasing scattering cross section $\sigma_{\rm s}$ and a corresponding increase in the average path length $\bar{L}$ for the scintillation light would merely broaden the distribution of arrival times of the scintillation photons at the detector (no photon losses). In a practical system, however, $\eta_{\rm i} < 1$ and $\alpha_{\rm c}(\lambda) > 0$, which lead to increasing loss of scintillation light with increasing $\sigma_{\rm s}$. A quantitative evaluation of the loss in this case not only depends on the microscopic material parameters but also on the geometry of the system (see Fig. 2.3) and would require a ray-tracing calculation. However, making the gross assumption that the combined loss from self-absorption and coating absorption simply scales linearly with $\bar{L}$, then the average fraction of scintillation light transmitted through the nanocomposite to the photodetector surface becomes

$$T = \exp\left\{-\sigma_{\rm s}\bar{L}N_{\rm p}\right\} = \exp\left\{-\frac{6\sigma_{\rm s}\bar{L}f_{\rm p}}{\pi d^3}\right\} \tag{2.6}$$

where T is the transmittance of scintillation light, $N_{\rm p}$ is the particle number density, and $f_{\rm p}$ is the volume fraction occupied by the particles in the medium. Combining Eqs. 2.4 and 2.6 and assuming the particles to be monodisperse then yields

$$T = \exp\left\{-4\pi^4 \frac{d^3\bar{L}f_p}{\lambda^4}\left(\frac{m^2-1}{m^2+2}\right)^2\right\} \tag{2.7}$$

Note that deviation from monodispersity, which is typical in actual nanocomposites, may be relevant because of the strong d^3 dependence. In that case both σ_s and N_p have to be weighted by the normalized particle size distribution $\rho(d)$, and the largest particles will tend to dominate the net scattering cross section. Particle agglomeration is another potential complication that is not captured within the framework of scattering by spheres. Furthermore, the scintillation light is not monochromatic but rather is comprised of a distribution of wavelengths. Therefore, σ_s has to be weighted by the normalized activator luminescence spectrum $g(\lambda)$, and the shorter wavelengths will tend to have a greater scattering cross section because of the strong λ^4 dependence.

The maximum particle size that is acceptable for a nanocomposite scintillator can be estimated from Eq. 2.7. Consider the example of spherical $LaBr_3$:5%Ce^{3+} particles in a polystyrene matrix, and assume the particles to be monodisperse, the scintillation light to be monochromatic, and multiple scattering to be negligible. The mean emission wavelength of Ce^{3+} in $LaBr_3$ is λ_0 = 370 nm [82], and the refractive indices of the particles and the matrix at λ_0 are estimated to be 2.305 [82] and 1.683 [39], respectively. Also note that $LaBr_3$ is birefringent, and the refractive index assumed here is the average of the ordinary (n_o) and extraordinary (n_e) indices. This yields $\lambda = \lambda_0/n_m$ = 220 nm and $m = n_p/n_m$ = 1.370, and Rayleigh scattering is therefore expected to be a good approximation ($m \cdot x < 1$) if the particle diameter is <50 nm. The transmittance calculated from Eq. 2.7 is shown in Fig. 2.5 for different particle volume fractions (f_p) and average path lengths ($\bar{L}$). The ranges of both of these parameters were chosen to be representative of a nanocomposite scintillator in a typical radiation detector application such as that shown in Fig. 2.3. It is evident that for this set of material parameters, particle diameters of <10 nm are required in order to achieve high transmittance of scintillation light in such a nanocomposite. A similar transmittance calculation in the Rayleigh approximation is shown in Fig. 2.6 for a nanocomposite of Ce^{3+}-doped $Y_3Al_5O_{12}$ (YAG:Ce^{3+}) in a polystyrene matrix. Transmittance of >80% is observed for nanoparticle volume fractions of up to 50% and particle sizes <6 nm. Optical scattering is less pronounced in this composite because (1) YAG:Ce^{3+} has a lower refractive index that is better matched to that of polystyrene (m = 1.081) than that of $CeBr_3$ (m = 1.370), and (2) the scintillation wavelength of YAG:Ce^{3+} is longer (550 nm) compared to that of $CeBr_3$ (370 nm), which enters the exponent of Eq. 2.7 to the 4th power. These two effects alone reduce scattering loss in a YAG:Ce^{3+} nanocomposite by a factor of ~77 compared to a $CeBr_3$ nanocomposite. The 5–10 nm desired particle size range estimated from such calculations should be viewed as a rough estimate, given the numerous and significant approximations that have been made above. Nevertheless, it provides a useful starting point that drives the design and fabrication of nanocomposite scintillator materials and also highlights the associated materials science challenges.

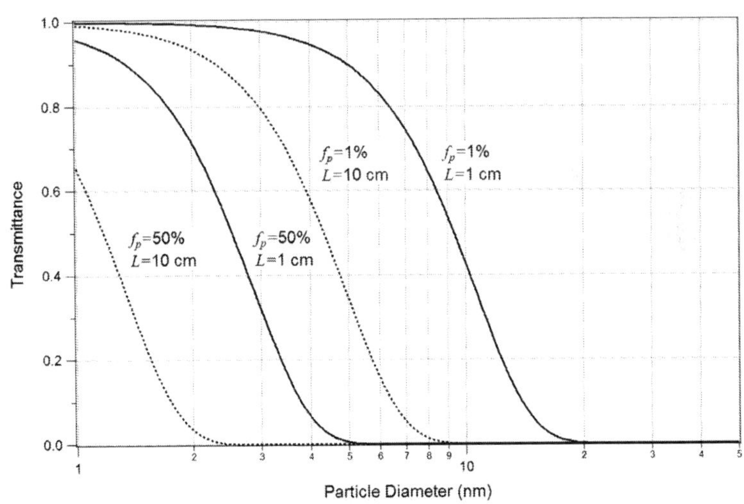

Figure 2.5 Calculated transmittance (Eq. 2.7) for monodisperse spherical $LaBr_3$:5%Ce^{3+} particles (assuming monochromatic emission at 370 nm) in a polystyrene matrix using the single-scattering approximation in the Rayleigh regime. Calculations for a particle volume fraction f_p of 1% and 50% as well as for an average path length $\overline{L}$ of 1 cm (solid curves) and 10 cm (dotted curves) are shown.

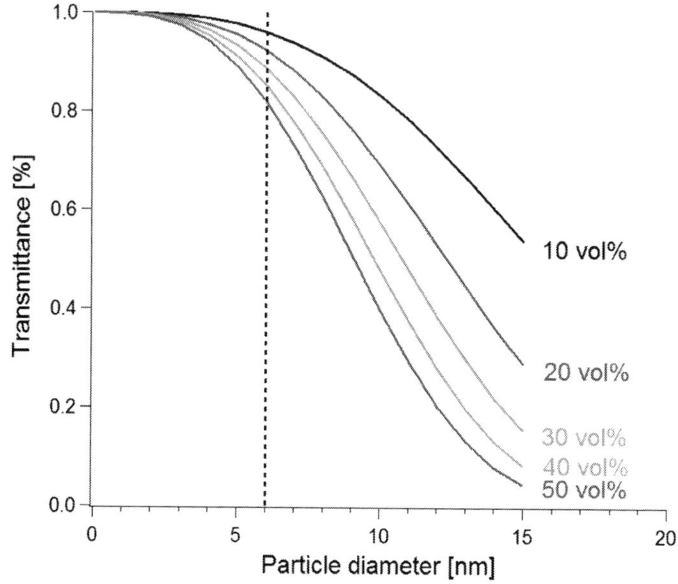

Figure 2.6 Calculated transmittance for monodisperse spherical particles of Ce^{3+}-doped $Y_3Al_5O_{12}$ (YAG:Ce^{3+}) in polystyrene assuming monochromatic emission at 550 nm and using the single-scattering approximation in the Rayleigh regime with a 1 cm optical path length.

2.2.2 Energy Deposition

In designing a nanocomposite scintillator, it is important to understand the underlying requirement for high performance as a spectroscopic γ-ray scintillator, which is different from the requirement for an imaging screen for example. For γ-ray spectroscopy, the energy deposition of individual γ-ray interactions are measured, and the respective scintillation pulse heights of these interactions are then used to deduce the incident γ-ray energy spectrum. At energies below the pair-production threshold, the energy deposition is dominated by two interactions: Compton scattering and the photoelectric effect. For Compton scattering, an incoming γ-ray interacts with an electron, producing an outgoing γ-ray of lower energy and an energetic electron in the material [43]. Because there are two incoming and outgoing particles, this interaction is possible even for free electrons, and binding energy does not have a major contribution to the scattering cross section. The Compton cross section therefore varies approximately linearly with atomic number and density. For the photoelectric effect, the incoming γ-ray is absorbed, and the full energy is transferred to the energetic electron [43]. Because the γ-ray photon is massless and the electron has mass, recoil against the nucleus is required to conserve energy and momentum. This causes the interaction to occur preferentially with the most tightly bound electrons, so that the 1S electrons dominate the scattering. This is the origin of the often-quoted $Z^{4.5}$ dependence of the photoelectric effect on atomic number. It becomes clear that for both Compton scattering and the photoelectric effect, the transduction process consists of converting energy carried by a γ-ray into energy carried by an energetic electron, followed by conversion of the energy carried by the energetic electron into excitations in the material which are then converted to optical photons. For a homogeneous system, the deposited energy can be derived from the NIST eStar database, which tabulates the energy loss per unit path length as a function of energy, dE/dx, for a wide range of materials [64].

An additional complication arises in nanocomposites where, in the basic case, the matrix does not scintillate. When a γ-ray is absorbed by the nanocomposite, the energy deposition occurs in both the non-scintillating matrix and the scintillating nanoparticles. This (i) causes a reduction in the overall photon yield compared to the respective scintillator crystal and (ii) introduces an additional variance due to partition of energy deposition in the matrix material and nanoparticles. In principle, the use of a scintillating matrix can provide some compensation for energy lost in the matrix, although likely with a lower conversion efficiency than the high Z nanoparticles in the composite. Li has carried out calculations of the fractional energy deposition in the nanoparticles relative to the total energy deposition in the material [49]. These calculations assumed scintillator nanoparticles dispersed in a scintillating (polystyrene) matrix, a polystyrene light yield equal to that of unloaded polystyrene, a nanoparticle light yield equal to that of the respective bulk crystal, and a negligible amount of ligands on the surface of the nanoparticles (often used in practice as surfactants, see Section 2.4.2). Respective calculations for $CeBr_3$ as well as YAG:Ce^{3+} scintillating nanoparticles in polystyrene are shown in Fig. 2.7.

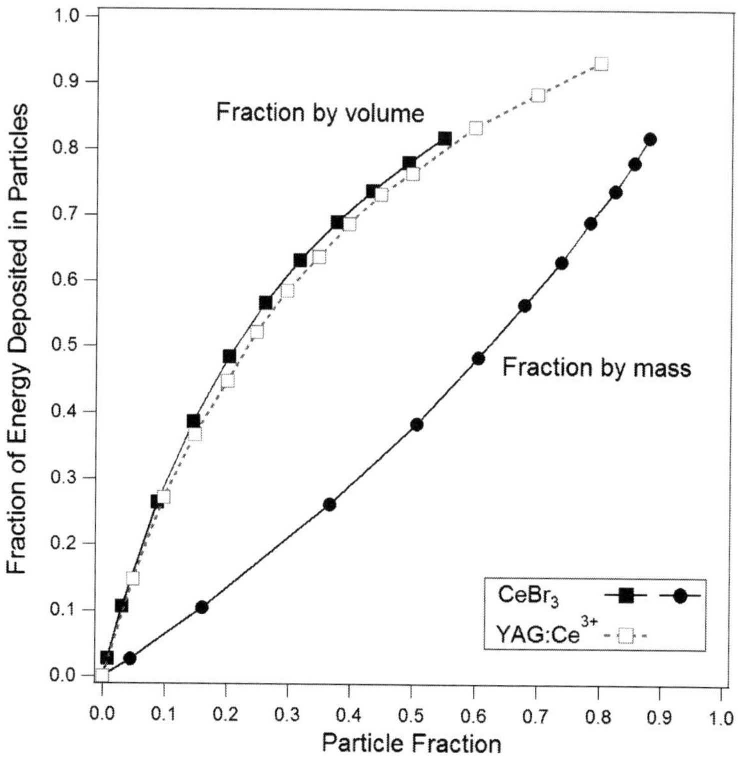

Figure 2.7 Calculated fractional energy deposition in nanocomposites of $CeBr_3$ particles (solid lines) and of YAG:Ce^{3+} (dotted line) in a polystyrene matrix.

Because the energy deposition of an energetic electron depends primarily on the electron density of the material through which it is propagating, a large fraction of energy is deposited in the nanoparticles, in comparison to their volume fraction. Note that for 50% volume fraction of nanoparticles, nearly 80% of the energy is deposited in the nanoparticles (see Fig. 2.7). This result highlights the requirement of high nanoparticle loading in order to achieve useful scintillation performance from a nanocomposite. Figure 2.8a shows the simulated energy spectrum of a ^{137}Cs source for different mass fractions of $CeBr_3$ nanoparticles in a polystyrene matrix. This calculation took into account the variance because of partition of energy deposition in the matrix material and nanoparticles. Note the shift from dominantly Compton scattering to balanced Compton and photoelectric interactions as the $CeBr_3$ nanoparticle loading increases. These calculations set the requirements for the nanoparticle loading of a composite for γ-ray spectroscopy applications. In this system, a loading of 20–50% by volume is required for good spectroscopic performance. A qualitatively similar behavior in the simulated energy spectrum of a ^{137}Cs source is obtained for a nanocomposite of $YAG:Ce^{3+}$ as shown in Fig. 2.8(b). The photon yield of $YAG:Ce^{3+}$ is ~4.5× lower than that of $CeBr_3$ (see Table 2.1), and the energy resolution that can be achieved by a $YAG:Ce^{3+}$ nanocomposite is lower than that for a $CeBr_3$ nanocomposite of equal loading. Only at very high volume fractions of $YAG:Ce^{3+}$ nanoparticles would one achieve a satisfactory detector performance. As will be shown in Section 2.4.2, achieving loadings of 20–50% by volume in a nanocomposite is technically very challenging, because the particles have to be dispersed uniformly without aggregation. Also note that 50% loading by volume of a matrix with monodisperse particles is approaching the theoretical density of close-packed monodisperse spheres (~74%). Nonetheless, transparent composites with >50% nanoparticles by volume have been fabricated [59] (see Section 2.4.2).

2.2.3 Energy Transport

The transport of energy deposited in a material by a γ-ray interaction involves the movement of both electrons (e) and holes (h) through the crystal as well as the final recombination at the luminescent center [67]. In the best case, the net transport efficiency is ~40% ($1/\beta$ term in Eq. 2.1), which is the key limiting factor for scintillator photon yield. Energy transport therefore is one of the fundamental aspects that determine the overall scintillator performance. Furthermore, energy-transport processes determine the response function of the scintillator, i.e., the proportionality of the number of scintillation photons produced per γ-ray energy absorbed. Despite its importance, the details of energy transport in scintillators are still not well understood [67]. The lack of understanding is largely due to the complex interplay of a variety of processes, including e–h densities along ion tracks [51, 52], trapping sites within the bandgap of the crystal [67], carrier dynamics within the crystal [51, 52], nonluminescent pathways for energy dissipation [45], and the luminescent center lifetime [67]. All of these factors have at least a theoretical dependence on the crystal dimensions, although little direct research and results on energy transport in nanocomposite scintillators is available in the open literature.

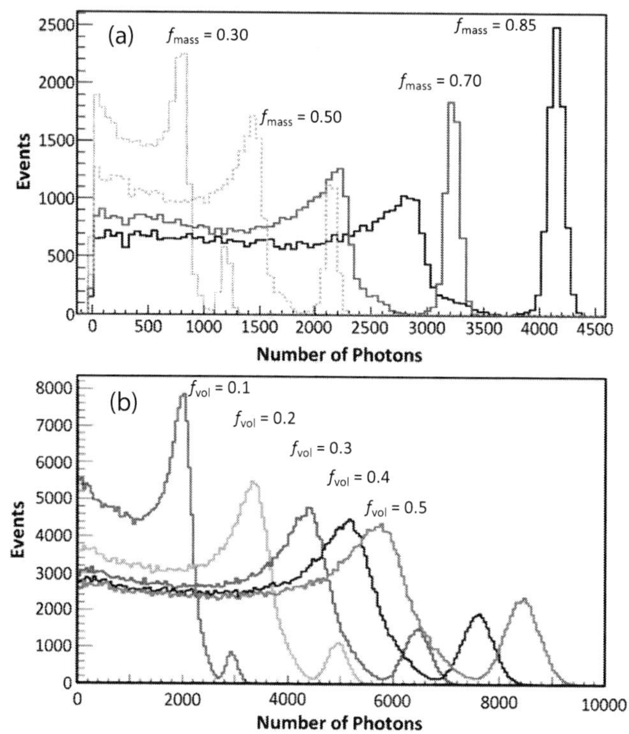

Figure 2.8 Simulated energy spectra of a ^{137}Cs source (dominated by a 662 keV γ-ray) obtained from nanocomposites of (a) $CeBr_3$ particles in a polystyrene matrix for different nanoparticle mass fractions and (b) CeF_3 particles in a polystyrene matrix for different nanoparticle volume fractions.

The origin of nonproportionality, i.e., the deviation from the ideal $n_p \propto E_\gamma$ (see Eq. 2.1), is still unclear and an area of active research [35, 40]. Nonproportionality is more prominent in insulating scintillators than semiconductors. A series of measurements of scintillator nonproportionality led Payne and coworkers at LLNL [70, 71] to describe the experimental data with two parameters: one for exciton–exciton (dipole–dipole) annihilation based on the

Birks equation [8] and one for the recombination branching ratio based on the Onsager mechanism [28]. On the basis of this theory, Fig. 2.9 illustrates some of the more important interconnections between the multitude of energy-transfer processes affecting the light yield inferred from the electron response curves. The orange boxes describe the physical states induced by high energy electron relaxation, green boxes describe receivers of the electron/exciton energy, and red and yellow boxes denote radiative and nonradiative relaxation processes, respectively. The branching ratio between free electrons and holes versus free exciton formation is denoted by $\eta_{e/h}$. Reducing the size of the crystal will eventually limit the distance free carriers can travel, which increases the local carrier density. This would favor the formation and excitation of the activators by free excitons. This is to illustrate that particle size may affect the relative rates of the processes shown in Fig. 2.9, which could be beneficial or detrimental to the scintillator photon yield depending on the actual rates. Unfortunately, these parameters currently elude measurement, limiting the ability to directly test the accompanying theory.

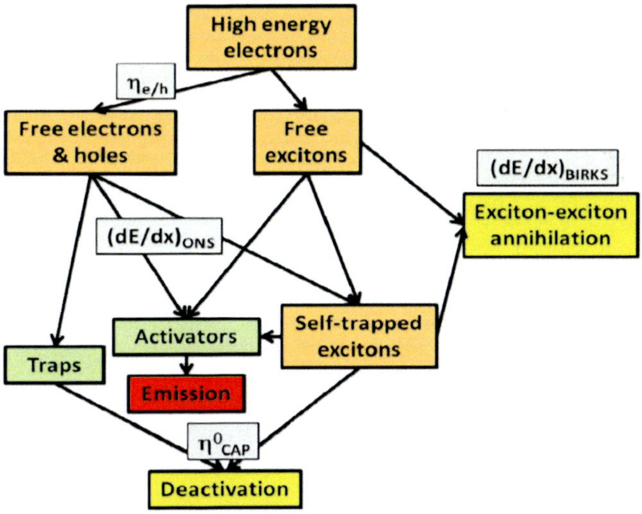

Figure 2.9 Relationship between fundamental processes occurring in scintillators with the parameters of the model in Payne et al. Reprinted from Ref. 71, Copyright 2011, with permission from IEEE.

In an attempt to find measurable parameters for nonproportionality, Li et al. [53] developed a model to measure nonproportionality in terms of the electron and hole mobilities, quantities which have traditionally not been measured for insulator materials. Nonproportionality was found to be due to different mobilities of electrons and holes, with larger discrepancies resulting in greater nonproportionality. Using this general approach, Williams et al. [87] found for CsI that the branching ratio between free carriers and excitons varied significantly as a function of initial e–h density. Excitons were readily formed at high e-h density despite slow hole mobility. However, at lower ionization densities, the disparity in the electron and hole mobilities led to free carriers. Figure 2.10 shows calculated carrier diffusion profiles for CsI:Tl [87], showing that the radial extent of the electron density increases from an initial 5–8 nm to ~50 nm within 9.5 ps. If the crystal size in this example were reduced to say 10 nm, the electron (hole) distribution at the 9.5 ps relaxation time would look much more similar to the initial distribution. As a result, the carriers would be confined to a smaller volume, exciton formation would be favored, and the scintillator photon yield could be affected. Summarizing work in the Williams group at Wake Forest [51–53, 87], Li et al. [53] identified two main causes for nonproportionality and absolute photon yield in scintillators: (1) competition between carrier diffusion and carrier concentration-dependent quenching, and (2) the difference between electron and hole mobilities. The summary depiction of these phenomena is shown in Fig. 2.11 [53]. It organizes the various physical processes driving energy transport into a logical sequence or "decision tree" and illustrates the hierarchy of energy-transport processes that affect light yield and proportionality. The Onsager energy loss mechanism $(dE/dx)_{\mathrm{ONS}}$ dominates in the limit of materials with low phonon energy (ω_{LO}), whereas the Birks energy loss mechanism $(dE/dx)_{\mathrm{BIRKS}}$ dominates in the limit of high ω_{LO}.

Bizarri and Dorenbos presented detailed experimental and theoretical studies of the charge carrier and exciton dynamics in LaX_3:Ce compounds (X = Cl or Br) [11]. They found a strong dependence of the Ce^{3+} decay on both temperature and Ce^{3+} concentration and postulated the competition of three processes: (1) prompt capture of primary charge carriers by Ce^{3+}, (2) thermally activated energy transfer from self-trapped excitons (STE) near Ce^{3+} ions, and (3) slow thermally activated migration of STEs towards Ce^{3+} ions followed by fast energy transfer. The authors were able to explain the single decay mode at room temperature (STE diffusion was faster than the Ce^{3+} lifetime) and the Ce^{3+} concentration dependencies. However, this model also implies that smaller crystal dimensions would tend to suppress long-range STE migration and would overall enhance the STE→Ce^{3+} energy transfer rate. Biswas and Du performed density functional calculations on Ce^{3+}-doped elpasolites (Cs_2YLiCl_6:Ce^{3+}) to gain insight on the energy transport and scintillation properties of this material [10]. They suggested that the higher electronegativity of Ce^{3+} compared to Cs^+ and Li^+ combined with the relatively large (7.4 Å) distance between rare earth ions to be the root cause for the slow carrier transport to Ce^{3+} ions in this material. Reduction of the crystal size should confine the carriers to a smaller volume, and it is conceivable that this would result in a higher Ce^{3+} excitation rate compared to bulk Cs_2YLiCl_6:Ce^{3+}.

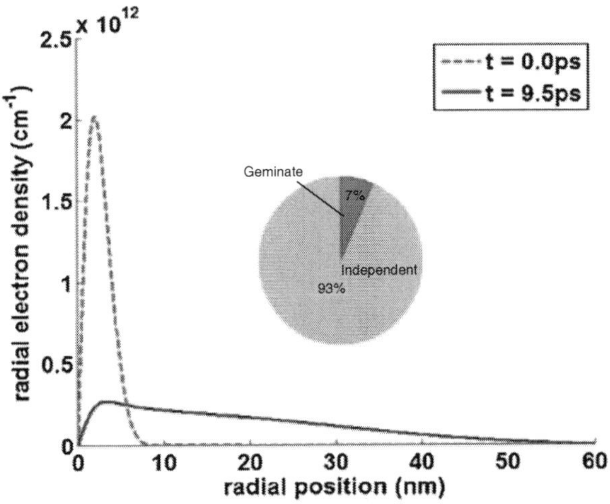

Figure 2.10 Calculated radial electron density representing the number of electrons (holes) in a cylindrical shell at radial distance r, per shell thickness Δr and unit length. The carrier diffusion profile was calculated for CsI:Tl at excitation densities of 2.5×10^{18} e–h/cm^3. Reprinted from Ref. 87, Copyright 2011, with permission from John Wiley & Sons.

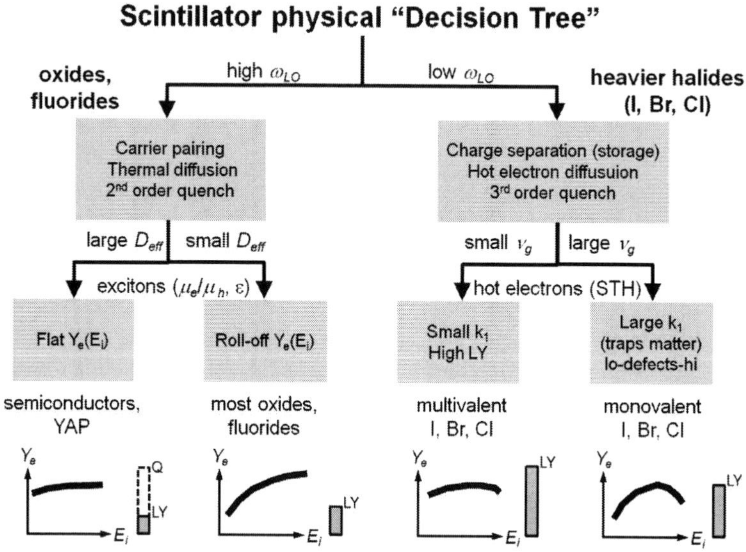

Figure 2.11 Qualitative influence of the phonon energy (ω_{LO}), band-edge diffusion coefficient (D_{eff}), and hot-electron velocity (v_g) on the electron energy response (curves) and light yield (LY), illustrated along the bottom. ω_{LO} determines the electron thermalization rate: fast on the left side with thermalized diffusion versus slow on the right side with hot free carriers. Subsequent branching in the thermalized case (left) is governed by D_{eff}, whereas the non-thermalized case (right) is controlled by v_g. Reprinted from Ref. 53, Copyright 2012, with permission from John Wiley & Sons.

The surface-to-volume ratio and thus the importance of surface phenomena greatly increases as the crystal size is reduced to the nanometer scale. The termination of the crystalline periodicity at the surface leads to a surface structure that differs from the bulk. The accompanying lower-energy surface state may facilitate nonradiative recombination. It may also affect the oscillator strength of electronic transitions of activators near the surface via an altered local electric field. Typical diffusion lengths of free carriers in ionic crystals [34] are on the order of 100 nm in alkali halides [5, 75] and tens of nm in oxides and oxysulfides [61]. For particle sizes in the 5–50 nm range, therefore, the surface is well within reach of free carriers and is expected to play a significant role in energy transport and relaxation processes. Jacobsohn et al. have prepared nanoparticles of LaF_3:Eu (4.4 ± 0.8 nm), CaF_2:Eu (10 ± 2 nm), and BaF_2:Ce (18 ± 3 nm) and found efficient scintillation despite the inherently large surface-to-volume ratio of these powders [34]. Other evidence also indicates that nanophosphors can be doped at a higher activator concentration before concentration quenching becomes important [63], which would enable more efficient energy transport in nanophosphors compared to the respective bulk material. This is an active area of research and more studies are needed to identify the type and relative importance of surface-induced processes.

2.3 Experimental Methods

Material characterization plays a key role in the development of new scintillator materials in general and nanocomposites in particular. The most important parameters include the optical absorption and luminescence spectrum of the activator, the activator excited-state lifetime, the scintillator stopping power, the photon yield, and ultimately the energy resolution. The overall efficiency of converting an incident γ-ray to scintillation photons, η, is given by a sequence of processes

$$\eta = \eta_a \times \eta_t \times \eta_c \times \eta_i \times \eta_e \tag{2.8}$$

where the individual efficiencies describe the absorption of the γ-ray by the material (η_a), the thermalization and transport of energy to create electron–hole pairs ($\eta_t = 1/\beta$), the capture of electron–hole pairs by the activator (η_c), the activator intrinsic quantum efficiency (η_i), and the efficiency of extracting the scintillation light from the bulk of the material (η_e). Different subsets of efficiency factors in Eq. 2.8 can be probed with different types of excitation. Direct optical excitation of the activator (photoluminescence) primarily probes $\eta_i \times \eta_e$, excitation with soft X-rays (radioluminescence) primarily probes $\eta_c \times \eta_i \times \eta_e$, and ultimately excitation with γ-rays provides information on all of the processes combined. Probing these subsets of efficiencies is also important for determining the causes of the overall good or bad response of a material to γ-radiation. It is impractical, however, to perform all these measurements for each material under consideration. One therefore usually employs a staged approach in which materials are first rapidly screened by optical spectroscopy, promising candidates are then characterized by radioluminescence spectroscopy, and finally larger samples of the best materials undergo more extensive characterization as scintillators using γ-ray excitation. This approach is particularly effective when large numbers of samples are under consideration, such as is often the case in the development of nanocomposite scintillators.

2.3.1 Optical Characterization

The most commonly performed measurements are photoluminescence (PL) spectroscopy, photoluminescence-excitation (PLE) spectroscopy, and time-resolved PL spectroscopy. In PL spectroscopy, the luminescence spectrum is measured with the material being excited by an optical source, i.e., an excitation that causes a local electronic excitation of the activator without ionizing the material. In PLE spectroscopy, the PL intensity of a selected radiative transition is measured as a function of the wavelength of the excitation light. The two measurements provide information on the optical absorption and emission transitions of the activator itself. An example is shown in Fig. 2.12 for Eu^{2+}-doped BaFCl crystals suspended in oleic acid. Electronic absorption and emission transitions between the 4f ground state and the 5d excited states can occur on the Eu^{2+} activator, and the respective excitation peak at 334 nm (29940 cm^{-1} or 3.71 eV in energy units) and emission peak at 399 nm (25060 cm^{-1} or 3.11 eV in energy units) are observed in the PLE and PL spectra. The energy difference between the maxima in PLE and PL is referred to as the Stokes shift, and it is 4880 cm^{-1} or 0.6 eV in this case. The Stokes shift, together with the spectral width of the optical transitions, is important because it determines the spectral overlap of the absorption $\alpha(\lambda)$ and luminescence $g(\lambda)$ spectra and thus the amount of self-absorption of scintillation light in the material [46, 83]. Self-absorption occurs at wavelengths for which the PL and PLE spectra overlap, which corresponds to the 350–390 nm range in the case of BaFCl:Eu^{2+} as shown in Fig. 2.12. The magnitude of self-absorption, together with the activator intrinsic quantum efficiency, determines the efficiency with which scintillation light can be extracted from a material. A large Stokes shift and correspondingly small self-absorption, combined with a high quantum efficiency, maximizes the amount of scintillation light that can emerge from a bulk sample. This is particularly important for nanocomposites where residual optical scattering increases the average path length $\overline{L}$ and thus self-absorption losses of the scintillation light compared to a single crystal with minimal scattering.

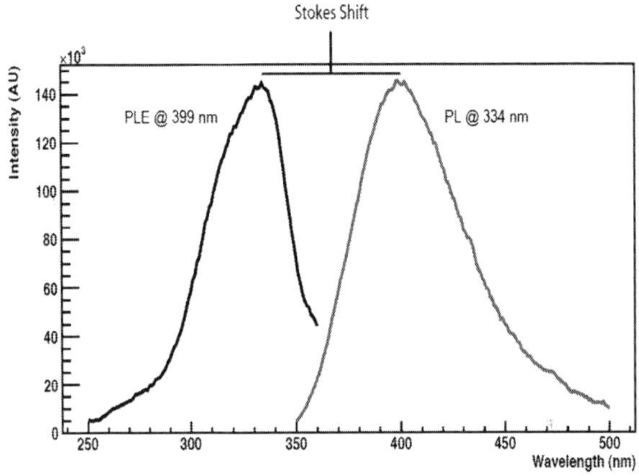

Figure 2.12 Photoluminescence (PL) and photoluminescence excitation (PLE) spectra of BaFCl:Eu crystals in oleic acid [49].

The measurement of the activator excited-state lifetime is another important optical characterization technique. Typical excited-state lifetimes of activators such as Ce^{3+}, Eu^{2+}, Pr^{3+}, Tl^{+}, and Bi^{3+} are in the range from tens to hundreds of nanoseconds [19] and are typical for parity-allowed electronic transitions such as the 5d→4f transitions in Ce^{3+}, Eu^{2+}, and Pr^{3+} considered here. The lifetime is measured by optically exciting the activator with a short (typically <10 ns) pulse and measuring the associated transient response of the activator luminescence. An example of such a transient is shown in Fig. 2.13 for the case of Ce^{3+}-doped LaF_3 nanoparticles [68]. In this case, a pulsed ultraviolet (UV) light emitting diode (LED) was used to excite the 4f→5d transition of Ce^{3+}. The resulting 5d→4f photoluminescence shows a rise time that is limited by the finite temporal response of the excitation source followed by a slower decay. In this sample, the 5d excited-state decay was found to be bi-exponential with time constants of 20.9 ns and 35.4 ns, respectively. Time resolved data such as this is important because it is usually the activator lifetime that determines the maximum count rate the scintillator/photodetector assembly can be used for.

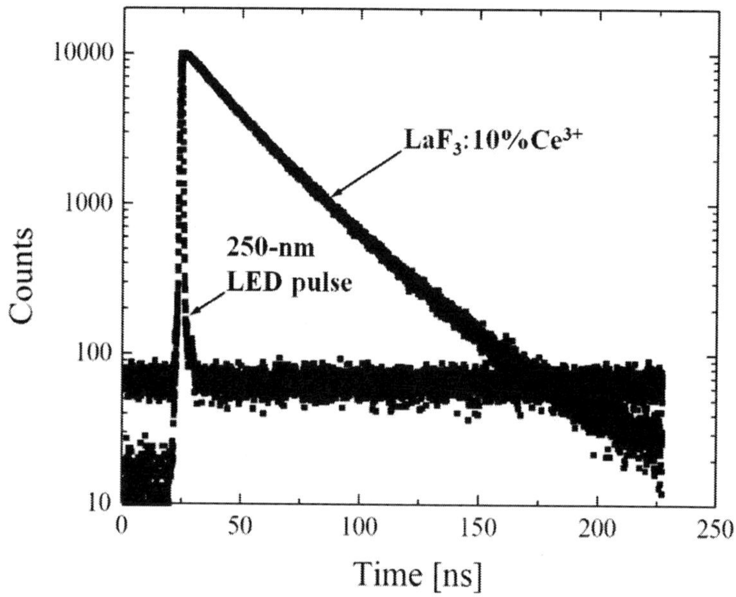

Figure 2.13 Excited-state decay transient of Ce^{3+} in 13±4.5 nm diameter nanoparticles of LaF_3. Also shown in the short pulse of a 250 nm UV LED that was used to excite the sample. Reprinted from Ref. 68, Copyright 2009, with permission from SPIE.

Using a scintillator in a mode that counts individual γ-ray energy deposition events requires complete decay of the system response before the next γ event is recorded. If the γ-ray absorption rate is comparable or higher than the activator decay rate then the pulses may start to "pile up" and will degrade the energy resolution of the system. It should also be noted that a lengthening of the measured lifetime is often observed in systems with significant self-absorption and high activator quantum yield. This is because each reabsorption/reemission event essentially restarts the decay and produces a new decay having the intrinsic decay constant. This effect can become particularly pronounced for long optical path lengths. For example, up to 40% lengthening of the observed excited-state lifetimes (relative to the intrinsic activator lifetime) were observed in Yb^{3+}-doped YAG crystals when the optical path length for the luminescence light was excessively long [27]. Similar effects are likely active in nanocomposites where scattering tends to increase the average path length for the scintillation light. However, no such measurements have been reported yet.

2.3.2 Radioluminescence Characterization

Radioluminescence (RL) spectroscopy measures the luminescence spectrum of a sample under X-ray irradiation. X-ray energies used for such experiments are typically in the tens of keV range and are obviously ionizing, however they are still 10–100 times lower than the γ-ray energies usually encountered in many scintillator applications. RL is one of the most frequently used and most important characterization techniques in scintillator development. Good RL response is a necessary, but not sufficient, indicator that a sample will scintillate in response to γ-rays or neutrons. In a typical RL experiment, X-rays generated by an X-ray tube are directed onto a sample, and the resulting scintillation light is collected by an optical fiber and analyzed in a spectrometer. Fig. 2.14 shows an example of an RL spectrum of BaFCl:Eu^{2+} nanocrystals mixed with tributyl phosphate and cyclohexane [49].

RL spectroscopy can also be used quantitatively to determine the photon yield of scintillators under X-ray excitation. Generally, the photon yield measured using γ-ray excitation is greater than using X-ray excitation [41]. An example of the energy dependence of the photon yield in several Ce^{3+}-doped bulk scintillators is shown in Fig. 2.15. These materials exhibit an overall decreasing photon yield as the energy of the incident ionizing radiation is decreased from the γ-ray range (>100 keV) to the X-ray range (0.1–100 keV). This behavior is observed for a wide range of scintillator materials [42]. The photon yield measured under X-ray excitation is therefore a useful first indicator for the performance of the material under γ-ray excitation. An RL spectroscopy setup such as the one described above may be used to determine the absolute X-ray photon yield of a scintillating sample. This can be done by comparing the total radioluminescence intensity of a test sample with that of a reference sample with known photon yield [46]. The RL counts recorded by the spectrometer, C, are given by

$$C = \Omega \times E_{\text{dep}} \times \Delta t \times Y \tag{2.9}$$

where Ω is the light collection efficiency of the spectrometer, E_{dep} is the total X-ray energy deposition rate in the sample, Δt is the X-ray exposure time, and Y is the light yield of the sample. Referring to Fig. 2.14, C corresponds to the total area under the curve. It should be noted that E_{dep} is a rate, so the total amount of energy deposited in the sample during the experiment is $E_{\text{dep}} \times \Delta t$. In essence, Eq. 2.9 assumes that all of the energy deposited in the scintillator will be converted to photons in proportion to a conversion factor Y, and that the photons are detected with efficiency Ω. Therefore, by comparing the RL intensities of two samples measured using the same system, the light collection efficiencies cancel, allowing the ratio of count rates to be expressed as a function of light yields and energy deposition:

$$\frac{C_1}{C_2} = \frac{Y_1 E_{\text{dep}}^{(1)}}{Y_2 E_{\text{dep}}^{(2)}} \tag{2.10}$$

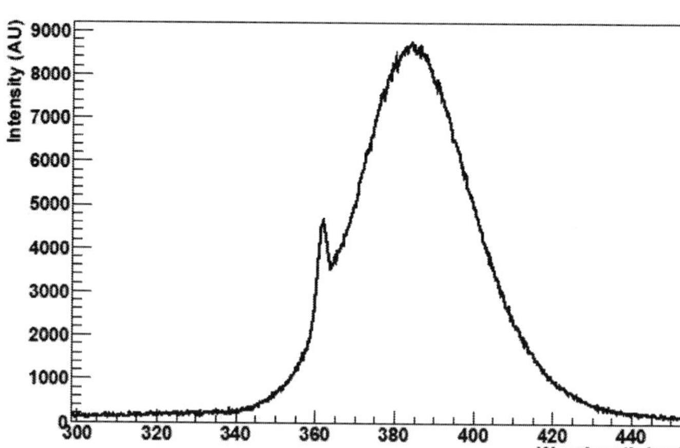

Figure 2.14 Radioluminescence (RL) spectrum of a sample containing BaFCl:Eu^{2+} nanocrystals mixed with tributyl phosphate and cyclohexane [49].

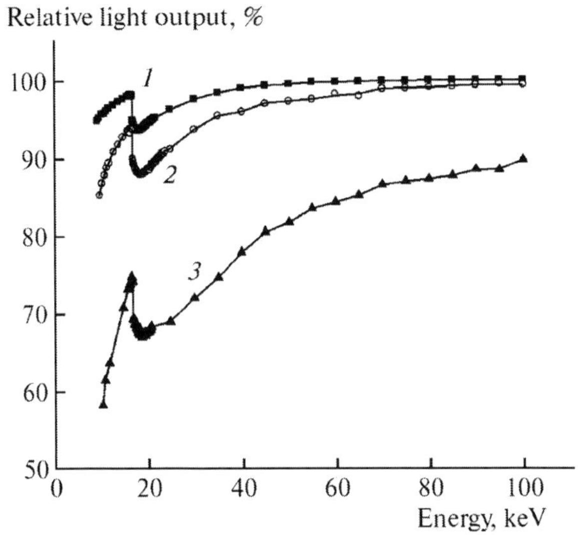

Figure 2.15 Scintillation light output of (1) YAP:Ce, (2) YSO:Ce, and (3) YPO:Ce bulk scintillators normalized to the scintillation light output at for irradiation with 662 keV γ-rays [41]. Reprinted from Ref. 41, Copyright 2012, with permission from Pleiades Publishing, Ltd.

The light yield of a test sample can then be determined using Eq. 2.10 and a reference sample. Inorganic scintillators such as NaI and $CeBr_3$ have high effective atomic numbers (Z_{eff}), and complete absorption of X-rays in the material and thus unity energy deposition rates can usually be assumed. With low Z materials such as liquid or plastic scintillators however, energy deposition is incomplete and has to be taken into account explicitly. This can be accomplished by performing Monte Carlo simulations of X-ray attenuation. Monte Carlo codes such as Geant4 [1, 3] and MCNP [88] have been used to calculate the energy deposition in low-Z scintillators. Such calculations require exact specification of the X-ray emission source spectrum, the RL spectrometer geometry, and the chemical composition of the scintillator. Once set up, Monte Carlo simulations can quickly provide accurate energy deposition values. Another option is to perform an X-ray attenuation experiment for the given sample. The respective experimental setup is relatively straightforward and requires three RL measurements as shown in Fig. 2.16: one with a reference sample (measurement 1), one with a test sample (measurement 2), and one with both, where the test sample is placed between the X-ray source and the reference sample (measurement 3). The energy deposition rate can be determined by comparing the RL intensities of measurements 1 and 3, provided that a reference sample with known photon yield is used. It is assumed that the difference in energy deposition rates between the two measurements is due to X-ray absorption in the test sample, where the relative energy deposition rate in the test sample is given by

$$\frac{E_{\text{dep}}^{\text{Test Sample}}}{E_{\text{dep}}^{\text{Reference Sample}}} = 1 - \frac{E_{\text{dep}}^{\text{Measurement 3}}}{E_{\text{dep}}^{\text{Measurement 1}}} = 1 - \frac{C_{\text{Measurement 3}}}{C_{\text{Measurement 1}}} \tag{2.11}$$

Using Eqs. 2.9 and 2.10, measurements 1 and 3 can then be directly compared to find the photon yield of the test sample.

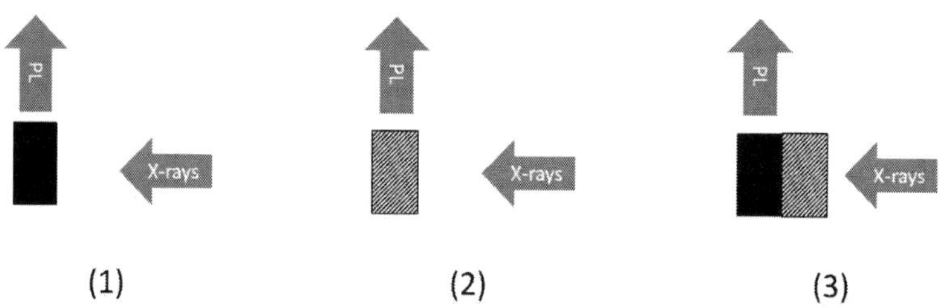

Figure 2.16 Schematic showing the three measurements that are performed to determine the X-ray attenuation of a test sample (shaded) using a reference sample (solid).

A similar approach can be used to measure the photon yield of nanocomposite scintillators. A common question that arises is how the photon yield of the nanoparticles in the composite compares to the photon yield of their bulk crystal analog. This introduces the additional complication of partition of the energy deposition between the scintillating nanoparticles, the matrix, and possibly the ligands used to functionalize the surface of the nanoparticles. Explicit quantification of the energy deposited in the nanoparticles is needed in this case. Li et al. have proposed and demonstrated a two-part approach [49, 50]. Their technique first uses Monte

Carlo simulations to determine the energy deposited in the bulk scintillator in an RL measurement, and it then partitions this energy into separate host, ligand, and crystals portions. The calculated amount of energy deposited in the nanoparticles can then be used with Eq. 2.11 to find the photon yield of the nanoparticles.

2.3.3 Gamma-Ray Response Characterization

The γ-ray response is measured by exposing the sample to a γ-ray source and recording a pulse height spectrum from the signal collected from the sample/photodetector assembly. ^{137}Cs, ^{60}Co, and ^{57}Co are sources that are typically used for such experiments. In most cases, the sample is mounted directly on a photomultiplier tube (PMT) using optical grease to ensure good optical coupling between the sample and the PMT. Data acquisition systems can take various forms; one example is shown in Fig. 2.17 (left). In this system, the signal from the PMT is split using a fan-in fan-out board. One output of the fan-in fan-out is used as a trigger signal for the charge to digital converter (QDC). A discriminator is used to set a minimum pulse height for triggering. A dual gate generator is used to delay the trigger signal and to set the length of the trigger. Another output of the fan-in fan-out is used as the signal input for the QDC. A long signal cable is used to synchronize the arrival time of the signal and the corresponding trigger. The pulse height spectrum recorded with such a setup can be used to determine the light yield and the energy resolution of a test sample. Figure 2.17 (right) shows the example of a pulse height spectrum recorded from a sample containing CeF_3 nanoparticles in a liquid scintillating matrix in response to γ-rays from a ^{137}Cs source (γ-ray energy of 661.7 keV) [78]. The curve in the figure is a fit to the Compton edge, and the "photopeak" was used to convert QDC channels to energy. The ^{137}Cs Compton edge is seen at 477.4 keV and a weak photopeak is observed at 661.7 keV. In this case, the matrix material itself was chosen to be scintillating and thus contributed to the pulse height spectrum of the nanocomposite. This makes it challenging to distinguish the individual contributions. The liquid scintillator matrix was likely responsible for most of the Compton continuum while the "photopeak" is a result of the high-Z nanoparticles in the composite.

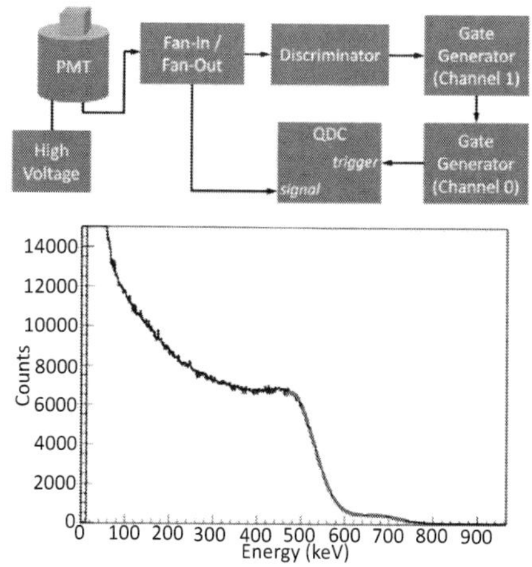

Figure 2.17 Above: Schematic of a setup for γ-ray response characterization of a sample (shown as cube). Below: Pulse height spectrum of CeF_3 nanoparticles in liquid scintillator in response to 661.7 keV γ-rays from a ^{137}Cs source [78].

2.4 Nanocomposite Materials

2.4.1 Organic–Organic Composites

Standard plastic scintillators typically consist of a base of polystyrene (PS) or polyvinlytoluene (PVT) doped with organic fluorophores such as POPOP (1,4-bis(5-phenyloxazol-2-yl) benzene) and PPO (2,5,-diphenyloxazole). They are produced from a solution of the appropriate organic dye mixture in a monomer, such as styrene, which is then polymerized [62]. Plastic scintillators are widely available commercially in numerous configurations of size, shape, and fluorophore composition. The fluorophores are dissolved in the matrix at the molecular level, and such materials therefore fall outside the definition of nanocomposites. Rather, they are the organic analog of an inorganic crystal that is doped with a luminescent metal ion (such as Ce^{3+} or Tl^{+}).

The class of materials consisting of organic nanoparticles in an organic matrix has not received much attention for scintillation applications because there is no obvious performance advantage of dispersing low-*Z* plastic scintillator nanoparticles in a low-*Z* matrix compared to using standard plastic scintillators. However, there are some reports of organic–organic composites at the millimeter scale. Karavaeva et al. incorporated 0.5–2.5 mm sized crystals of stilbene or *p*-terphenyl into a two-component organosilicon scintillating matrix [33, 38]. The fluorophore crystals were doped with a 0.1% of 1,4-diphenyl-1,3-butadiene, and the resulting composites were found to be effective scintillators for fast neutron detection. While these materials are composites, the micrometer-scale size of the single crystals dispersed in the matrix excludes them from the class of nanocomposites. Another example was reported by Brown et al. who used PPO and POPOP doped polystyrene and polyvinyltoluene nanoparticles in scintillation counting as colloidal suspensions [13]. But also this material does not cleanly fit in the organic–organic composite category since the colloids were dispersed in water.

2.4.2 Inorganic–Organic Composites

The exploration of nanocomposite materials containing inorganic materials in an organic polymeric matrix has gained considerable attention during the last decade. Over 500 reviews alone have been published on the topic of "polymer nanocomposites". Solid materials include carbon nanotubes, mineral nanoparticles, metal and alloy nanoparticles, and numerous luminescent nanoparticles such as quantum dots and rare earth–doped oxides, oxyhalides, and halides. Such materials are being pursued for energy storage and harvesting, for solid state lighting applications, and as structural materials. The appeal of this concept is not hard to imagine; nearly any desirable material property of a polymeric system can be influenced by the incorporation of a second phase of a dispersed material. A review of some general concepts regarding nanocomposite production including optical considerations can be found in Ref. [54].

Potdevin et al. developed a nanocomposite comprised of Tb^{3+}-doped $Y_3Ga_5O_{12}$ (YGG) and poly-N-vinylpyrrolidine [72]. Their focus was on solid state lighting applications, but they discuss the concept of high loading fractions in an optically transparent film. As shown in Fig. 2.18, a highly transparent nanocomposite loaded at 33 wt% (Fig. 2.18a) is shown to have good emission response when exposed to UV excitation (Fig. 2.18b). The cast films were made to a thickness of ~350 μm and were quite flexible. Furthermore, the emission of the nanocomposite was identical to the emission of the neat YGG:Tb^{3+} powder, indicating that the incorporation of the nanoparticles into the polymeric film had no measurable impact on the local environment of the Tb^{3+} activators.

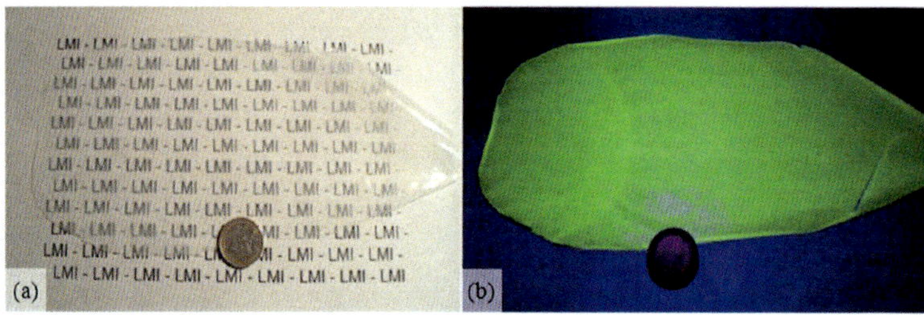

Figure 2.18 (a) As-cast film of YGG:Tb-loaded PVP in ambient light, and (b) the same film exposed to UV excitation [72]. Reprinted from Ref. 72, Copyright 2012, with permission from American Chemical Society.

Tan et al. developed a method of casting thin films of highly loaded CeF_3:Er^{3+},Yb^{3+} into both polymethyl-methacrylate (PMMA) and polystyrene (PS) [80]. This work very clearly illustrates the importance of refractive index matching for improving optical transparency. As shown in Section 2.2.1.2 (see Eq. 2.7), the ratio of refractive index of the particles and matrix, $m = n_p/n_m$, is an important factor that determines the scattering cross section and thus the overall optical transmission through a given specimen. From Ref. [80], a ratio of refractive indices (at 500 nm) between the particles and the matrix of $m = 1.016$ for PS and $m = 1.087$ for PMMA is found. Thus one would expect that equivalent volume loadings of the CeF_3:Er^{3+},Yb^{3+} nanoparticles into the respective polymers yielded different optical transparencies. As shown in Fig. 2.19, this was indeed observed; the CeF_3/PS nanocomposite had significantly less scattering than the CeF_3/PMMA nanocomposite at loading fractions of up to 20 vol%.

The necessity to achieve high (ideally 10–50%) volume fractions of particles in nanocomposites for scintillation applications has been highlighted in Section 2.2. The challenges associated with loading a nanocomposite to such levels are many and varied. While there have been numerous articles published with the concept of optically active nanocomposites, there are relatively few research articles published on this particular topic with regards to radiation detection. McKigney et al. developed a method for manufacturing a highly loaded LaF_3:Ce nanocomposite that showed spectroscopic response to ^{137}Cs gamma rays [59, 60]. The method afforded a ~1 cm^3 transparent composite loaded to about 10–12 vol% that was then measured for energy response. X-ray diffraction measurements confirmed phase-pure LaF_3 and CeF_3 with no indication of any crystalline oxides. As shown in Fig. 2.20, the measured response to γ-irradiation with a ^{137}Cs source showed a characteristic photopeak with an estimated energy resolution of 16.5%, which is comparable to the energy resolution of single crystal CeF_3. Figure 2.20 also shows the energy response of an unloaded polymer for comparison, clearly demonstrating that the photopeak at 662 keV is due to the CeF_3 nanoparticles in the composite. Using these methods, McKigney et al. have fabricated a nanocomposite containing <10 nm diameter CeF_3 particles with >50 vol% in an organic matrix [23, 59, 60]. As shown in Fig. 2.21, this nanocomposite had a high degree of transparency. Furthermore, it produced clear photopeaks upon irradiation with ^{241}Am and ^{57}Co sources. The respective energy resolutions achieved in this experiment were in part limited by the small sample size (incomplete energy deposition) as well as the rather low intrinsic photon yield of CeF_3.

Figure 2.19 As-cast films of a CeF_3:Er^{3+},Yb^{3+} loaded into (a) PPMA and (b) PS at volume fractions ranging from 1% to 20% [80]. It is evident that the PS system is far more transparent that the PMMA system, consistent with the predictions of the Rayleigh scattering theory. Reprinted from Ref. 80, Copyright 2010, with permission from American Chemical Society.

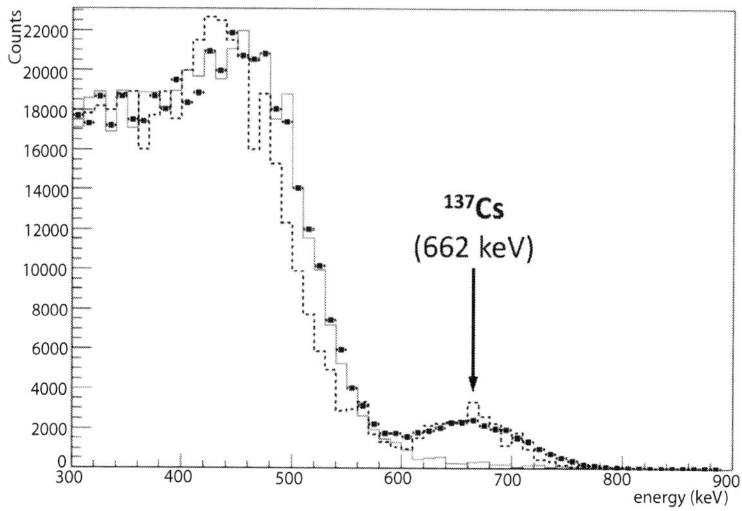

Figure 2.20 Gamma response of a nanocomposite consisting of 10 vol% LaF_3:20%Ce^{3+} in oleic acid (dotted line) and a similar sample without nanoparticle loading (solid line) [60]. Both samples show a clear Compton edge but the nanoparticle-loaded sample also shows a photopeak. Reprinted from Ref. 60, Copyright 2007, with permission from SPIE.

There are a number of other interesting reports on inorganic-organic nanocomposites for radiation detection. Cai et al. dispersed Gd_2O_3 nanocrystals (plates of 9 nm edge length and 1 nm thickness) in vinyltoluene and which was then thermally polymerized to polyvinyltoluene (PVT) using an aliphatic peroxide initiator [14]. The resulting nanocomposite contained 4.6 vol% Gd_2O_3 nanoparticles and had had good optical transparency with a 3 mm thick sample having a transmittance of 69.6% at 550 nm compared to 76% for pure PVT. The material showed good response to β radiation and also achieved an 11.4% energy resolution at 662 keV. DiMaio et al. created a nanocomposite by first functionalizing the surface of LaF_3 nanoparticles (5–10 nm size) with trifluorovinyl ether (TFVE) and then cross-linking the TFVE ligands [20]. In this case, the cross-linked ligands themselves form the organic matrix of the composite. The corresponding small amount of organic matrix allowed for high nanoparticle loadings approaching 80 wt% while maintaining good optical transparency. The authors described such materials as omnicomposites.

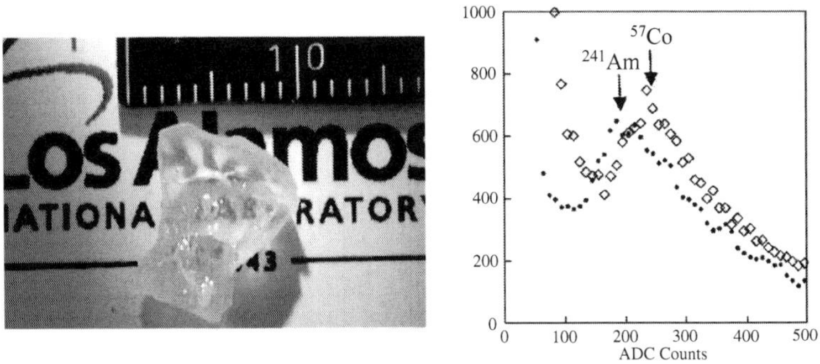

Figure 2.21 Left: Image of a nanocomposite consisting of CeF_3 nanoparticles in an organic matrix [59]. The particle size is <10 nm and the composite consists of >50 vol% of CeF_3. Right: Measured photopeaks obtained by irradiating a small piece of CeF_3 nanocomposite with ^{241}Am (dots) and ^{57}Co (diamonds). Reprinted from Ref. 59, Copyright 2007, with permission from Elsevier.

A number of studies have pursued semiconductor quantum dots (QDs) as the scintillating component. For example, Chen's group used a radiative energy-transfer approach to create potential nanocomposite scintillators by coupling LaF_3:Ce nanoparticles with CdTe quantum dots [31, 89]. CdTe shows little X-ray luminescence response on its own, but when coupled with LaF_3:Ce, a significant increase in the X-ray induced luminescence was observed as a result of energy transfer from LaF_3:Ce to CdTe (see Fig. 2.22). The authors achieved similar results with BaFBr:Eu/CdTe nanocomposites [16]. Letant et al. loaded a mesoporous silica glass with CdSe/ZnS core–shell QDs to produce a transparent 1/16 inch thick nanocomposite that showed an energy resolution of 30% for the ^{241}Am γ-ray line [47] as well as a response to α radiation [48]. Wagner et al. explored numerous nanocomposite systems, including semiconducting QDs, lanthanide halides, and glass ceramics [84]. This group also addressed the optical considerations regarding scattering and absorption. Highly loaded QD composites suffered from self-absorption as the volume loading was increased. Lawrence et al. found that co-doping a polymer matrix with CdTe and CdSe quantum dots and PPO resulted in a higher luminescence output than composites that used either of the dopants individually due to energy transfer from PPO to the QDs [44]. Zhong et al. found that energy transfer from a blue-emitting conjugated polymer matrix, HEH-PF (poly[9-hexyl-9-(2-ethylhexyl)-fluorene-2,7-diyl]), to BiI_3 nanoparticles resulted in photoluminescence quenching [92]. To utilize this phenomenon, oleylamine was added to the nanocomposite, forming an interfacial layer between the two phases that suppressed the charge transfer. Upon γ-ray exposure, the oleylamine degraded and allowed photoluminescence quenching in a manner that was linearly proportional to dose. The material appears to be a sensitive radiation dosimeter but has not been fully tested.

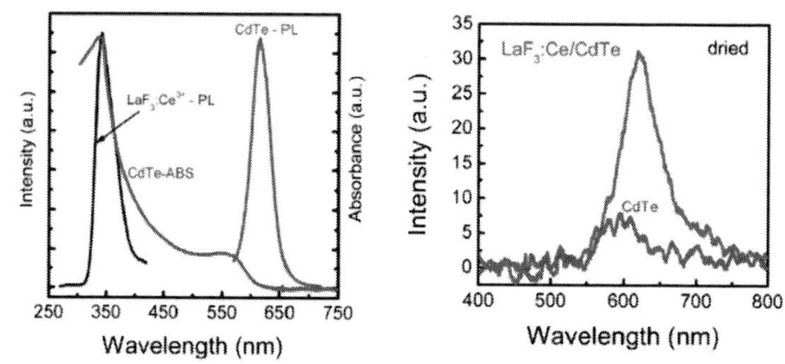

Figure 2.22 Left: Spectral overlap of the LaF_3:Ce^{3+} emission (PL) and CdTe absorption (ABS) enables a strong orange CdTe emission in the coupled system. Right: X-ray luminescence response of neat CdTe and the mixed composite. Reprinted from Ref. 31, Copyright 2012, with permission from AIP Publishing LLC.

2.4.3 Inorganic–Inorganic Composites

Nanocomposites containing scintillating inorganic nanoparticles embedded in an inorganic matrix belong to the larger family of materials known as glass ceramics. Glass ceramics are fabricated on an industrial scale for a wide range of thermal, structural, and electrical applications and find use, for example, in cooktops, kitchenware, telescope mirrors, machinable glass, electrical insulators, building walls, dental implants, and armor [90]. The common approach to fabricating glass ceramics consists of first preparing a solid amorphous precursor (glass or gel) followed by heat treatment of the precursor in order to induce the nucleation and growth of one or more crystalline phases within the solid in a process known as controlled crystallization. Crystallinities of up to 99.5% are possible, with typical glass ceramics having a crystallinity in the 30–70 vol% range [90]. The crystallization process is usually enhanced and optimized by the addition of nucleating agents such as TiO_2 and ZrO_2. Optical transparency is not of concern in many commercial applications, and crystallite grain sizes in the tens of micrometer range are typical. Transparent glass ceramics have been fabricated for optical amplifiers, up-converters, solar concentrators, harmonic generators, and gratings. Here, the crystalline phase is often a silicate having a refractive index that is comparable to the glass matrix. This allows for excellent transparency even for relatively large crystallite sizes up to 200 nm [90]. Application of this concept to nanocomposite scintillators however is challenging because efficient scintillator materials all have refractive indices near or above 1.8 (see Table 2.1), i.e., much greater than the silica glass matrix ($n \approx 1.5$). This in turn necessitates much smaller particle sizes in order to maintain acceptable transparency (see Section 2.2.1.2.).

Nanocomposite glass ceramics are being pursued for application in X-ray imaging and slow neutron detection. The latter methods typically use nuclear reactions between the incident neutron and a target nucleus in the scintillator. A popular nuclear reaction uses the large cross section of ^{6}Li to capture slow neutrons according to ${}^6_3\mathrm{Li} + {}^1_0\mathrm{n} \rightarrow {}^3_1\mathrm{H} + {}^4_2\alpha$. The resulting α particle (${}^4_2\alpha$) and triton (${}^3_1\mathrm{H}$) deposit 2.05 and 2.73 MeV of energy into the scintillator, respectively (assuming negligible energy of the incident neutron) [43], and produce electron–hole pairs along their trajectories that subsequently produce scintillation light in analogy to γ-ray scintillators. The smaller effective charge of the triton gives it a mean free path within the scintillator that is ~10× greater than that of the α particle, and it therefore provides the main contribution to the light output of the scintillator [66]. Nikitin et al. have fabricated a nanostructured glass–ceramic neutron scintillator by first synthesizing $LiAlSi_4O_{10}$ nanoparticles via a solution precipitation method, mixing the nanoparticles with the other SiO_2, Li_2CO_3 (~85% ^{6}Li enriched), Al_2O_3, MgO, and CeO_2, melting the mixture for a period of time such that the nanoparticles were not completely dissolved in the glass melt, and finally rapid quenching of the melt [66]. The "as quenched" glass contained a high density of crystal seeds that could be grown during a subsequent heat treatment near the glass transition temperature, resulting in a transparent glass ceramic. As shown in Fig. 2.23, the nanocomposite material had a photon yield that was 240% of the photon yield of commercial GS20 scintillator glass.

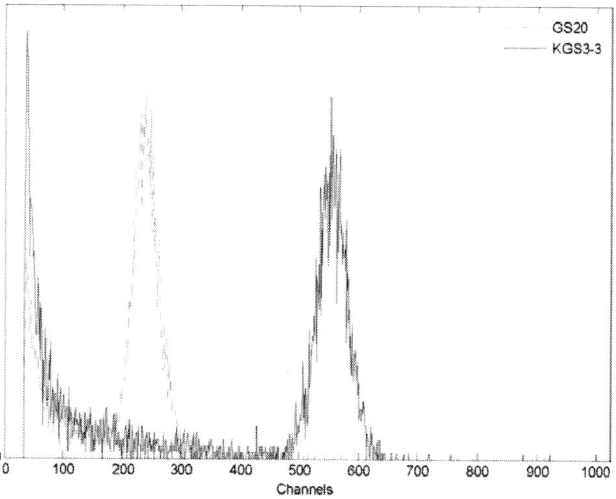

Figure 2.23 Pulse height spectra produced by detectors using commercial glass (GS20) and nanocomposite glass–ceramic (KGS3-3) as scintillator under thermal neutron exposure. The photon yield of the glass ceramic is 240% that of the commercial glass. Reprinted from Ref. 66, Copyright 2013, with permission from IEEE.

Chen, Johnson, and coworkers produced a transparent glass ceramic consisting of fluorozirconate glass (ZBLAN) containing a high density of hexagonal $BaCl_2:Eu^{2+}$ particles with an average crystallite size of <20 nm [15, 36, 76]. The glass ceramic was fabricated by first melting the constituents of ZBLAN glass (ZrF_4, BaF_2, LaF_3, AlF_3, NaF) along with $EuCl_2$ at 750 °C and quenching of the melt to form a glass. The glass was then heated to 200 °C to induce the formation of crystallization nuclei that were subsequently grown to ~20 nm size at 235 °C over periods ranging from 5 to 30 hours. The glass ceramic showed good transparency over the 1 mm sample thickness, and the authors used it successfully to record X-ray computed tomography images in a benchmark against $CdWO_4$. Appleby et al. used a melt-quench-treat method to produce $BaCl_2:Eu^{2+}$ nanoparticles in a B_2O_3–Li_2O glass [4]. The glass was enriched in ^{10}B and ^{6}Li to maximize the neutron absorption cross section and enable use in thermal neutron imaging. In a similar melt-quench-treat method, Fu et al. produced a glass ceramic consisting of a SiO_2–Al_2O_3–CaO glass containing nanocrystals of $CaF_2:Eu^{2+}$ with ~35 nm size [24]. The photon yield of the transparent glass ceramic under X-ray excitation was estimated to be ~30% of that of a $CaF_2:Eu^{2+}$ single crystal. The stopping power for X-rays and neutrons is generally much higher than for γ-rays and, combined with the geometry of an imaging application, relatively thin sheets of scintillator are desired. This relaxes the challenging transparency requirements for γ-ray scintillators to some degree.

Sol–gel chemistry offers alternative fabrication methods for nanostructured glass ceramics, including attractive low-temperature routes to oxide-based scintillator materials such as LSO, $LuBO_3$, LuAG, $LuPO_4$, $SrHfO_3$, and $CdWO_4$. A review on this topic by Nedelec highlights the versatility of sol–gel reactions [65] but also illustrates that the majority of sol–gel studies to date involve the preparation of scintillator materials in thin-film or powder form. The fabrication of larger and transparent nanocomposite monoliths has not been reported yet. Most noteworthy in this context is the work by Bescher et al., who have reported a transparent glass ceramic containing Ce^{3+}-doped lutetium pyrosilicate ($Lu_2Si_2O_7$) nanocrystals in a silica glass matrix [6, 7]. They used a sol–gel process starting from TEOS (tetraethyl orthosilicate: $Si(OC_2H_5)_4$), lutetium-isopropoxide (prepared in situ from lutetium metal and isopropanol in a $HgCl_2$ catalyzed reaction), and cerium nitrate in isopropanol to form a transparent silica gel under ambient conditions after several days. Subsequent heating of the gel to 1200°C for several hours led to the growth of crystallites (~50 nm diameter) within the SiO_2 glass (Fig. 2.24). X-ray diffraction indicated that the nanocrystals were not the orthosilicate (Lu_2SiO_5, LSO) but rather the pyrosilicate ($Lu_2Si_2O_7$) phase [7]. Nevertheless, the light output of the glass ceramic was 65–70% of that of a LSO:Ce^{3+} single crystal for irradiation with ^{137}Cs and ^{241}Am. The authors also measured the energy resolution for both the glass ceramic and the single-crystal LSO:Ce^{3+} using a ^{241}Am source and found both samples to have similar performance (~35%, see Fig. 2.24). Both samples were relatively small (<10 mm^3) which limited the energy resolution due to incomplete absorption of γ-rays. This is an example of a nanocomposite scintillator rivaling the performance of a traditional single-crystal scintillator. It also highlights the significant cost savings that may be possible: while Czochralski crystal growth of LSO:Ce^{3+} requires furnace temperatures of 2000°C over periods of many days in expensive iridium crucibles, the fabrication of the glass ceramic sample involved only low temperatures and one short heat treatment at 1200 °C. Mansuy et al. reported a similar synthesis of rare earth–doped lutetium silicate that avoided the mercury catalyst and used a metathesis reaction between anhydrous $LuCl_3$ and potassium isopropoxide (prepared by reacting metallic potassium with anhydrous 2-propanol) to obtain the lutetium-isopropoxide precursor [56, 57]. Their process yielded phase-pure LSO, and a second pyrosilicate phase was only observed for lower heat-treatment temperatures [57]. The resulting nanostructured LSO:Ce showed a good scintillation yield of ~20,000 photons/MeV (compared to ~33,000 photons/MeV for bulk LSO:Ce, see Table 2.1) [56].

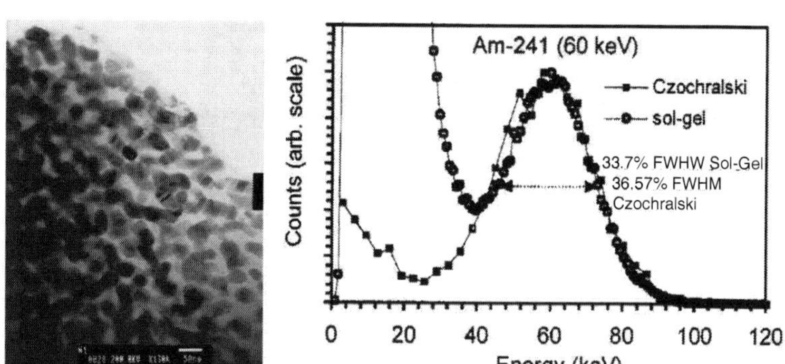

Figure 2.24 Left: Transmission electron microscopy image of lutetium pyrosilicate ($Lu_2Si_2O_7$) crystals in a transparent silica matrix [6]. Right: Energy spectrum of a ^{241}Am source recorded from the nanocomposite scintillator produced by the sol–gel method (2.5 mm^3) and from a LSO:Ce^{3+} single crystal (8.5 mm^3). Reprinted from Ref. 6, Copyright 2000, with permission from Springer.

2.5 Summary and Outlook

The detection and characterization of ionizing radiation continues to gain importance in many fields of science. In the past decade, this growth has been further fueled by an increased focus on detecting and tracking the global flow of benign and illicit nuclear materials. During the same time, new and highly efficient inorganic scintillators have emerged and are making their way into a diverse range of radiation sensing equipment. The desired widespread deployment of these scintillators however has been hampered by the challenges in growing large halide crystals and the associated high cost. This has created an urgent need for materials with sufficiently good scintillation performance that can be fabricated at significantly lower cost than single crystal scintillators. The main focus of recent research has been on developing materials that incorporate nanocrystals of known scintillators into a matrix with the goal of creating optically transparent composites using inexpensive fabrication methods. Scintillating nanoparticles have been incorporated into both organic and inorganic matrices. Fabrication of such nanocomposites continues to face significant challenges that fall into two general categories. First, the best performing scintillators are rare earth-doped chlorides and bromides that are highly hygroscopic and demand special care during processing. This is further exacerbated by the large surface area of nanoparticles. Halides readily react with oxygen, and many desired host materials such as silicates are often chemically incompatible. Second, the need for high optical transparency and energy deposition primarily in the nanoparticles requires materials with a high concentration of <10 nm sized nanoparticles. Creating such highly loaded nanocomposites with minimal particle aggregation has proven very difficult, particularly for inorganic nanoparticles in organic matrices. Some of these difficulties are avoided in the class of inorganic–inorganic composites, where the in situ growth processes are more amenable to creating monodisperse scintillating nanoparticles. Recent efforts have produced nanocomposite scintillators that approach the performance of their bulk counterparts, at least on the small scale. The many avenues for fabricating nanocomposite scintillators combined with the diverse needs of radiation detection applications will continue to fuel this field of research. The resulting materials may not only replace some of the currently expensive scintillator crystals but also enable entirely new applications in sensor networks and medical imaging that are currently cost prohibitive.

Acknowledgments

This work was supported in part by the Los Alamos National Laboratory LDRD Program (LDRD 20120246ER).

References

1. Agostinelli, S., Allison, J., Amako, K., Apostolakis, J., Araujo, H., Arce, P., Asai, M., Axen, D., Banerjee, S., Barr, G., Behner, F., Bellagamba, L., Boudreau, J., Broglia, L., Brunengo, A., Burkhardt, H., Chauvie, S., Chuma, J., Chytracek, R., and Cooperman, G. (2003). G4: A simulation toolkit, *Nuc. Instr. Meth. Phys. Res. A*, **506**, pp. 250–303.
2. Alaribe, L., Disch, C., Fauler, A., Engels, R., Keller, E., Cecilia, A., dos Santos Rolo, T., Hamann, E., and Fiederle, M. (2012). $SrI_2:Eu^{2+}$-scintillators for spectroscopy and X-ray imaging applications, *IEEE Trans. Nucl. Sci.*, **59**(5), pp. 2193–2195.
3. Allison, J., Amako, K., Apostolakis, J., Araujo, H., Arce Dubois, P., Asai, M., Barrand, G., Capra, R., Chauvie, S., Chytracek, R., Cirrone, G. A. P. , Cooperman, G., Cosmo, G., Cuttone, G., Daquino, G. G., Donszelmann, M., Dressel, M., Folger, G., Foppiano, F., Generowicz, J., Grichine, V., Guatelli, S., Gumplinger, P., Heikkinen, A., Hrivnacova, I., Howard, A., Incerti, S., Ivanchenko, V., Johnson, T., Jones, F., Koi, T., Kokoulin, R., Kossov, M., Kurashige, H., Lara, V., Larsson, S., Lei, F., Link, O., Longo, F., Maire, M., Mantero, A., Mascialino, B., McLaren, I., Mendez Lorenzo, P., Minamimoto, K., Murakami, K., Nieminen, P., Pandola, L., Parlati, S., Peralta, L., Perl, J., Pfeiffer, A., Pia, M. G. , Ribon, A., Rodrigues, P., Russo, G., Sadilov, S., Santin, G., Sasaki, T., Smith, D., Starkov, N., Tanaka, S., Tcherniaev, E., Tomé, B., Trindade, A., Truscott, P., Urban, L., Verderi, M., Walkden, A., Wellisch, J. P., Williams, D. C., Wright, D., and Yoshida, H. (2006). Geant4 developments and applications, *IEEE Trans. Nuc. Sci.*, **53**(1), 270–278.
4. Appleby, G. A. and Vontobel, P. (2008). Optimisation of lithium borate – barium chloride glass-ceramic thermal neutron imaging plates, *Nuc. Instr. Meth. Phys. Res. A*, **594**, pp. 253–256.
5. Benitez, E. L., Husk, D. E., Schnatterly, S. E., and Tarrio, C. (1991). A surface recombination model applied to large features in inorganic

phosphor efficiency measurements in the soft X-ray region, *J. Appl. Phys.*, **70**(6), pp. 3256–3260.

6. Bescher, E., Robson, S. R., Mackenzie, J. D., Patt, B., Iwanczyk, J., and Hoffman, E. J. (2000). New lutetium silicate scintillators, *J. Sol-Gel Sci. Tech.*, **19**, pp. 325–328.
7. Bescher, E., Robson, S. R., Mackenzie, J. D., Patt, B., Iwanczyk, J., and Hoffman, E. J. (2000). New transparent, polycrystalline silicate scintillators, *Proc. SPIE*, **3943**, 139–146.
8. Birks, J. B. (1964). *The Theory and Practice of Scintillation Counting*, Pergamon Press, New York.
9. Birowosuto, M. D. and Dorenbos, P. (2009). Novel γ- and X-ray scintillator research: on the emission wavelength, light yield and time response of Ce^{3+} doped halide scintillators, *Phys. Status Solidi A*, **206**(1), pp. 9–20.
10. Biswas, K. and Du, M.-H. (2012). Energy transport and scintillation of cerium-doped elpasolite Cs_2LiYCl_6: Hybrid density functional calculations. *Phys. Rev. B*, **86**, 014102.
11. Bizarri, G. and Dorenbos, P. (2007). Charge carrier and exciton dynamics in $LaBr_3$:Ce^{3+} scintillators: Experiment and model. *Phys. Rev. B*, **75**, 184302.
12. Bohren, C. F. and Huffmann, D. R. (2010). *Absorption and Scattering of Light by Small Particles*, Wiley-Interscience, New York.
13. Brown, S. S., Rondinone, A. J., and Dai, S. (2007). Applications of nanoparticles in scintillation detectors, in *Antiterrorism and Homeland Defense* (eds. Reynolds, J. G., Lawson, G. E. and Koester, C. J.), ACS Symposium Series 980, America Chemical Society, Washington, DC, pp 117–129.
14. Cai, W., Chen, Q., Cherepy, N., Dooraghi, A., Kishpaugh, D., Chatziioannou, A., Payne, S., Xiang, W., Pei, Q. (2013). Synthesis of bulk-size transparent gadolinium oxide – polymer nanocomposites for gamma-ray spectroscopy, *J. Mater. Chem. C*, **1**, pp. 1970–1976.
15. Chen, G., Johnson, J., Schweizer, S., Woodford, J., Newman, P., and MacFarlane, D. (2006). Transparent $BaCl_2$:Eu^{2+} glass-ceramic scintillator, *Proc. SPIE*, **6142**, 61422X.
16. Chen, W., Zhang, J., Westcott, S. L., Joly, A. G., Malm, J.-O., and Bovin, J.-O. (2006). The origin of X-ray luminescence from CdTe nanoparticles in CdTe/BaFBr:Eu^{2+} nanocomposite phosphors, *J. Appl. Phys.*, **99**, 034302.

17. Creager, A. N. H. (2009). Phosphorus-32 in the Phage Group: radioisotopes as historical tracers of molecular biology, *Stud. Hist. Philos. Biol. Biomed. Sci.*, **40**, pp. 29–42.

18. Del Sordo, S., Abbene, L., Caroli, E., Mancini, A. M., Zappettini, A., and Ubertini, P. (2009). Progress in the development of CdTe and CdZnTe semiconductor radiation detectors for astrophysical and medical applications, *Sensors*, **9**, pp. 3491–3526.

19. Derenzo, S., Boswell, M., Weber, M., and Brennan, K. *Scintillation Properties*, [online database] http://scintillator.lbl.gov

20. DiMaio, J. R., Kokuoz, B., and Ballato, J. (2008). Omni-composites: A new strategy for forming bulk nanostructured materials, *J. Am. Chem. Soc.*, **130**, pp. 5628–5629.

21. Dorenbos, P. (2002). Light output and energy resolution of Ce^{3+}-doped scintillators, *Nuc. Instr. Meth. Phys. Res. A*, **486**, pp. 208–213.

22. Doyle, J. (2011). *Nuclear Safeguards, Security and Nonproliferation: Achieving Security with Technology and Policy*, Elsevier Science, Burlington.

23. Feller, R. K., Purdy, G. M., Ortiz-Acosta, D., Stange, S., Li, A., McKigney, E. A., Esch, E. I., Muenchausen, R. E., Gilbertson, R., Bacrania, M., Bennett, B. L., Ott, K. C., Brown, L., Macomber, C. S., Scott, B. L., and Del Sesto R. E. (2011). Large-scale synthesis of $Ce_xLa_{1-x}F_3$ nanocomposite scintillator materials, *J. Mater. Chem.*, **21**, pp. 5716–5722.

24. Fu, J., Kobayashi, M., Sugimoto, S., and Parker, J. M. (2009). Scintillation from Eu^{2+} in Nanocrystallized Glass, *J. Am. Ceram. Soc.*, **92**(9), pp. 2119–2121.

25. Glodo, J., van Loef, E. V. D., Higgins, W. M., and Shah, K. S. (2008). Mixed lutetium iodide compounds, *IEEE Trans. Nuc. Sci.*, **55**(3), pp. 1496–1500.

26. Halzen, F. and Klein, S. R. (2010). IceCube: An instrument for neutrino astronomy. *Rev. Sci. Instrum.*, **81**, 081101.

27. Hehlen, M. P. (1997). Reabsorption artifacts in measured excited-state lifetimes of solids, *J. Opt. Soc. Am. B*, **14**(6), pp. 1312–1318.

28. Hoffman, G. J. and Albrecht, A. C. (1991). Near-threshold photoionization spectra in nonpolar liquids – A germinate pair based model, *J. Phys. Chem.*, **95**(6), pp. 2231–2241.

29. Hofstadter, R. (1948). Alkali halide scintillation counters, *Phys. Rev.*, **74**, pp. 100–101.

30. Hofstadter, R. and McIntyre, J. A. (1950). The measurement of gamma-ray energies with single crystals of NaI(Tl), *Phys. Rev.*, **80**(4), pp. 631–637.

31. Hossu, M., Liu, Z., Yao, M., Ma, L., and Chen, W. (2012). X-ray luminescence of CdTe quantum dots in LaF_3:Ce/CdTe nanocomposites, *Appl. Phys. Lett.*, **100**, 013109.

32. Iltis, A., Mayhugh, M. R., Menge, P., Rozsa, C. M., Selles, O., and Solovyev, V. (2006). Lanthanum halide scintillators: Properties and applications, *Nuc. Instr. Meth. Phys. Res. A*, **563**, pp. 359–363.

33. Iwanowska, J., Swiderski, L., Moszynski, M., Szczesniak, T., Sibczynski, P., Galunov, N. Z., and Karavaeva, N. L. (2011). Neutron/gamma discrimination properties of composite scintillation detectors, *J. Instr.*, **6**, P07007.

34. Jacobsohn, L. G., Sprinkle, K. B., Roberts, S. A., Kucera, C. J., James, T. L., Yukihara, E. G., DeVol, T. A., and Ballato, J. (2011). Fluoride nanoscintillators, *J. Nanomat.*, **2011**, pp. 1–6.

35. Jaffe, J. E., Jordan, D. V., and Peurrung, A. J. (2007). Energy nonlinearity in radiation detection materials: Causes and consequences, ." (2007), *Nuc. Instr. Meth. Phys. Res. A*, **570**(1), pp. 72–83.

36. Johnson, J. A., Schweizer, S., Henke, B., Chen, G., Woodford, J., Newman, P. J., and MacFarlane, D. R. (2006). Eu-activated fluorozirconate glass-ceramic scintillators, *J. Appl. Phys.*, **100**, 034701.

37. Johnson, T. E. and Birky, B. K. (eds) (1998). *Health Physics and Radiological Health* (1995). Lippincott Williams & Wilkins, Baltimore.

38. Karavaeva, N. L. and Tarasenko, O. A. (2009). Development of new composite scintillators based on organic single crystal grains, *Funct. Mat.*, **16**(1), pp. 92–96.

39. Kasarova, S. N., Sultanova, N. G., Ivanov, C. D., and Nikolov, I. D. (2007). Analysis of the dispersion of optical plastic materials, *Opt. Mater.*, **29**, pp. 1481–1490.

40. Khodyuk, I. V. and Dorenbos, P. (2010). Nonproportional response of $LaBr_3$:Ce and $LaCl_3$:Ce scintillators to synchrotron X-ray irradiation, *J. Phys.: Condens. Matter*, **22**, 485402.

41. Khodyuk, I. V., Rodnyi, P. A., and Dorenbos, P. (2012). Energy dependence of the relative light output of $YAlO_3$:Ce, Y_2SiO_5:Ce, and YPO_4:Ce scintillators, *Instr. Exp. Tech.*, **55**(2), pp. 187–197.

42. Khodyuk, I. V. (2013). *Nonproportionality of Inorganic Scintillators*, Ph.D. Thesis, Delft University of Technology, Delft, The Netherlands.

43. Knoll, G. F. (2010). *Radiation Detection and Measurement*, 4th ed., John Wiley & Sons, Hoboken.

44. Lawrence, W. G., Thacker, S., Palamakumbura, S., Riley, K. J. and Nagarkar, V. V. (2012). Quantum dot-organic polymer composite materials for radiation detection and imaging, *IEEE Trans. Nuc. Sci.*, **59**(1), pp. 215–221.

45. Lecoq, P., Annenkov, A., Gektin, A., Korzhik, M., and Pedrini, C. (2006). *Inorganic Scintillators for Detector Systems. Physical Principles and Crystal Engineering*, Springer, Berlin.

46. Lempicki, A., Brecher, C., Szupryczynski, P., Lingertat, H., Nagarkar, V. V., Tipnis, S. V., and Miller, S. (2002). A new lutetia-based ceramic scintillator for X-ray imaging, *Nuc. Instr. Meth. Phys. Res. A*, **488**(3), pp. 579–590.

47. Létant, S. E. and Wang, T.-F. (2006). Semiconductor quantum dot scintillation under γ-ray irradiation, *Nano Lett.*, **6**(12), pp. 2877–2880.

48. Létant, S. E. and Wang, T.-F. (2006). Study of porous glass doped with quantum dots or laser dyes under alpha irradiation, *Appl. Phys. Lett.*, **88**, 103110.

49. Li, A. (2011). *Development and Evaluation of Novel Structured Scintillation Detectors*, Ph.D. Thesis, North Carolina State University, Raleigh, NC.

50. Li, A., Smith, N., Hehlen, M. P., McKigney, E. A., and Gardner, R. (2012). Light yield measurement method for milled nanosize inorganic crystals, *Appl. Rad. Isotop.*, **70**, pp. 1219–1222.

51. Li, Q., Grim, J. Q., Williams, R. T., Bizarri, G. A., and Moses, W. W. (2011). A transport-based model of material trends in nonproportionality of scintillators, *J. Appl. Phys.*, **109**(12), 123716.

52. Li, Q., Grim, J. Q., Williams, R. T., Bizarri, G. A., and Moses, W. W. (2011). The role of hole mobility in scintillator proportionality, *Nuc. Instr. Meth. Phys. Res. A*, **652**(1), pp. 288–291.

53. Li, Q., Grim, J. Q., Ucer, K. B., Burger, A., Bizarri, G. A., Moses, W. W., and Williams, R. T. (2012). Host structure dependence of light yield and proportionality in scintillators in terms of hot and thermalized carrier transport, *Phys. Status Solidi RRL*, **6**(8), pp. 346–348.

54. Li, S., Lin, M., Toprak, M., Kim, D., and Muhammed, M. (2010). Nanocomposites of polymer and inorganic nanoparticles for optical and magnetic applications, *Nano Reviews*, North America, **1**, 5214.

55. Macdougall, D. (2008). Nature's clocks: How scientists measure the age of almost everything. University of California Press, Berkeley, CA

56. Mansuy, C., Mahiou, R., and Nedelec, J. M. (2003). A new sol-gel route to Lu_2SiO_5 (LSO) scintillator: powders and thin films, *Chem. Mater.*, **15**(17), pp. 3242–3244.

57. Mansuy, C., Leroux, F., Mahiou, R., and Nedelec, J. M. (2005). Preferential site substitution in sol-gel derived Eu^{3+} doped Lu_2SiO_5: a combined study by X-ray absorption and luminescence spectroscopies, *J. Mater. Chem.*, **15**(38), pp. 4129–4135.

58. Mares, J. A., Nikl, M., Solovieva, N., D'Ambrosio, C., de Notaristefani, F., Blazek, K., Maly, P., Nejezchleb, K., Fabeni, P., Pazzi, G. P., de Haas, J. T. M., van Eijk, C. W. E., and Dorenbos, P. (2003). Scintillation and spectroscopic properties of Ce^{3+}-doped $YAlO_3$ and $Lu_x(RE)_{1-x}AlO_3$ ($RE=Y^{3+}$ and Gd^{3+}) scintillators, *Nuc. Instr. Meth. Phys. Res. A*, **498**, pp. 312–327.

59. McKigney, E. A., Del Sesto, R. E., Jacobsohn, L. G., Santi, P. A., Muenchausen, R. E., Ott, K. C., McCleskey, T. M., Bennett, B. L., Smith, J. F., and Cooke, D. W. (2007). Nanocomposite scintillators for radiation detection and nuclear spectroscopy, *Nuc. Instr. Meth. Phys. Res. A*, **579**, pp. 15–18.

60. McKigney, E. A., Muenchausen, R. E., Cooke, D. W., Del Sesto, R. E., Gilbertson, R. D., Bacrania, M. K., Bennett, B. L., Jacobsohn, L. G., McCleskey, Ott, K. C., Sitarz, S. C., Smith, J. F., and Stange, S. (2007). LaF_3:Ce nanocomposite scintillator for gamma-ray detection, *Proc. SPIE*, **6706**, 67061A.

61. Mikhailik, V. B., Kraus, H., Miller, G., Mykhaylyk, M. S., and Wahl, D. (2005). Luminescence of $CaWO_4$, $CaMoO_4$, and $ZnWO_4$ scintillating crystals under different excitations, *J. Appl. Phys.*, **97**(8), 083523.

62. Moser, S. W., Harder, W. F., Hurlbut, C. R., and Kusner, M. R. (1993). Principles and practice of plastic scintillation design, *Radiat. Phys. Chem.*, **41**(1/2), pp. 31–36.

63. Muenchausen, R. E., McKigney, E. A., Jacobsohn, L. G., Blair, M. W., Bennett, B. L., and Cooke, D. W. (2008). Science and application of oxyorthosilicate nanophosphors, *IEEE Trans. Nuc. Sci.*, **55**(3), pp. 1532–1535.

64. National Institute of Science and Technology (NIST). *ESTAR: Stopping-Power and Range Tables for Electrons*, [online database] http://physics.nist.gov/PhysRefData/Star/Text/ESTAR.html

65. Nedelec, J. M. (2007). Sol-gel processing of nanostructured inorganic scintillating materials, *J. Nanomat.*, **2007**, 36392.

66. Nikitin, A., Fedorov, A., and Korjik, M. (2013). Novel glass ceramic scintillator for detection of slow neutrons in well logging applications, *IEEE Trans. Nuc. Sci.*, **60**(2), pp. 1044–1048.

67. Nikl, M., Laguta, V. V., and Vedda, A. (2008). Complex oxide scintillators: Material defects and scintillation performance, *Phys. Status Solidi RRL (b)*, **245**(9), pp. 1701–1722.

68. Osiński, M., Pumley, J. B., Withers, N. J., Rivera, A. C., Akins, B. A., Sankar, K., and Smolyakov, G. A. (2009). Lanthanide-halide-based nanoscintillators for portable radiological detectors, *Proc. SPIE*, **7306**, 730617.

69. Owens, A., Bos, A. J. J., Brandenburg, S., Dathy, C., Dorenbos, P., Kraft, S., Ostendorf, R. W., Ouspenski, V., and Quarati, F. (2007). γ-ray performance of a 1242 cm^3 $LaCl_3$:Ce scintillation spectrometer, *Nuc. Instr. Meth. Phys. Res. A*, **574**, pp. 110–114.

70. Payne, S. A., Cherepy, N. J., Hull, G., Valentine, J. D., Moses, W. W., and Choong, W.-S. (2009). Nonproportionality of scintillator detectors: Theory and Experiment, *IEEE Trans. Nuc. Sci.*, **56**(4), pp. 2506–2512.

71. Payne, S. A., Moses, W. W., Sheets, S., Ahle, L., Cherepy, N. J., Sturm, B., Dazeley, S., Bizarri, G., and Choong, W.-S. (2011). Nonproportionality of scintillator detectors: Theory and Experiment. II, *IEEE Trans. Nuc. Sci.*, **58**(6), pp. 3392–3402.

72. Potdevin, A., Chadeyron, G., Thérias, S., Mahiou, R. (2012). Luminescent nanocomposites made from finely dispersed $Y_3Ga_5O_{12}$:Tb powder in a polymer matrix: Promising candidates for optical devices, *Langmuir*, **28**, 13526–13535.

73. Quinten, M. (2011). Mie's theory for single spherical particles, in *Optical Properties of Nanoparticle Systems: Mie and Beyond*, Wiley-VCH Verlag GmbH & Co. KGaA, Weinheim, Germany.

74. Ronda, C. R. and Srivastava, A. M. (2008). Chapter 5: Scintillators in *Luminescence – From Theory to Applications* (ed. Ronda, C. R.), Wiley, Weinheim.

75. Schwartz, K. (1997). *Atomic Physics Methods in Modern Research*, Springer, Berlin.

76. Schweizer, S., Henke, B., Miclea, P. T., Ahrens, B., and Johnson, J. A. (2010). Multi-functionality of fluorescent nanocrystals in glass ceramics, *Rad. Meas.*, **45**, pp. 485–489.

77. Siciliano, E. R., Ely, J. H., Kouzes, R. T., Milbrath, B. D., Schweppe, J. E., and Stromswold, D. C. (2005). Comparison of PVT and NaI(Tl) scintillators

for vehicle portal monitor applications, *Nuc. Instr. Meth. Phys. Res. A*, **550**(3), pp. 647–674.

78. Stange, S. (2010). *Nanocomposite Scintillators for Neutron Capture Measurements*, Ph.D. Thesis, University of Michigan, Ann Arbor, MI.

79. Stock, S. R. (2008). Recent advances in X-ray microtomography applied to materials. *Int. Mat. Rev.*, 53(3), pp. 129–181.

80. Tan, M. C., Patil, S. D., and Riman, R. E. (2010). Transparent infrared-emitting CeF_3:Yb-Er polymer nanocomposites for optical applications, *Appl. Mat. Int.*, **2**(7), 1884–1891.

81. Tavernier, S., Gektin, A., Grinyov, B., and Moses, W.W. (eds) (2006) *Proceedings of the NATO Advanced Research Workshop on Radiation Detectors for Medical Applications*, Alushta, Crimea, Ukraine.

82. van Dam, H. T., Seifert, S., Drozdowski, W., Dorenbos, P., and Schaart, D. R. (2012). Optical absorption length, scattering length, and refractive index of $LaBr_3$:Ce^{3+}, *IEEE Trans. Nuc. Sci.*, **59**(3), pp. 656–664.

83. van Loef, E. V., Higgins, W. M., Glodo, J., Brecher, C., Lempicki, A., Venkataramani, V., Moses, W. W., Derenzo, S. E., and Shah, K. S. (2006). Scintillation properties of $SrHfO_3$:Ce^{3+} and $BaHfO_3$:Ce^{3+} ceramics, *IEEE Nuc. Sci. Symp. Conf. Record*, **N40-3**, pp. 1538–1540.

84. Wagner, B. K., Kang, Z., Nadler, J., Rosson, R., and Kahn, B. (2012). Nanocomposites for radiation sensing, *Proc. SPIE*, **8373**, 83730K.

85. Weber, M. J. (2002). Inorganic scintillators: today and tomorrow, *J. Lumin.*, **100**, pp. 35–45.

86. Wilkinson, F. (2007). Methodology for packaging reliable scintillation detectors, *Proc. SPIE*, **6707**, 670702.

87. Williams, R. T., Grim, J. Q., Li, Q., Ucer, K. B., and Moses, W. W. (2011). Excitation density, diffusion drift, and proportionality in scintillators, *phys. stat. sol (b)*, **248**(2), pp. 426–438.

88. X-5 Monte Carlo Team (2008). *MCNP - A General Monte Carlo N-Particle Transport code, Version 5*, Los Alamos National Laboratory, Los Alamos.

89. Yao, M., Zhang, X., Ma, L., Chen, W., Joly, A. G., Huang, J., and Wang, Q. (2010). Luminescence enhancement of CdTe nanostructures in LaF_3:Ce/CdTe nanocomposites, *J. Appl. Phys.*, **108**, 103104.

90. Zannoto, E. D. (2010). A bright future for glass-ceramics, *Am. Ceram. Soc. Bull.*, **89**(8), pp. 19–27.

91. Zemel, B. (1995). *Tracers in the Oil Field*, Elsevier Science, Amsterdam.

92. Zhong, H., Zhao, Y., and Pei, Q. (2008). Photoluminescence quenching of conjugated polymer nanocomposites for gamma ray detection, *Nanotechnology*, **19**(50), 505503.

Chapter 3

Glass–Ceramic Scintillator

Jacqueline A. Johnson,[a] Russell L. Leonard,[a] Carlos Alvarez,[b] Brooke Barta,[c] and Stefan Schweizer[d,e]

[a]*Department of Mechanical, Aerospace and Biomedical Engineering, University of Tennessee Space Institute, Tullahoma, TN 37388, USA*

[b]*Department of Materials Science and Engineering, Northwestern University, Evanston, IL 60209, USA*

[c]*Electro-Optical Systems Laboratory, Georgia Tech Research Institute, Atlanta, GA 30332, USA*

[d]*Fraunhofer Application Center for Inorganic Phosphors, Branch Lab of Fraunhofer Institute for Microstructure of Materials and Systems IMWS, Lübecker Ring 2, 59494 Soest, Germany*

[e]*Department of Electrical Engineering, South Westphalia University of Applied Sciences, 59494 Soest, Germany*

jjohnson@utsi.edu

Glass–ceramic scintillators are a field rich in research opportunities and potential applications. The chapter leads the reader in the direction of potential exploration routes to improve such materials so that they may be used commercially. The work described here encompasses novel X-ray, γ-ray, α-particle, and neutron scintillator materials based on glass ceramics. Synthesis methods, characterization, and applications, including medical and

Nanocomposite, Ceramic, and Thin Film Scintillators
Edited by Martin Nikl

ISBN 978-981-4745-22-2 (Hardcover), 978-981-4745-23-9 (eBook)
www.panstanford.com

homeland security, are discussed. These materials have enormous potential owing to their ability to accept a broad range of elemental compositions, protect normally hygroscopic crystalline materials, and be cast in almost any shape or size. Consequently, the materials have a great future in radiation devices and are also attractive for fundamental science studies.

3.1 Introduction

Glass ceramics are composite materials consisting of small crystallites embedded within a glass matrix. In general, they are fabricated like a glass but have some of the special properties of ceramics.

The conventional synthesis of glass ceramic materials is usually a 2-step process in which the constituent powders are melted together for some period of time followed by rapid cooling, creating a glass. After the glass solidifies and cools, it is heated once more and undergoes partial crystallization to become a glass ceramic. The crystallites developed are generally in the nanometer range and so may exhibit all the properties of nanomaterials, which are attributed to the large ratio of surface area to volume. The properties of the materials can be tailored by changing the composition of the base glass and the heat-treatment protocol. It is also possible to separately synthesize nanoparticles and disperse them in a glass or polymer matrix, which has the advantage that nanoparticles do not have to survive the matrix processing conditions.

Glass ceramics can also be made via a sol–gel route. Sol–gel processing is a low-temperature synthesis, in which the hydrolysis and condensation of the precursors result in the formation of a dispersion of colloidal nanoparticles in a liquid, otherwise known as the sol. The subsequent polymerization and condensation reactions turn the sol into a gel that has an interconnected, rigid network with pores of submicron dimensions. Nanocrystallites are precipitated in the same way as conventional glass ceramics, i.e., by heat treatments.

The overall principle of X-ray scintillation in glass–ceramic materials is that incident radiation generates multiple electron–hole (e–h) pairs and recombination of the e–h pair releases energy. This recombination energy is transferred to a doped activator, often but not always, a lanthanide ion, which then emits a photon in the visible or near-visible spectrum. This allows such a material to be used as a detector, either passively or actively. Scintillation detectors have applications in medicine, nondestructive evaluation, high-energy physics, and homeland security. Conversion efficiency, scintillation decay time, density, chemical stability, and afterglow are important properties along with stopping power and resolution (both spatial and energy). The majority of these properties are highly dependent on nanoparticle size, which is a direct result of the heat-treatment temperature and time.

For glass ceramics to be successfully used as scintillators, several material properties must first be optimized. For example, the nucleation and growth of crystallites must be readily controllable by varying aging temperature or duration, and the lanthanide activator must be chosen to produce an emission wavelength that matches the peak spectral sensitivity of the photodetector used.

In this chapter, we discuss glass–ceramic materials predominantly, but not exclusively, containing lanthanides. The trivalent lanthanide ions have a rich energy spectrum from the near infrared to the ultraviolet spectral region, which makes them prominent for many optical applications. In Section 3.2 we describe the materials themselves, how these materials are synthesized, the principal of scintillation, and characterization methods. In Section 3.3 we describe applications in medicine, particularly image plates and the intricate requirements for their use pertaining to human subjects. In Section 3.4 the use of glass ceramics in homeland security will be discussed, where it is desirable to minimize the time for identification of any contraband material. Conclusions are given in Section 3.5.

3.2 Glass Ceramics

3.2.1 Materials Systems

Typical melt-derived glass ceramics can be composed of many different systems; those pertaining to scintillators can be broadly classified as oxides [1, 2] and fluorides [3–6] or a mixture of the two, oxyfluorides [7–11]. The sol–gel synthesis method is also an option for making glass ceramics [12–14], as is dispersing nanoparticles in a polymer or glass matrix [15].

One recent oxide material is vanadium-doped zinc borosilicates [2], which is a superfast scintillator. The base glass is formed from $ZnO\text{-}SiO_2\text{-}B_2O_3$. Various amounts of V_2O_5 were added to enhance the optical properties. The different nanocrystalline phases seen depend on heat treatment temperatures and times. For details see Hamnabard et al. [2].

Two other recent examples of oxide materials are yttrium and lutetium orthoborates, synthesized in thin layers [1]. Orthoborates are well known as effective phosphors and scintillators. Such materials are generally fabricated by reacting lanthanide oxide powders with boron anhydride B_2O_3 melt at elevated temperatures. Kedrov et al. used lutetium and yttrium oxides, activated by Ce^{3+} (~1 at.%) mixed with glass-forming media and glass components, which were heated in a furnace at 650 °C. The resulting vitreous mass is ground with alcohol to a fine powder and deposited on either a fused silica or sapphire substrate. After annealing, a glass–ceramic transparent film was formed containing crystallites of ~9 nm of either yttrium or lutetium orthoborate.

A class of fluorochlorozirconate (FCZ) glasses has been developed for X-ray image plates [3–5]. These plates are based on the ZBLAN (zirconium, barium, lanthanum, aluminum, and sodium fluoride) formulation with additional chlorine and europium doping and a small amount of indium fluoride to maintain zirconium in the 4^+-valence state. The constituent chemicals are melted at ~750 °C for 1 hour then poured into a preheated brass mold at 200 °C and cooled to room temperature over 4 hours. This system has a glass transition temperature marginally above 210 °C and so is easy to work with. Heat-treating at ~270 °C brings about the precipitation of $BaCl_2$ nanocrystallites, which contain europium dopants. Scintillation arises upon irradiation with ionizing radiation such as X-rays.

A similar material system has been used as a neutron image plate. The material also contains europium-doped $BaCl_2$ nanocrystallites but the glass matrix is now composed of lithium borate; neutron absorbing isotopes, ^{10}B and ^{6}Li were used to enrich the material [6–9]. The mixture of oxide and halide components makes this material an oxyhalide glass ceramic. Eu^{2+}-doped $SiO_2\text{-}Al_2O_3\text{-}CaO\text{-}CaF_2$ precipitates CaF_2 nanocrystallites after synthesis and heat treatment but requires much higher processing temperatures due to the oxide component. Ce^{3+}-doped GdF_3 and $GdBr_3$ were precipitated in aluminosilicate glasses to produce transparent samples suitable for γ-ray detection, although energy resolution was poor in comparison to NaI:Tl.

The development of new glass–ceramic materials is currently a busy field, especially in the sol–gel area due to the room temperature options for synthesis and ability to produce uniform dispersion of nanoparticles. Sol–gels can also be oxides or oxyhalides. One example of an oxide sol–gel is a Ce-doped lutetium silicate [12], which has properties that compare favorably to single crystal lutetium oxyorthosilicate (LSO). A more intricate material is LaF_3:Ce-doped organically modified silica (ORMOSIL). Incorporating organics into sol–gel-derived silica allows for flexibility, permeability, and porosity. The ORMOSIL was synthesized using a co-polymerization of tetramethoxysilane (TMOS) and (3-aminopropyl)triethoxysilane (APTES) [16, 17]. Yao et al. synthesized both water- and ethanol-soluble LaF_3:Ce to incorporate into ORMOSOL [14]. These samples are shown on the left and right of Fig. 3.1, respectively.

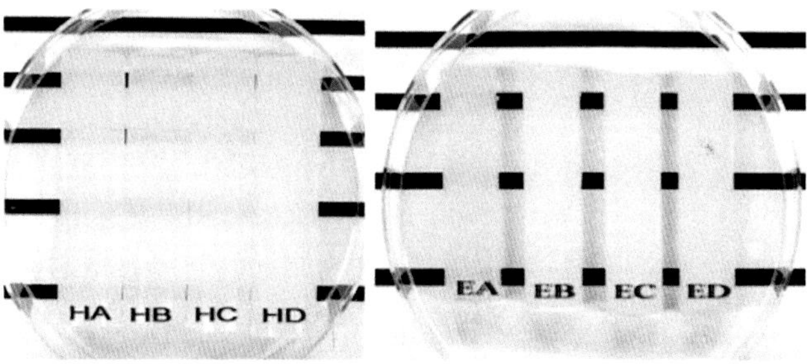

Figure 3.1 Left: Water-based LaF_3:Ce-doped ORMOSILS with loading levels of HA-3.63%, HB-7.01%, HC-10.17%, and HD-13.11%. Right: Ethanol-based LaF_3:Ce-doped ORMOSILS with loading levels of EA-15.66%, EB-27.09%, EC-35.78%, and ED-42.63%. Reprinted from Ref. 14, with the permission of AIP Publishing.

The final class of materials to be discussed is nanocomposites made by the separate creation of nanoparticles, which are then dispersed into a polymer, epoxy, or glass matrix [15]. CdTe and CdHgTe quantum dots (QDs) were incorporated into poly-methylmethacrylate (PMMA) matrices. The samples are shown in Fig. 3.2. These are promising systems for γ-ray spectroscopy.

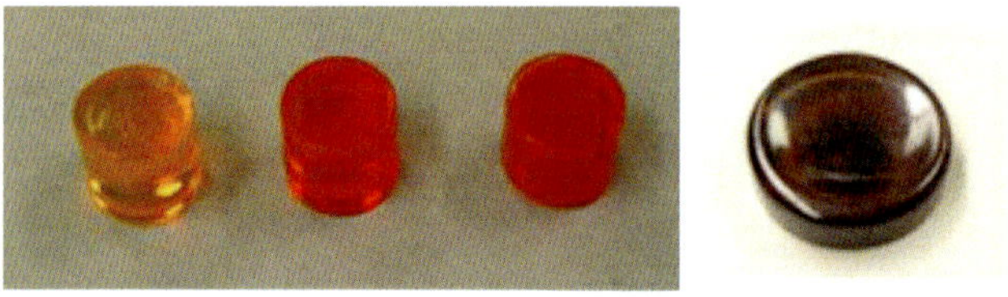

Figure 3.2 PMMA-QD composites (left CdTe, right CdHgTe). Reprinted with permission from Ref. 15.

3.2.2 Characterization

Samples are typically analyzed before the annealing step by differential scanning calorimetry (DSC) in order to ascertain the glass transition and crystallization temperatures, which in turn determines the heat-treatment protocol for precipitating nanocrystallites within the glass matrix. Figure 3.3 shows a typical DSC curve [18]; the important points of interest are marked for an FCZ glass with an Eu^{2+} to Eu^{3+} fraction of 60%, measured with a heating rate of 10 K/min. T_g is the glass transition temperature, i.e., the point where glass viscosity and specific heat capacity drop significantly. The crystallization onset, T_x, and the peak temperature of the first crystallization, T_{p1}, are attributed to the crystallization of hexagonal $BaCl_2$, while T_{p2} is attributed to the phase transition from hexagonal to orthorhombic $BaCl_2$. DSC measurements with different heating rates, α, enable the determination of the $BaCl_2$ crystallization activation energy, E_a, by Kissinger's method [19]:

$$\frac{\mathrm{d}\left[\ln(\alpha / T_\mathrm{p}^2)\right]}{\mathrm{d}(1/T_\mathrm{p})} = -\frac{E_{\mathrm{a,\,Kissinger}}}{R} \tag{3.1}$$

where R is the gas constant. With higher heating rates the crystallizations are shifted to higher temperatures; their peak intensities are increased due to an enhanced heat flow $\mathrm{d}H_c/\mathrm{d}T$ into the samples, which is given by the product

$$\mathrm{d}H_c/\mathrm{d}T = m \cdot c_\mathrm{p} \cdot \alpha \tag{3.2}$$

with the sample mass, m, the sample specific heat capacity, c_p, and the constant heating rate, α. When plotting $\ln(\alpha/T_p{}^2)$ vs. $(1/T_p)$ as shown in Fig. 3.4, E_a can be obtained from the slope of the linear fit. As Eq. 3.1 suggests, lower E_a values correspond to more stable glasses and a larger separation between T_p and T_g.

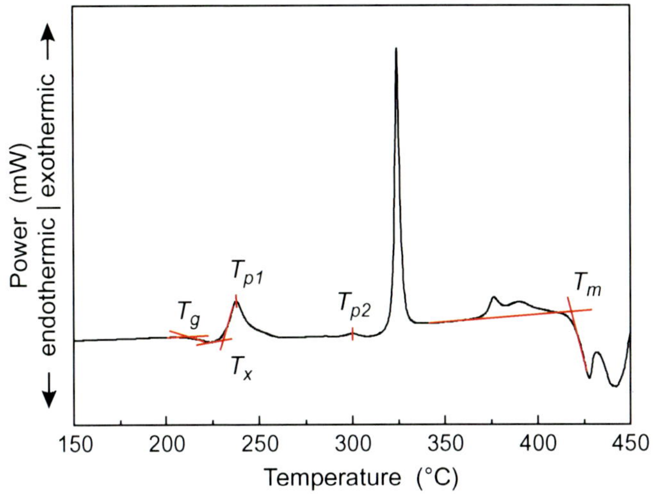

Figure 3.3 Plot of the DSC data for the FCZ glass with a Eu^{2+} to Eu^{3+} fraction of 60%. The heating rate was 10 K/min. Reprinted from Ref. 18, Copyright 2013, with permission from Elsevier.

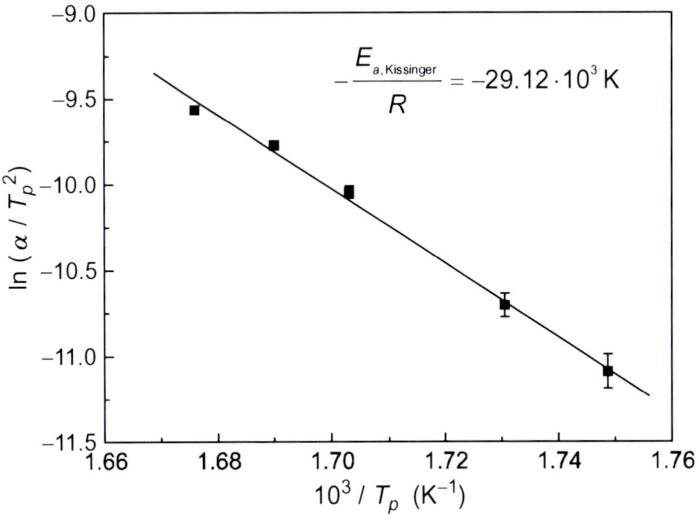

Figure 3.4 Fit for $EuCl_3$-doped FCZ glass. The slope gives the apparent activation energy, E_a. Reprinted from Ref. 18, Copyright 2013, with permission from Elsevier.

The structural phase of the nanocrystallites, determined by X-ray diffraction (XRD), is important as it can change the properties of the material. For example, Fig. 3.5 shows XRD data for a Eu-doped FCZ glass ceramic held for 20 minutes at temperatures of 250, 260, 270, 280, 285, and 292 °C [20]. The as-poured glass does not show any crystalline phases, whereas annealing at or below 270 °C leads to the formation of hexagonal phase $BaCl_2$ nanocrystals (space group P2m (189), see top bar pattern). A phase transition to the orthorhombic $BaCl_2$ structure (space group Pnam (62), see bottom bar pattern) can be observed upon annealing at 285 °C and 292 °C.

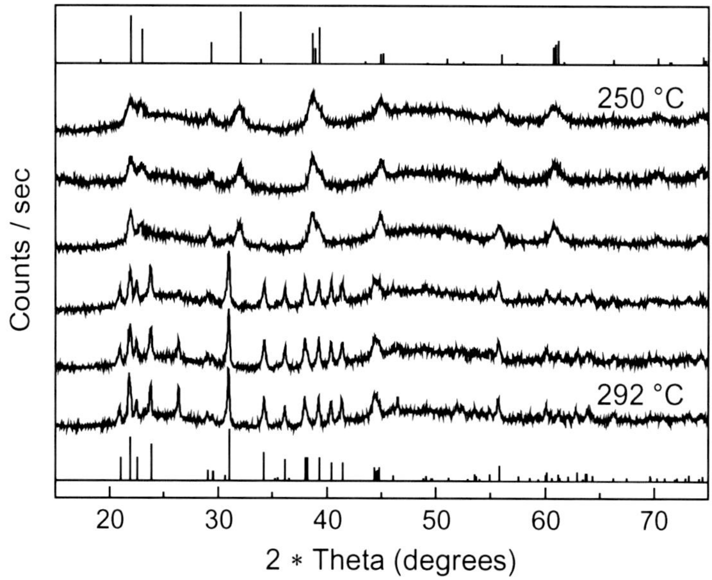

Figure 3.5 XRD data for a Eu-doped FCZ glass ceramic held for 20 minutes at 250, 260, 270, 280, 285, and 292 °C with PDF 24-0094, orthorhomic $BaCl_2$ and PDF 45-1313, hexagonal $BaCl_2$. Reprinted from Ref. 20, with the permission of AIP Publishing.

XRD can be used to give a rough estimate of nanocrystallite size using the Scherrer equation if applied correctly. The Scherrer equation was published in 1918 [21]:

$$L = \frac{K\lambda}{\beta(2\theta)\cos\theta} \tag{3.3}$$

where L is the average nanocrystallite size, β is the FWHM at any θ, λ is the X-ray wavelength, and K is the constant of proportionality. Therefore, peak width is inversely proportional to crystallite size. However, contributions to peak broadening are not only due to the crystallites but will certainly have a contribution from instrumental broadening and likely from some micro-strain in the crystallite as well; the latter contributions must be accounted for. A further uncertainty is in the determination of K. K depends on how the width of the crystallites is determined, along with the shape and size distribution. For spherical crystals with cubic symmetry, when the size is determined from FWHM, $K = 0.94$. When the crystallite size is determined from the integral breadth, i.e., the total area under the peak divided by the height of the peak, $K = 0.89$. For a detailed discussion on the evaluation of K, see Langford and Wilson [22].

A further important characterization method for glass–ceramic scintillators is photoluminescence (PL) as it can give information about the lanthanide environment. For example, in FCZ glasses, PL easily distinguishes the hexagonal from the orthorhombic form of $BaCl_2$ crystallites [20]. This is very important for these materials as hexagonal structure leads to scintillation, i.e., spontaneous recombination of electron–hole pairs, and orthorhombic structure leads to storage of electron–hole pairs, which can later be released by stimulation with a laser. The heat-treated glass is generally excited in the middle or near ultraviolet portion of the electromagnetic spectrum; with heat treatment temperatures of 280 °C and below, a double emission spectrum is seen with peaks at ~407 and ~485 nm, typical of the hexagonal structured $BaCl_2$. At ~290 °C the structure transforms to the orthorhombic phase of $BaCl_2$, resulting in an altered emission spectrum [20] (Fig. 3.6).

While a lot can be learned from studies on bulk materials, the growth of nanoparticles inside a glass matrix can be seen directly from transmission electron microscopy (TEM), in particular with in situ heat treatments. Figure 3.7 shows a low magnification dark field TEM image of a sample prepared from an as-poured modified ZBLAN glass (left), together with a selected area diffraction pattern (right). The lack of contrast in the dark field image, and the broad rings in the diffraction pattern, show that the region is amorphous.

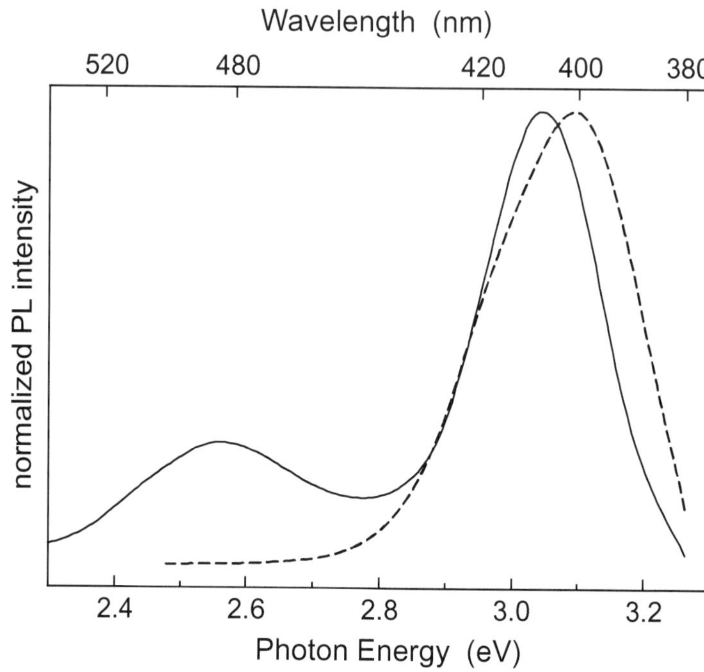

Figure 3.6 Normalized PL spectra of Eu-doped FCZ glasses processed at 260 °C (solid curve) and 290 °C (dashed curve) for 20 min. The PL was excited at 285 nm and recorded at room temperature. Reprinted from Ref. 20, with the permission of AIP Publishing.

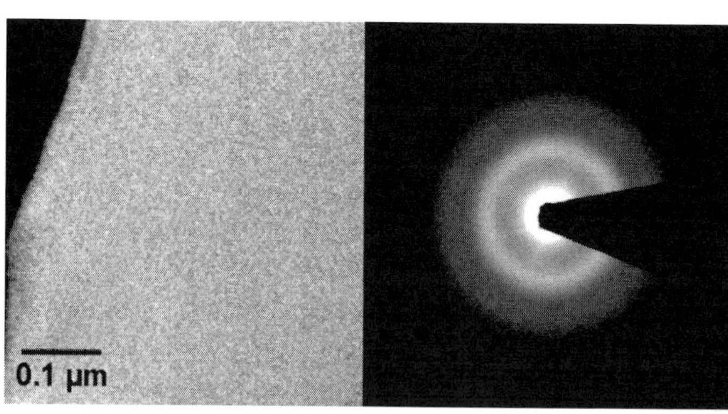

Figure 3.7 Low magnification dark field TEM image of ZBLAN (left) and selected area diffraction pattern (right). Reprinted from Ref. 23, Copyright 2011, with permission from Cambridge University Press.

Figure 3.8 shows a bright field image (left) and corresponding electron diffraction pattern of the same area (right) after the sample was annealed in the TEM. The contrast of the small regions and the polycrystalline diffraction indicate nanocrystallite formation during annealing; the size of the formed nanocrystals however, varies from a few nanometers to tens of nanometers. Figure 3.9 shows in situ high-resolution TEM images of hexagonal $BaCl_2$ nanocrystallites [23]. This work is in its early stages, but further details will soon be available in the literature to show how glass ceramics behave on the atomic scale.

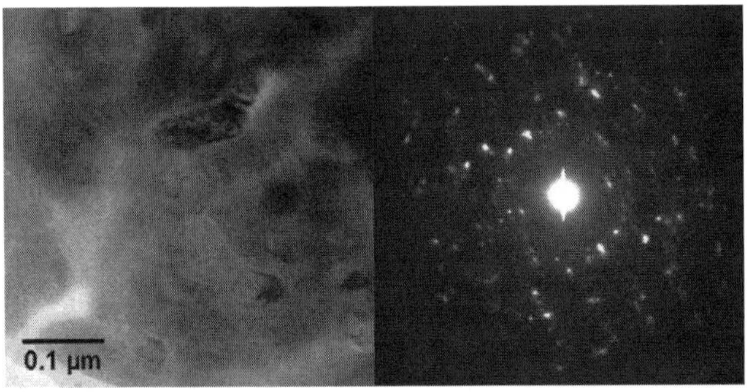

Figure 3.8 Bright field TEM image (left) and selected area diffraction pattern (right) of partially crystallized ZBLAN. Reprinted from Ref. 23, Copyright 2011, with permission from Cambridge University Press.

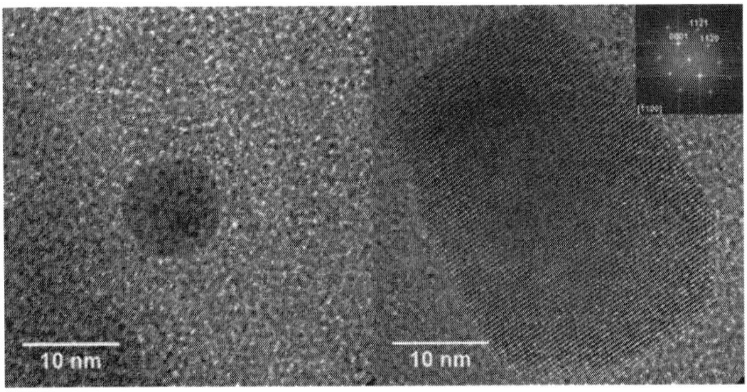

Figure 3.9 High-resolution TEM of $BaCl_2$ nanoparticles (right) with corresponding FFT of a hexagonal $BaCl_2$ nanocrystal at the $[\bar{1}100]$ zone axis. Reprinted from Ref. 23, Copyright 2011, with permission from Cambridge University Press.

3.3 Applications in Medicine

Scintillators are used in many aspects of medicine, for example, planar X-ray imaging, positron emission tomography (PET), X-ray computed tomography (X-ray CT), and single-photon emission tomography (SPECT). Such applications rely on single crystal or transparent ceramic technology. Table 3.1 shows the annual demand for scintillators used in medical imaging instruments, amounting to about 175 metric tons [24]. Ideally, these materials should have high density, high luminous efficiency, short decay time, no afterglow, good spectral match to photodetectors, and low cost. However, compromises nearly always have to be made.

Table 3.1 Annual worldwide volume of scintillator required by medical imaging instruments

Modality	Annual Production	Scintillator Content	Annual Scintillator Volume
Planar X-ray	1,000,000 screens	50 cc/screen	50,000 liters
X-ray CT	2,000 scanners	75 cc/scanner	150 liters
SPECT	2,000 scanners	3,000 cc/scanner	6,000 liters
PET	50 scanners	10,000 cc/scanner	500 liters

Traditional inorganic scintillators are CsI:Na, CsI:Tl, NaI:Tl, $CdWO_4$, and $Bi_4Ge_3O_{12}$ (BGO) [25, 26]. They have light yields larger than 10,000 photons per MeV but decay times slower than 200 ns. In comparison, the undoped halides BaF_2, CsF, CeF_3, and CsI, which have decay times in the range from 1 to 30 ns, have light yields of only a few thousand photons per MeV for the fast component. Other materials include cerium doped inorganics, Lu_2SiO_5:Ce, Gd_2SiO_5:Ce, $YAlO_3$:Ce, $LuAlO_3$:Ce, $Lu_2Si_2O_7$:Ce, and cerium-doped ^{6}Li glass [25, 26]. Plastic scintillators [27], anthracene and stilbene, are used almost exclusively as guard shields for low level counting due to their poor efficiency in detecting low energy X-rays. As no existing scintillator material has all the desired properties, there is an ongoing search for alternate materials.

Glass ceramics have the properties of both glass and ceramics. In addition, they can be produced in any shape or size, including large-area detectors with very high resolution and adequate efficiency, and can be tailored to produce individual desirable properties for specific medical applications. The material can be adapted to detect γ-rays as well as X-rays with the substitution of appropriate glass formers and additional dopants.

Two of the authors, Johnson and Schweizer, tested an FCZ plate in an X-ray imaging system built at the 2-BM beamline of the Advanced Photon Source (APS), Argonne National Laboratory [3, 4]. A 5.5 mm (horizontal) $\times$ 3.8 mm (vertical) monochromatic X-ray beam was used as an imaging source. Although a synchrotron may seem to differ too greatly from standard medical X-rays for this setup to be useful, synchrotron beams are being increasingly used for medical applications [3, 4]. The X-ray beam passes through a phantom and projects its image on the FCZ plate. Many phantoms have been used to test the resolution of the glass–ceramic plates, in particular, the precision test pattern MA0647, which contains a series of parallel gold grids with spatial frequencies up to 20 line pairs/mm. The X-ray images of the test patterns are formed instantly by focusing the light emitted from the FCZ plate through a 1.25× objective lens onto a cooled charge-coupled device (CCD) camera.

Figure 3.10 shows the scintillation intensity obtained for Eu-doped FCZ glass ceramics as a function of X-ray intensity. These measurements were performed at 16 keV, and the X-ray intensity was reduced by a series of aluminum filters. Acquisition times varied from 3 s to 30 min, depending on the X-ray intensity. At low X-ray intensity longer acquisition times were needed to overcome a problem with signal-to-noise ratio. The scintillation efficiency varies linearly with X-ray intensity over the range measured. The deviation of the 0.2% sample from the slope of the CWO and 2% sample is likely due to the poor signal-to-noise ratio at low intensity.

Figure 3.11 shows the modulation transfer function (MTF) versus spatial frequency for CWO and the 2% Eu-doped FCZ glass ceramic. The graph gives a measure of the image quality from the relative contrast of an imaging system as a function of spatial frequency. The resolution is defined as the value of the spatial frequency corresponding to a certain MTF value. Here, the value of resolution is taken at an MTF of 0.2. The resolution of the scintillator system is ~110 lp/mm (at MTF = 0.2). In other words, this system is able to resolve features as small as 4–5 μm. The resolution of CWO, which is the current standard scintillator for X-ray tomography, is ~130 lp/mm, which implies it can detect features of just less than 4 μm.

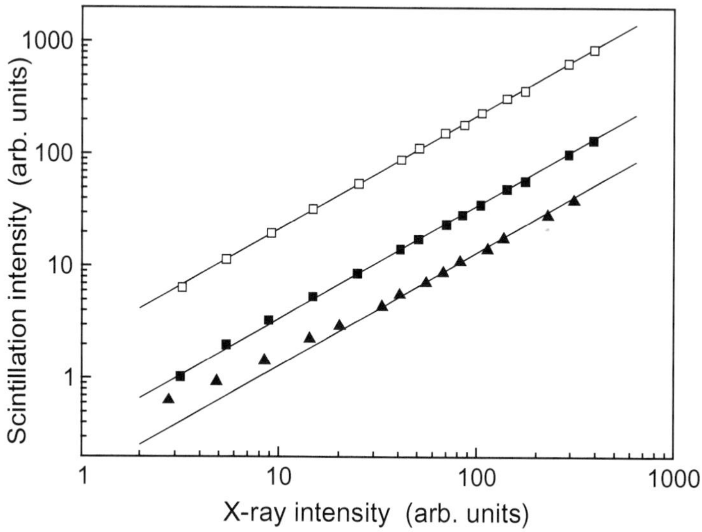

Figure 3.10 Scintillation intensity vs. X-ray dose of Eu-doped, FCZ glass ceramics (full symbols). The Eu-doping level was 0.2% (triangles) and 2% (squares). The scintillation efficiency of cadmium tungstate is shown for comparison (open squares). The lines for the CWO and 2% Eu data are a linear fit; the 0.2% data are believed to be inaccurate at low X-ray intensity, and so the line is made parallel to the other two lines. Reprinted from Ref. 3, with the permission of AIP Publishing.

In the glass–ceramic samples presented here as scintillators, the $BaCl_2$ crystallites are always in the hexagonal phase and show a typical emission at about 410 nm under X-ray excitation (see Fig. 3.12). Interestingly, there is a shift in the 407 nm peak to 412 nm when the doping level is increased from 0.2% to 2%. This shift could be due to overlapping bands, self-absorption by europium, or a perturbed crystal field. The band near 485 nm is tentatively attributed to Eu^{2+} ions associated with an impurity in the hexagonal phase; this band scales linearly with europium doping.

The efficiency of the glass–ceramic plate is an extremely important parameter for medical applications; a patient should be exposed to the minimum possible dose. The efficiency of the samples discussed here increases with respect to europium doping amount but levels off rather quickly. This result implies that little would be gained in efficiency by increasing the europium doping level beyond 2%. There is also a limit to how much europium can be incorporated into the material as this element creates nucleation sites, which lead to crystallization.

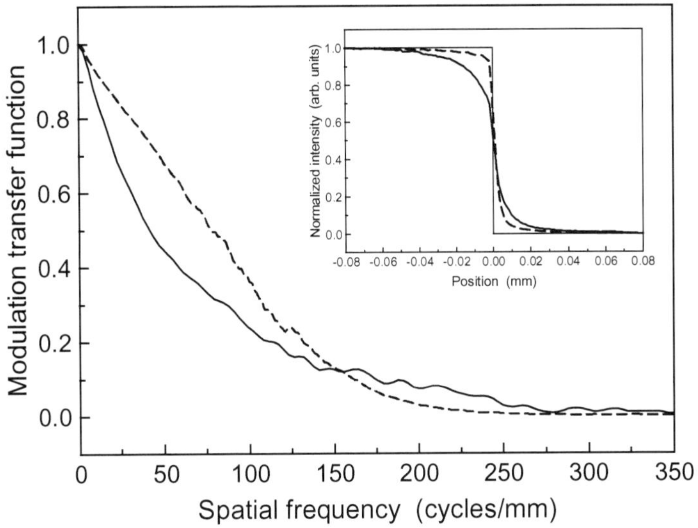

Figure 3.11 The MTF of a 2% Eu-doped, FCZ glass–ceramic scintillator (solid curve). The MTF of CWO (dashed curve) is shown for comparison. The inset shows the scintillation intensity profile for the FCZ plate (solid curve) and the CWO single crystal (dashed curve). The ideal profile (thin solid curve) is shown for comparison. Reprinted from Ref. 3, with the permission of AIP Publishing.

The current maximum efficiency obtained for these glass–ceramic plates is just 16% of CWO's efficiency. However, the glass–ceramic plate and the readout system should show improvement with optimization. The price and versatility of a glass ceramic is a significant advancement over current systems. The average crystallite size, determined from XRD is ~14 nm, much smaller than the wavelength of visible light. Therefore, light scattering due to particle size is negligible, and the glass ceramics are fully transparent. This behavior is important to resolution; resolution is soon lost once the particle size approaches the emission wavelength. Clarity of the image is crucial. The scintillation efficiency is linear over almost three orders of magnitude. At low X-ray intensity the signal-to-noise ratio is higher for equal time intervals; therefore, scan length was increased to compensate for this.

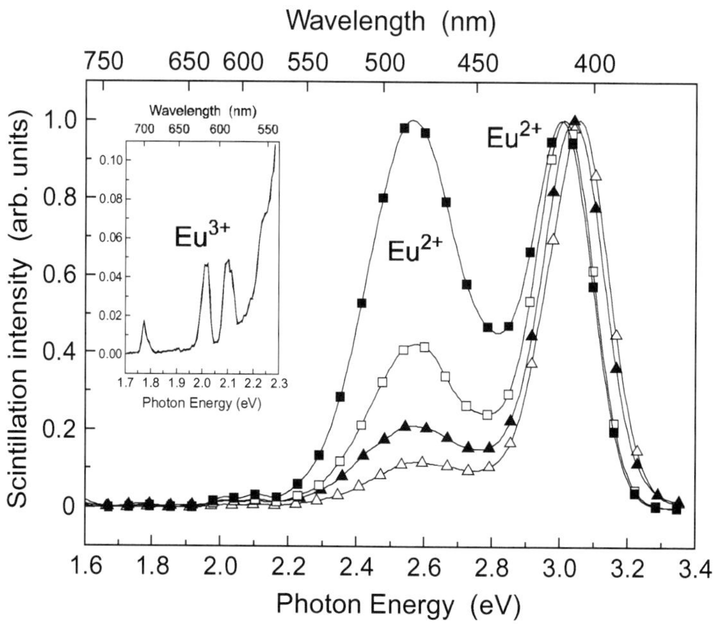

Figure 3.12 X-ray excited luminescence spectra of double-annealed FCZ glass ceramics. The Eu doping level was 0.2% (open triangles), 0.5% (full triangles), 1% (open squares), and 2% (full squares). The inset shows the Eu^{3+} bands in the 2% sample before smoothing the spectrum. Reprinted from Ref. 3, with the permission of AIP Publishing.

Figure 3.13 shows that the scintillation intensities of the glass ceramics with different Eu-doping levels follow the same curve. It increases quasi exponentially with X-ray energy and is proportional to the inverse mass absorption coefficient [3], i.e., with more photoelectric absorption there is less scintillation. The zirconium K-edge is sharp in the 1% and 2% doped sample, less sharp in the 0.5% sample, and hardly noticeable in the 0.2% sample. The fluorozirconate glass matrix obviously has an influence on the scintillation intensity of the glass ceramic. However, the inset in Fig. 3.13 shows that the intensity drop at 18 keV is not as absolute as one would expect from the (inverse) mass attenuation coefficient but varies from sample to sample; this variation is probably caused by the different lifetimes of the ionized Zr. It seems that the lifetime of the ionized Zr is influenced by the size, structure, and composition of the embedded nanoparticles.

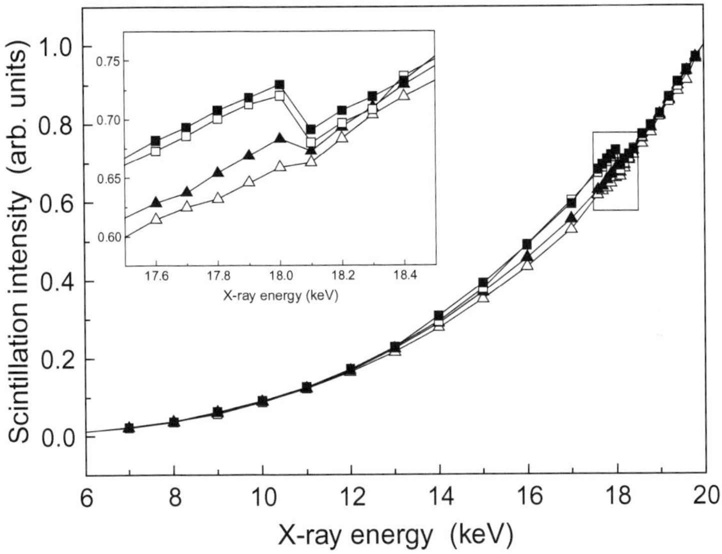

Figure 3.13 Scintillation intensity vs. X-ray energy of Eu-doped, FCZ glass ceramics. The Eu-doping level was 0.2% (open triangles), 0.5% (full triangles), 1% (open squares), and 2% (full squares). The lines are a guide to the eye, whereas the symbols represent the experimental data points. The inset shows the Zr edge at 18 keV. Reprinted from Ref. 3, with the permission of AIP Publishing.

The MTF curves shown in Fig. 3.11 indicate that in terms of resolution the glass-ceramic plate is almost state of the art but would only cost a fraction of the single-crystal CWO. This statement is supported by the image of a mouse leg in Fig. 3.14, where the detailed structure of the bones is clearly resolved in both images—the Eu^{2+}-doped scintillator is on the left, and CWO is on the right.

In general, optically active glass-ceramic materials have enormous potential as detectors in multiple medical imaging devices. However, optimization of efficiency proves to be the limiting factor. With further advancements, such materials could replace single crystal scintillators in many applications.

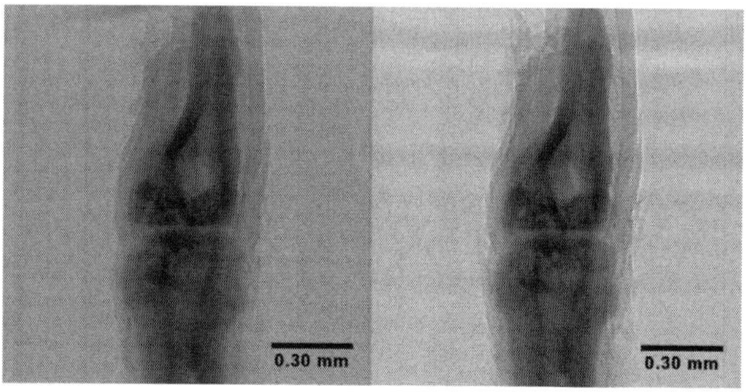

Figure 3.14 Scintillation image of a mouse joint, recorded with a double-annealed 2% Eu^{2+}-doped FCZ plate (left) and with a CWO single crystal (right). Reprinted from Ref. 3, with the permission of AIP Publishing.

3.4 Homeland Security Applications

In recent years, the United States Departments of Homeland Security (DHS) and Customs and Border Protection (CBP) have been charged with scanning every cargo container crossing domestic borders for illicit radioactive material [28]. This is accomplished by using γ-ray detection systems capable of discriminating between non-threatening radioisotopes, such as ^{137}Cs, which is often used in nuclear medicine, and fissile material, such as ^{235}U, that can be used to make nuclear weapons or "dirty" bombs. Scintillation detector systems, specifically thallium-doped sodium iodide (NaI(Tl)) single crystals, are by far the most popular for this purpose because they are inexpensive relative to other types of detectors, but are still able to identify isotopes with reasonable accuracy. However, increased demand for these systems has served as a catalyst for the research and development of new scintillator materials to surpass NaI(Tl), such as glass ceramics.

A schematic of the configuration of a standard radiation detection system is shown in Fig. 3.15 [29]. Scintillators (A) are coupled to a photomultiplier tube (PMT) (B) that converts the photon pulses produced by absorbed radiation energy into electrical pulses with amplitude proportional to the number of photons in the pulse. The peak spectral sensitivity wavelength will vary with the type of PMT used, though the most common models for γ-ray spectroscopy usually exhibit peak sensitivity around 350 nm. The electrical pulses are then passed to an array of signal amplifiers (C) and multi-channel analyzers (D) that sort the pulses, according to their amplitude, into channels that correspond to a particular energy. As each isotope emits a characteristic series of decay product energies, the source isotope can be identified by its pulse-height spectrum [30, 31]. The accuracy with which the source isotope may be identified depends on the energy resolution of the peaks in the pulse-height spectrum [32], which is defined by the full-width at half the maximum peak energy divided by the peak energy, i.e.

$$R = \frac{\Delta E}{E_{\text{peak}}} \tag{3.4}$$

Peak resolution increases with the number of photons produced per pulse by the scintillator [33–35], known as the light yield, and the number of photons produced by photoelectric interactions between the incident radiation particles and the scintillator material. NaI(Tl) scintillators generate 50,000 photons per MeV of radiation energy absorbed, resulting in an energy resolution of 6–7% for a 662 keV incident γ-ray [36]. Greater energy resolution improves the accuracy of isotope discrimination [37], as shown by the distinct spectra for ^{137}Cs and ^{60}Co shown in Fig. 3.16 [29].

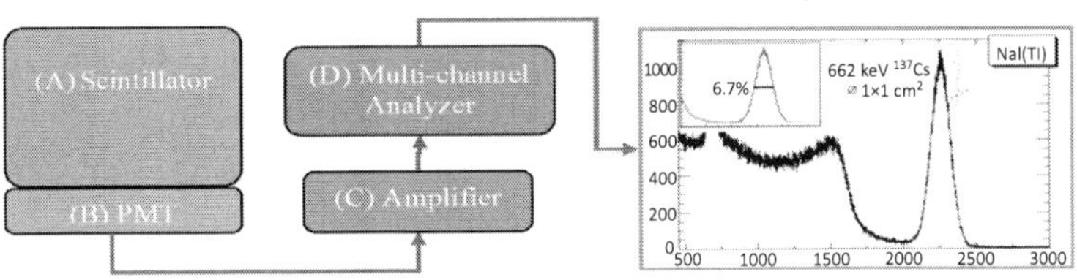

Figure 3.15 Schematic of γ-ray detection system. γ-ray spectra is reprinted from Ref. 29, Copyright 2002, with permission from Elsevier.

What is needed in a high-performance scintillator is efficient conversion of the absorbed incident radiation energy to detectable photons (high scintillation efficiency), with the light yield remaining proportional to the energy deposited over a wide energy range; a scintillator transparent to the emitted photon wavelengths to minimize absorption; the time required to convert radiation energy to photons (decay time) on the order of a few hundred nanoseconds or less; and a material fabricated to volumes of at least a few cubic centimeters [38]. No single scintillator fully meets all of these criteria; some properties must be sacrificed for the sake of improved performance elsewhere.

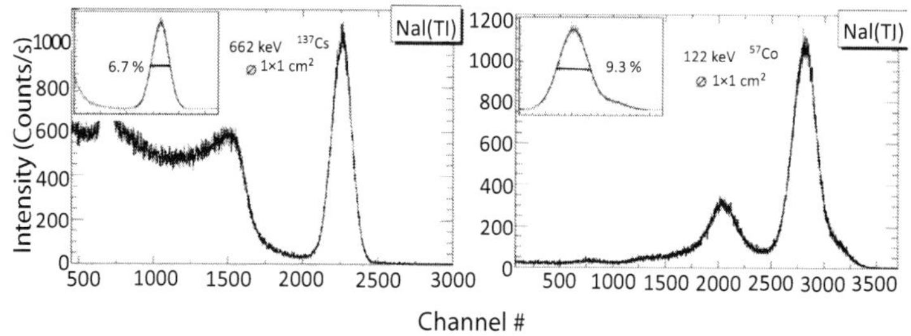

Figure 3.16 Energy spectra for ^{137}Cs (left) and ^{57}Co (right) generated by a NaI(Tl) single crystal. The inset indicates the scintillator's energy resolution at the photopeak. Reprinted from Ref. 29, Copyright 2002, with permission from Elsevier.

NaI(Tl) single crystals fulfill many of these requirements, though new single crystal compositions such as $LaBr_3$(Ce) and SrI_2(Ce) developed in the last decade suggest that energy resolution can be improved with the new compositions [39–42]. However, growing large volume crystals from these compounds is difficult because their low-symmetry crystal structures ($LaBr_3$(Ce) is monoclinic) cause anisotropic thermal expansion during cooling, which causes larger crystals to fracture [43]. Research efforts are also limited to cylindrical geometry, which eliminates investigation of the effect of varied geometries on the scintillator performance. Finally, most of these ionic salts are extremely hygroscopic, and so must be encased in stainless steel or another isolation medium to prevent degradation due to atmospheric moisture. Alpha particles cannot penetrate the cover and so these hygroscopic scintillators are useful only for gamma spectroscopy.

Glass–ceramic nanocomposite scintillators are a promising alternative to single crystals because new scintillator compositions, shapes, and sizes may be developed [3, 44–47]. Glass-ceramic nanocomposites contain nanoscale crystallites precipitated from an amorphous host glass. Using nanoscale scintillating crystallites rather than a single crystal monolith allows scintillator compositions such as $LaBr_3$(Ce) to be fabricated to larger (45 cm^3) volumes, and the spectroscopy performance to be directly compared to NaI(Tl) [38]. Also, because glass–ceramic synthesis begins by casting a homogeneous glass melt, new geometries may be fabricated and cast to near-net shape [48]. Finally, the glass matrix ensures environmental isolation of the hygroscopic scintillating crystals, and so glass–ceramic scintillators show potential to serve as viable detectors in alpha- and neutron-spectroscopy in addition to γ-rays [3, 49–51].

Currently there are no glass ceramics for neutron detection. However, this is an area of development for the authors. Glass-based detectors for thermal neutrons have already been made by researchers at Pacific Northwest National Laboratory [52] based on the interaction between incident neutrons, ^6Li, and Ce^{3+}. The reaction $^1n^0 + {}^6Li^{3+} \rightarrow {}^3H^+ + {}^4He^{2+}$ yields a tritium ion and α-particle or helium nucleus. The tritium ion then transfers energy to a Ce^{3+} ion, causing it to emit a visible photon.

There are several potential glass–ceramic systems that could serve as radiation detectors, including fluorozirconate glasses containing europium-doped barium chloride nanocrystals, sodium aluminosilicate (NAS) sodium borosilicate (NBS) and alumino-borosilicate (ABS) glasses containing cerium-doped gadolinium fluoride. Barta et al. [53] studied the silicates in detail.

Fluorochlorozirconate glass ceramics could be modified to include the components necessary for detection of neutrons. These formulations contain up to 3.5% LaF_3, which can be replaced in part or completely with CeF_3; they also contain up to 20% NaX (where X is either F or Cl), which can be partially replaced with ^6LiF. Thus, they could be doped with ^6Li and Ce^{3+}; some measurements with fluorochlorozirconate glass ceramics doped with Ce^{3+} have been done [54]. To moderate fast neutrons, it will be necessary to coat the glass with a polymer or some other neutron-absorbing material (i.e., ^{10}B, ^6LiF, or Gd).

Discrimination of neutrons from photons needs to be done for the detection of contraband materials. Neutrons are detected indirectly, through producing knocked-on charged particles, which then cause scintillation. In some materials the scintillation light from knocked-on protons has a different decay time than scintillation caused by photon-induced electrons. In these cases, good separation of neutrons and γ-rays can be achieved by pulse-shape analysis with preset counters like an NE213. On average, the neutron–γ-ray separation improves with the energy of the incident particles, as the pulses are bigger. However, for neutron-induced interactions of equivalent energy to photons, in organic scintillators, the pulses are smaller (~20%). Digital pulses can be captured and the difference between photon-induced and neutron-induced interactions quantified, using simultaneous time-of-flight measurements and pulse shape analysis. The time of flight tags the particles event-by-event, as photons travel at the speed of light while neutrons have a time of flight determined by their energy.

A comparison of the γ-ray and α-particle spectroscopy performance from the three different glass systems is shown in Fig. 3.17, left and right respectively, with the corresponding energy resolutions given in Table 3.2.

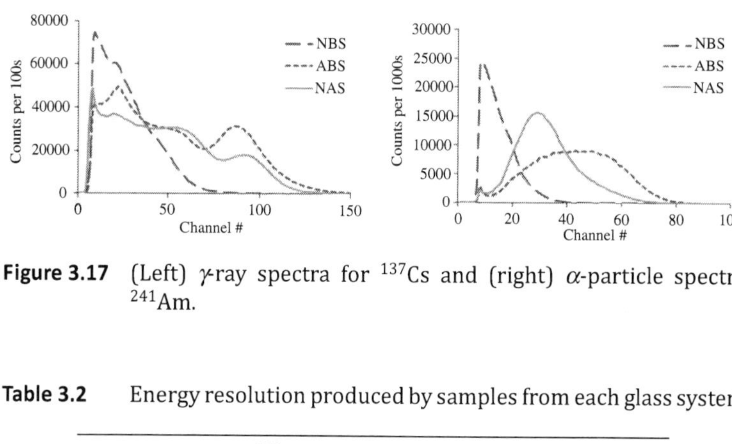

Figure 3.17 (Left) γ-ray spectra for ^{137}Cs and (right) α-particle spectra ^{241}Am.

Table 3.2 Energy resolution produced by samples from each glass system

Glass system	^{241}Am Resolution	^{137}Cs Resolution
NBS	No peak visible	No peak visible
ABS	89%	30%
NAS	32%	34%

The $GdBr_3$(Ce) doping concentrations were 15 mol% in the NBS matrix, 19 mol% in the NAS matrix, and 30 mol% in the ABS matrix. The ABS system produced the highest light yield in both α-particle and γ-ray spectroscopy, though the resolution at the 662 keV photopeak is finer in the NAS system (34%) compared to the

ABS system (38%). However, there is a narrow separation between the photopeak and the Compton edge in the ABS γ-ray spectrum, indicating that scattering and incomplete collection of photons were significant [37]. Overall, the energy resolution is poor, but further work on this and other glass–ceramic system could lead to significant improvements.

3.5 Conclusions

Glass ceramics have enormous potential in a number of applications but particularly as scintillators in medical and homeland security devices. This potential is largely due to their versatility. Numerous oxides, fluorides and oxyfluorides can be synthesized by a conventional melt-quench method and doped with optically active rare earths. The potential combination of glass formers and optically active components is vast and therefore, so are the possible properties of these materials. With the option of sol–gel synthesis, the net of applications is cast even wider. It is therefore surprising that very little work is going on in the area of scintillation and glass ceramics so an overview is a difficult task. The essence of the chapter is such that the reader can be encouraged to look at the possibilities of glass ceramics as scintillators for medical imaging, in particular, radiography, PET, SPECT, and CT. In addition, glass ceramics can be developed as homeland security devices for the detection of α and β particles along with γ-rays and neutrons or any combination. The future is thus bright for this field.

Acknowledgments

The characterization of FCZ glasses by TEM was supported by the National Science Foundation under Grant Number DMR1001381. The medical imaging work was supported by the National Institutes of Health under Grant Number 5R01EB006145—National Institute of Biomedical Imaging and Bioengineering. The content is solely the responsibility of the authors and does not necessarily represent the official views of the National Science Foundation, the National Institute of Biomedical Imaging and Bioengineering, or the National Institutes of Health.

The electron microscopy was accomplished at the Electron Microscopy Center at Argonne National Laboratory, a U.S. Department of Energy Office of Science Laboratory operated under Contract No. DE-AC02-06CH11357 by UChicago Argonne, LLC. This document was prepared by University of Tennessee Space Institute as a result of the use of facilities of the U.S. Department of Energy (DOE), which are managed by UChicago Argonne, LLC, acting under Contract No. DE-AC02-06CH11357. Neither UChicago Argonne, DOE, the U.S. Government, nor any person acting on their behalf: (a) make any warranty or representation, express or implied, with respect to the information contained in this document; or (b) assume any liabilities with respect to the use of, or damages resulting from the use of any information contained in the document.

One of the authors (S. Schweizer) wants to thank the "Ministerium für Innovation, Wissenschaft und Forschung des Landes Nordrhein-Westfalen" for its financial support to the Fraunhofer Application Center for Inorganic Phosphors in Soest.

References

1. Kedrov V.V., Shmytko I.M., Shmurak S.Z., Kudrenko E.A., Klassen N.V. (2012) Structural and morphological characteristics of yttrium and lutetium orthoborates synthesized in thin layers of borate oxide glasses. *Journal of Materials Research* **27**, pp. 2116–2121.
2. Hamnabard Z., Khalkhali Z., Qazvini S.S.A., Baghshahi S., Maghsoudipour A. (2012) Preparation, heat treatment and photoluminescence properties of V-doped ZnO-SiO_2-B_2O_3 glasses. *Journal of Luminescence* **132**, pp. 1126–1132.
3. Johnson J.A., Schweizer S., Henke B., Chen G., Woodford J., Newman P.J., *et al.* (2006) Eu-activated fluorochlorozirconate glass-ceramic scintillators. *Journal of Applied Physics* **100**, 034701.
4. Chen G., Johnson J., Weber R., Nishikawa R., Schweizer S., Newman P., *et al.* (2006) Fluorozirconate-based nanophase glass ceramics for high-resolution medical X-ray imaging. *Journal of Non-Crystalline Solids* **352**, pp. 610–614.
5. Schweizer S., Johnson J.A. (2007) Fluorozirconate-based glass ceramic X-ray detectors for digital radiography. *Radiation Measurements* **42**, pp. 632–637.

6. Appleby G.A., Vontobel P. (2008) Optimisation of lithium borate-barium chloride glass-ceramic thermal neutron imaging plates. *Nuclear Instruments & Methods in Physics Research Section a-Accelerators Spectrometers Detectors and Associated Equipment* **594**, pp. 253–256.
7. Appleby G.A., Edgar A., Williams G.V.M., Vontobel P. (2006) Structure and neutron imaging characteristics of lithium borate-barium chloride glass-ceramics. *Nuclear Instruments & Methods in Physics Research Section a-Accelerators Spectrometers Detectors and Associated Equipment* **564**, pp. 424–430.
8. Appleby G.A., Edgar A., Williams G.V.M., Bos A.J.J. (2006) Photostimulated luminescence from $BaCl_2 : Eu^{2+}$ nanocrystals in lithium borate glasses following neutron irradiation. *Applied Physics Letters*, **89**, 101902.
9. Appleby G.A., Bartle C.M., Williams G.V.M., Edgar A. (2006) Lithium borate glass ceramics as thermal neutron imaging plates. *Current Applied Physics* **6**, pp. 389–392.
10. Han C.L., Barta M., Dorn M., Nadler J., Rosson R., Wagner B., *et al.* (2011) Transparent oxyhalide glass and glass ceramics for gamma-ray detection. *Hard X-Ray, Gamma-Ray, and Neutron Detector Physics XIII*, **8142**.
11. Fu J., Kobayashi M., Sugimoto S., Parker J.M. (2009) Scintillation from Eu^{2+} in nanocrystallized glass. *Journal of the American Ceramic Society* **92**, pp. 2119–2121.
12. Bescher E., Robson S.R., Mackenzie J.D., Patt B., Iwanczyk J., Hoffman E.J. (2000) New lutetium silicate scintillators. *Journal of Sol-Gel Science and Technology* **19**, pp. 325–328.
13. Wallace S.A., Stephan A.C., Cooper R., Im H.J., Dai S. (2007) Lithiated sol-gel based neutron scintillators. *Nuclear Instruments & Methods in Physics Research Section a-Accelerators Spectrometers Detectors and Associated Equipment* **579**, pp. 84–187.
14. Yao M.Z., Hall R., Chen W., Mohite D.P., Leventis N., Lu N., *et al.* (2013) Luminescent LaF3:Ce-doped organically modified nanoporous silica xerogels. *Journal of Applied Physics* **113**, 013111.
15. Wagner B.K., Kang Z.T., Nadler J., Rosson R., Kahn B. (2012) Nanocomposites for radiation sensing. *Micro- and Nanotechnology Sensors, Systems, and Applications IV*, **8373**, doi:10.1117/12.920858.
16. Katti A., Shimpi N., Roy S., Lu H.B., Fabrizio E.F., Dass A., *et al.* (2006) Chemical, physical, and mechanical characterization of isocyanate cross-linked amine-modified silica aerogels. *Chemistry of Materials* **18**, pp. 285–296.

17. Leventis N. (2007) Three-dimensional core-shell superstructures: Mechanically strong aerogels. *Accounts of Chemical Research* **40**, pp. 874–884.
18. Paßlick C., Johnson J.A., Schweizer S. (2013) Crystallization studies on rare-earth co-doped fluorozirconate-based glasses. *J. Non. Cryst. Sol.* **371–372**, pp. 33–36.
19. Kissinger H.E.. (1957) Reaction kinetics in differential thermal analysis. *Anal. Chem* **29**, pp. 1702–1706.
20. Schweizer S., Hobbs L.W., Secu M., Spaeth J.M., Edgar A., Williams G.V.M., *et al.* (2005) Photostimulated luminescence from fluorochlorozirconate glass ceramics and the effect of crystallite size. *Journal of Applied Physics* **97**.
21. Scherrer P. (1918) Bestimmung der Grösse und der inneren Struktur von Kolloidteilchen mittels Röntgenstrahlen. *Nachr. Ges. Wiss. Göttingen* **26**, pp. 98–100.
22. Langford J.I., Wilson A.J.C. (1978) Scherrer after 60 years - Survey and some new results in determination of crystallite size. *Journal of Applied Crystallography* **11**, pp. 102–113.
23. Vu M., Alvarez C., Liu Y., Petford-Long A.K., Johnson J.A. (2011) In situ TEM studies of nanoparticle growth in a fluorozirconate (ZBLAN) glass matrix. *Microsc. Microanal.* **17**, pp. 496–497.
24. Moses W.W. (1999) Scintillator requirements for medical imaging. In: *International Conference on Inorganic Scintillators and Their Applications: SCINT99*. Moscow, Russia.
25. van Eijk C.W.E. (2001) Inorganic-scintillator development. *Nuclear Instruments & Methods in Physics Research Section a-Accelerators Spectrometers Detectors and Associated Equipment* **460**, pp. 1–14.
26. van Eijk C.W.E. (2002) Inorganic scintillators in medical imaging. *Physics in Medicine and Biology* **47**, pp. R85–R106.
27. Schorr M.G., Torney F.L. (1950) Solid non-crystalline scintillation phosphors. *Physical Review* **80**, pp. 474–474.
28. Congress US. Security and accountability for every port act Washington, D.C. 2006.
29. Moszynski M., Zalipska J., Balcerzyk M., Kapusta M., Mengesha W., Valentine J.D. (2002) Intrinsic energy resolution of NaI(T1). *Nuclear Instruments & Methods in Physics Research Section a-Accelerators Spectrometers Detectors and Associated Equipment* **484**, pp. 259–269.

30. Milbrath B.D., Peurrung A.J., Bliss M., Weber W.J. (2008) Radiation detector materials: An overview. *Journal of Materials Research* **23**, pp. 2561–2581.

31. Klann R.T., Shergur J., Mattesich G. (2009) Current state of commercial radiation detection equipment for homeland security applications, *Nucl. Technol.* **168**, pp. 79–88.

32. Drozdowski W., Dorenbos P., Bos A.J.J., de Haas J.T.M., Kraft S., Maddox E., *et al.* (2007) Effect of proton dose, crystal size, and cerium concentration on scintillation yield and energy resolution of LaBr3 : Ce. *Ieee Transactions on Nuclear Science* **54**, pp. 736–740.

33. Nassalski A., Kapusta M., Batsch T., Wolski D., Mockel D., Enghardt W., *et al.* (2007) Comparative study of scintillators for PET/CT detectors. *Ieee Transactions on Nuclear Science* **54**, pp. 3–10.

34. Moszynski M., Balcerzyk M., Kapusta M., Syntfeld A., Wolski D., Pausch G., *et al.* (2005) $CdWO_4$ crystal in gamma-ray spectrometry. *Ieee Transactions on Nuclear Science* **52**, pp. 3124–3128.

35. Darambara D.G.. (2006) State-of-the-art radiation detectors for medical imaging: Demands and trends. *Nuclear Instruments & Methods in Physics Research Section a-Accelerators Spectrometers Detectors and Associated Equipment* **569**, pp. 153–158.

36. Rodnyi P.A. (1997) *Physical processes in inorganic scintillators*. Boca Raton: CRC Press.

37. Price W.J. (1958) *Nuclear radiation detection*. New York,: McGraw-Hill.

38. Knoll G.F. (2010) *Radiation detection and measurement*. 4th ed. Hoboken, N.J.: John Wiley.

39. Milbrath B.D., Choate B.J., Fast J.E., Hensley W.K., Kouzes R.T., Schweppe J.E. (2007) Comparison of LaBr3 : Ce and NAI(Tl) scintillators for radio-isotope identification devices. *Nuclear Instruments & Methods in Physics Research Section a-Accelerators Spectrometers Detectors and Associated Equipment* **572**, pp. 774–784.

40. Wilson C.M., Van Loef E.V., Glodo J., Cherepy N., Hull G., Payne S., *et al.* (2008) Strontium iodide scintillators for high energy resolution gamma ray spectroscopy. In: *Conference on Hard X-Ray, Gamma-Ray and Neutron Detector Physics X*. San Diego, CA.

41. Higgins W.M., Churilov A., van Loef E., Glodo J., Squillante M., Shah K. (2008) Crystal growth of large diameter LaBr3 : Ce and CeBr3. *Journal of Crystal Growth* **310**, pp. 2085–2089.

42. Cherepy N.J., Hull G., Drobshoff A.D., Payne S.A., Van Loef E., Wilson C.M., *et al.* (2008) Strontium and barium iodide high light yield scintillators. *Applied Physics Letters* **92**, 083508

43. Findley K.O., Johnson J., Bahr D.F., Doty F.P., Frey J. (2007) Fracture and deformation behavior of common and novel scintillating single crystals - art. no. 670706. In: *Conference on Penetrating Radiation Systems and Applications VIII*. San Diego, CA, pp. 70706–70706.

44. Yu Y.L., Chen D.Q., Wang Y.S., Liu F., Ma E. (2007) A new transparent oxyfluoride glass ceramic with improved luminescence. *Journal of Non-Crystalline Solids* **353**, pp. 405–409.

45. Qiao X.S., Fan X.P., Wang M.Q., Zhang X.H. (2008) Spectroscopic properties of Er^{3+}-Yb^{3+} co-doped glass ceramics containing BaF_2 nanocrystals. *Journal of Non-Crystalline Solids* **354**, pp. 3273–3277.

46. Petzoldt J., Pannhorst W. (1991) Chemistry and structure of glass-ceramic materials for high-precision optical applications. *Journal of Non-Crystalline Solids* **129**, pp. 191–198.

47. Mortier M., Bensalah A., Dantelle G., Patriarche G., Vivien D. (2007) Rare-earth doped oxyfluoride glass-ceramics and fluoride ceramics: Synthesis and optical properties. *Optical Materials* **29**, pp. 1263–1270.

48. Cherepy N.J., Payne S.A., Asztalos S.J., Hull G., Kuntz J.D., Niedermayr T., *et al.* (2009) Scintillators with potential to supersede lanthanum bromide. *Ieee Transactions on Nuclear Science* **56**, pp. 873–880.

49. Fu J., Parker J.M., Brown R.M., Flower P.S. (2003) Compositional dependence of scintillation yield of glasses with high Gd_2O_3 concentrations. *Journal of Non-Crystalline Solids* **326**, pp. 335–338.

50. Chewpraditkul W., Chen D., Yu B., Zhang Q., Shen Y., Nikl M., *et al.* (2011) Luminescence and scintillation of Eu^{2+}-doped high silica glass. *Physica Status Solidi-Rapid Research Letters* **5**, pp. 40–42.

51. Schweizer S., Henke B., Miclea P.T., Ahrens B., Johnson J.A. (2010) Multi-functionality of fluorescent nanocrystals in glass ceramics. *Radiation Measurements* **45**, pp. 485–489.

52. Seymour R.S., Richardson B., Morichi M., Bliss M., Craig R.A., Sunberg D.S. (2000) Scintillating-glass-fiber neutron sensors, their application and performance for plutonium detection and monitoring. *Journal of Radioanalytical and Nuclear Chemistry* **243**, pp. 387–388.

53. Barta M., Brooke. (2012) Nanocomposite glass-ceramic scintillators for radiation spectroscopy, *Georgia Institute of Technology*.

54. Selling J., Schweizer S., Spaeth J.M., Corradi G., Edgar A., Williams G.V.M. (2005) *Radiation Defects in Ce-Doped $BaCl_2$ and Fluorochlorozirconate Glass-Ceramic X-ray Storage Phosphors*, WILEY-VCH Verlag.

Chapter 4

Transparent High-Density Oxide Ceramics Prepared by Spark Plasma Sintering

Takashi Goto,[a] Akihiko Ito,[a] Akira Yoshikawa,[a] Shunsuke Kurosawa,[a] and Jan Pejchal[a,b]

[a]*Institute for Materials Research, Tohoku University, Sendai 980-8577, Japan*

[b]*Institute of Physics ASCR, Prague 16351, Czech Republic*

kurosawa@imr.tohoku.ac.jp

Transparent ceramics are promising optical materials for solid state lasers, scintillators, high-power phosphors for solid state lighting, and various optical elements. Spark plasma sintering (SPS) is a versatile technique for obtaining fully dense ceramics with fine grains and high transparency because of its fast densification. By combining optimum sintering conditions, such as two-step sintering and subsequent annealing, SPS can be used to synthesize various transparent ceramics. This chapter describes the preparation of several material groups by SPS and provides in detail their characteristics which are suitable for scintillators.

Nanocomposite, Ceramic, and Thin Film Scintillators
Edited by Martin Nikl

ISBN 978-981-4745-22-2 (Hardcover), 978-981-4745-23-9 (eBook)
www.panstanford.com

4.1 Introduction

4.1.1 Transparent Polycrystalline Ceramics

Novel materials and processing techniques can bring about rapid advancements in optical devices such as solid state lasers and scintillators. Optical materials are commonly used in the form of a single crystal grown from a melt because of its high crystallinity and transparency. However, some materials with promising characteristics cannot be synthesized because of their high melting points and phase transformations during cooling. Transparent polycrystalline ceramics are an alternative to transparent materials because of their lower processing temperatures, their ability to be formed into large and complicated shapes, and their ability to be uniformly doped with a large concentration of active ions.

Translucent polycrystalline Al_2O_3 ceramics were used in lamps and windows [1]. Polycrystalline ceramics are generally opaque because of the presence of various scattering sources in these materials, such as grain boundaries, secondary phases, impurities, defects, and residual pores [2]. Therefore, the preparation of starting powders and subsequent sintering processing have been carefully designed to obtain transparent polycrystalline ceramics. Highly transparent neodymium-doped yttrium aluminum garnet (Nd:YAG) ceramics were synthesized by Ikesue et al. [3], who also demonstrated their laser oscillation. Kintaka et al. [4] have prepared Ba-based complex oxide ceramics for optical imaging systems.

In conventional sintering techniques, specific homemade nanopowders are often required, and high temperatures and prolonged sintering times result in materials with low transmittance and poor mechanical properties due to grain coarsening. A method for the preparation of fine-grained transparent ceramics with good mechanical properties using commercially available common powders would be more useful in practical applications.

4.1.2 Spark Plasma Sintering

Spark plasma sintering (SPS), also known as plasma-activated sintering (PAS), pulsed electric current sintering (PECS), and field-assisted sintering technique (FAST), is a versatile technique for consolidating various kinds of materials [5–7]. A wide spectrum of materials, including metals and alloys, ceramics, and composites, has been consolidated by SPS. Compared to pressureless sintering and hot pressing, SPS is advantageous for the consolidation of ceramics in a short time and at low temperatures. SPS can be used to prepare ceramics with limited grain coarsening and with good mechanical properties.

4.1.2.1 A schematic of the SPS method

SPS is a fast densification process in which sintering is activated by a combination of uniaxial pressure and electric current (Fig. 4.1). A powder is poured into a conductive die. A uniaxial pressure is applied to powder compact by punches. Graphite has generally been used as a die and punch material at pressures lesser than 100 MPa, whereas WC and SiC are used at higher pressures. A large pulsed direct current is supplied to a conductive graphite die through the punches and through the powder compact itself if it is conductive. The powder compact is heated rapidly, primarily by electric current (Joule heating). The sintering temperature during SPS is monitored by a thermocouple and/or an optical pyrometer. The graphite die often contains a small hole to accommodate the thermocouple and to allow measurement of the temperature by an optical pyrometer under black-body conditions. The heating rate can reach 1000 K min^{-1}.

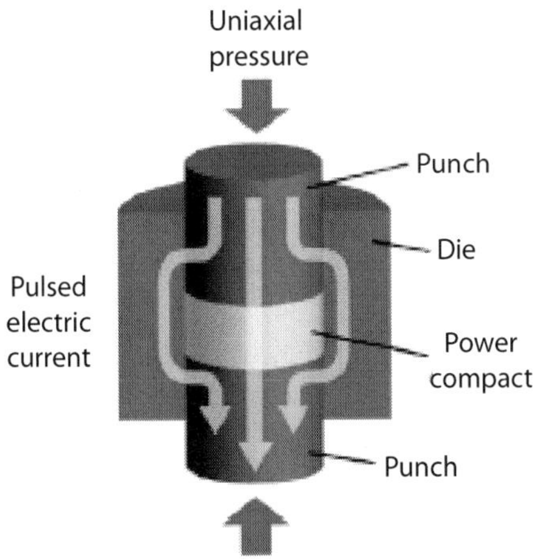

Figure 4.1 Schematic of an SPS setup.

4.1.2.2 Densification in the SPS process

The sintering is generally categorized into solid state sintering and liquid sintering and additionally into viscous-flow sintering and transient-liquid sintering. SPS usually proceeds via solid state sintering, but sometimes liquid or transient-liquid phase sintering is implemented in SPS. From the shrinkage behavior in SPS (Fig. 4.2a), the sintering (densification) process can be divided into three stages: (i) packing and rearrangement of powder under applied pressure, (ii) grain sliding and neck formation, and (iii) isolated pore removal by grain boundary diffusion. The densification occurs dominantly during stage (ii). The grain sliding is promoted mainly by pressure, depending on the shape and size of the particles and the particle size distribution. The neck formation can be dominated by diffusion: (i) surface diffusion, (ii) grain boundary diffusion, (iii) volume diffusion, and (iv) gas-phase diffusion (Fig. 4.2b). The diffusion is accelerated predominantly by temperature. Therefore, the main parameters related to densification by SPS are the pressure, temperature, and heating rate (time).

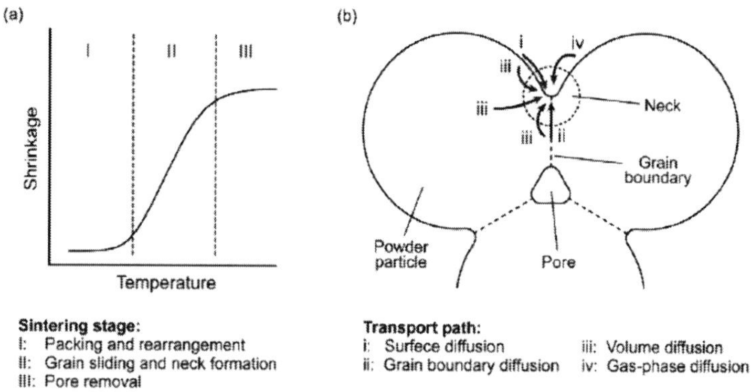

Figure 4.2 A typical shrinkage curve obtained during SPS (a) and the mass transport paths (b).

In SPS, a fixed high pressure (10–300 MPa) and a high heating rate are used, and a higher pressure and heating rate usually lead to more extensive and faster densification, thereby minimizing grain growth. However, An et al. have proposed that a two-step sintering process for SPS, i.e., a low pressure and high heating rate at low temperatures, and a high pressure and low heating rate at high temperatures. This approach has yielded materials with greater transparency than that achieved via single-step sintering [8]. The neck formation in stage (ii) is the main densification process; however, it would also retard the rearrangement of powder and grain sliding. Isolated pores at triple junctions of grain boundaries are difficult to eliminate by diffusion. An appropriate neck evolution, the rearrangement of powder, and grain sliding can promote densification. Therefore, the pressure, heating rate, and temperature should be optimized in the first and second stages. Less pressure applied in the first stage and a lower heating rate in the later stages would promote further densification.

Although a high heating rate greater than 100 K min^{-1} is widely used in SPS, a low heating rate might be beneficial for preparing highly transparent ceramics [8–10]. During the second step in SPS, pores can diffuse through grain boundaries. The high pressure and low heating rate in the second step facilitate the removal of pores without significant grain coarsening.

Many researchers have discussed the contributions of the electric field. Although early reports assumed that spark or plasma is generated between powder particles, no clear experimental evidence has demonstrated their formation or contribution to densification [11]. Electromigration under a dc field through a powder compact and high local temperature gradients may enhance mass transport, leading to rapid densification during SPS [12,13].

4.1.3 Transparent Ceramics Prepared by SPS

SPS is a promising process for preparing transparent ceramics. The combination of a short holding time and a low sintering temperature prevents grain coarsening while providing high transparency.

In 2002 Wu et al. prepared the transparent $PbZrO_3$–$PbTiO_3$–$Pb(Zn_{1/3}Nb_{2/3})_3$ (PZ–PT–PZN) ceramics by SPS [14]. Subsequently, more than 100 papers have been published on the preparation of transparent ceramics by SPS. Table 4.1 summarizes the literature on the preparation of transparent ceramics by SPS and selected materials are described in more details below.

Table 4.1 Transparent ceramics prepared by SPS technique

Material	Temperature (K)	Pressure (MPa)	Heating Rate (K min^{-1})	Grain Size (µm)	Transmittance (%)	Ref.
α-Al_2O_3	1273	500(Ø10)	100	0.2	65(645,0.8t)	31
Al_2O_3-SiO_2	1573	40(Ø15)	100	0.035	80(2100,1t)	69
AlN	2173	30(Ø12)	200	4–5	15(2500,0.5t)	70
(Ba,Sr)TiO_3	1273	50(Ø10)	100	0.8–1	74(650,0.3t)	71
β-TCP	1273	60(Ø8)	25	1.5	n.r.	72
Eu_2O_3	1823	55(Ø15)	5	4.5	35(800,0.8t)	42
$La_2Zr_2O_7$	1673	100(Ø10)	100	1.5	54(2000,1t)	73
$Lu_2Hf_2O_7$	1723	100(Ø10)	100	0.02	78(2000,1t)	41
Lu_2O_3	1723	100(Ø10)	100	0.75	60(550,1t)	8
$Lu_3Al_5O_{12}$	1773	100(Ø10)	100	0.52	78(2000,1t)	60
$MgAl_2O_4$	1593	72(Ø20)	1	0.25	74(550,2.3t)	54
MgO	2173	100(Ø10)	100	116	70(300,1t)	36
Sc_2O_3	1923	100(Ø10)	100	0.3–3.6	27(600,1t)	40
SiO_2	1573	100(Ø10)	100	n.r.	75(200,1t)	62
Y_2O_3	1323	300(Ø10)	20	0.4	68(700,1t)	37
Y_2O_3-ZrO_2	1373	400(Ø10)	10	0.095	57(600,1t)	75
$Y_3Al_5O_{12}$	1598	74(Ø20)	8	2	66(600,1t)	58
Zeolite	1573	50(Ø12)	100	n.r.	n.r.	63

In the Transmittance column in parentheses, there is the wavelength in nanometers at which transmission was measured, and sample thickness in mm designated by "t." "n.r." stands for "not reported," and Ø for sample diameter in mm.

4.1.3.1 α-Al_2O_3

SPS has been extensively employed to fabricate translucent/transparent α-Al_2O_3 ceramics [15–34] because of many applications, such as in lamp windows. Kim et al. [15] first prepared transparent α-Al_2O_3 ceramics by SPS. They reported that the resulting α-Al_2O_3 ceramics exhibited high transparency when prepared using a high pressure (500 MPa) without pretreatment of the powder or the use of additives. Chakravarty et al. [19] and Dang et al. [23] reported that high applied pressures resulted in α-Al_2O_3 with greater transparency. Most researchers who have published results related to SPS of α-Al_2O_3 ceramics reported that a low heating rate favored high transmittance [15,16,25,27,32]. Makino et al. [29] reported that a high heating rate could be feasible for preparing α-Al_2O_3 ceramics with high transparency. Lallemant et al. [25,27] studied the effect of dopants (Mg, Zr, Y, and La) and prepared highly transparent α-Al_2O_3 ceramics using La-doped nanopowders [32]. Nanko et al. [34] prepared transparent α-Al_2O_3 ceramics using a two-step sintering process. The effects of SPS parameters on the densification and transparency have been discussed by Aman et al. [24] and Santanach et al. [26].

4.1.3.2 MgO

Chaim et al. [35] prepared nanocrystalline MgO ceramics with an average grain size of 54 nm, and Wakahara et al. [36] prepared microcrystalline MgO ceramics with an average grain size of 116 μm. In the ultraviolet range, the microcrystalline MgO ceramics exhibited high transmittance, whereas the nanocrystalline MgO ceramics were opaque.

4.1.3.3 Sesquioxides (Y_2O_3, Sc_2O_3, Eu_2O_3, Lu_2O_3)

Zhang et al. [37] prepared transparent Y_2O_3 ceramics under a high pressure (300 MPa). An et al. [38] prepared transparent Y_2O_3 ceramics using a two-step sintering process, and its transmittance was comparable to that of Y_2O_3 prepared under a high pressure [37] and was higher than that of Y_2O_3 prepared at a low heating rate [39]. Preparation and optical properties of other sesquioxide transparent ceramics, namely, Sc_2O_3 [40, 41], Eu_2O_3 [42] and high density Lu_2O_3 [43, 44] have been reported as well.

4.1.3.4 Spinel ($MgAl_2O_4$, $ZnAl_2O_4$)

The SPS of transparent $MgAl_2O_4$ ceramics has been widely studied [9, 45–53]. Transparent $MgAl_2O_4$ ceramics were prepared at a low heating rate (less than 10 K min^{-1}). Bonnefont et al. [54] applied a three-step sintering process that involved a decreasing heating rate (from 100 to 10 to 1 K min^{-1}) with increasing sintering temperature to obtain highly transparent $MgAl_2O_4$ ceramics. The preparation of transparent $ZnAl_2O_4$ ceramics using nanopowders with a sintering aid has been also reported [55].

4.1.3.5 Garnets ($Y_3Al_5O_{12}$, $Lu_3Al_5O_{12}$)

Transparent $Y_3Al_5O_{12}$ (YAG) ceramics can be used in a wide variety of applications as optical and structural materials [56–59]. The preparation and pretreatment of starting powders affects the densification of transparent YAG ceramics [58]. Transparent $Lu_3Al_5O_{12}$ (LuAG) ceramics have also been prepared by reactive SPS using a two-step sintering process [60].

4.1.3.6 Glasses

Transparent amorphous SiO_2 glass ceramics have been prepared by SPS [61, 62]. After being annealed in air, these glass ceramics exhibited high ultraviolet transmittance [62]. Transparent zeolite glass ceramics have also been prepared by SPS [63], as well as transparent chalcogenide [64, 65] and tellurite [66] glass ceramics for use as infrared windows.

4.1.4 Fundamental Properties of Lu-Based Oxide Ceramics

The fundamental properties of typical Lu-based oxide ceramics are listed in Table 4.2. Phase diagrams for Lu_2O_3–TiO_2 [67, 80] and Lu_2O_3–Al_2O_3 [68, 81] are illustrated in Fig. 4.3. Lu_2O_3 and $Lu_3Al_5O_{12}$ have widely been studied as practical optical materials, whereas the literature contains few reports on the synthesis and properties of $Lu_2Ti_2O_7$, Lu_3NbO_7, and $Lu_2Hf_2O_7$.

Table 4.2 Fundamental properties of typical Lu-based oxides

	Lu_2O_3	$Lu_2Hf_2O_7$	$Lu_3Al_5O_{12}$
Crystal structure	Bixbyite	Defect-fluorite	Garnet
Space group	$Ia\bar{3}$	$Fm\bar{3}m$	$Ia\bar{3}d$
JCPDS#	65-3172	-	73-1368
Lattice parameter (nm)	1.0396	0.515	1.1906
Density (Mg m^{-3})	9.426	9.95	6.695
Melting point (K)	2643	3173[c]	2327
Hardness (GPa)	12.5[a]	-	17.2
Fracture toughness (MPa $m^{1/2}$)	4.1[a]	-	1.1
Thermal conductivity (W m^{-1} K^{-1})	12.6	-	8.3
Bandgap (eV)	5.5	-	7.0
Refractive index	1.8997[b]	-	1.8184[b]

a: Nd:Lu_2O_3, b: λ = 589.3 nm, c: estimated from Lu_2O_3–HfO_2 phase diagram.

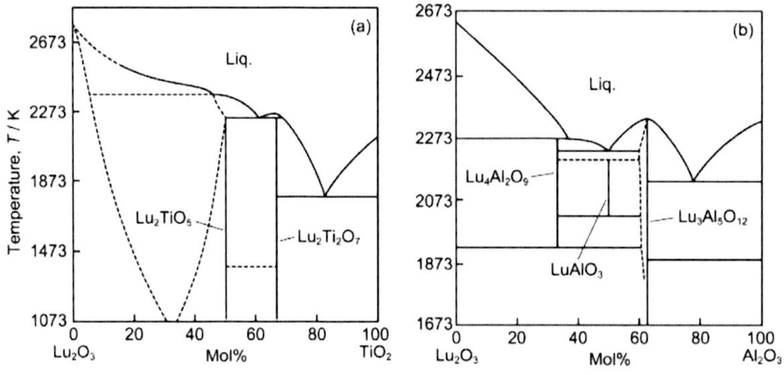

Figure 4.3 Phase diagrams for Lu_2O_3–TiO_2 (a) and Lu_2O_3–Al_2O_3 (b) systems.

4.2 Preparation of Transparent Lu_2O_3 Ceramics

Lu_2O_3 has a cubic bixbyite structure (space group: $Ia\overline{3}$; lattice parameter: a = 1.039 nm) (Fig. 4.4) [69]. Lu cations have an fcc sublattice and occupy the 24d and 8b sites. An O anion occupies the 48e site. The bixbyite structure is derived from a simple fluorite lattice by doubling the lattice parameter and leaving one-fourth of the O anions orderly, as illustrated in Fig. 4.4 b,c [70, 71].

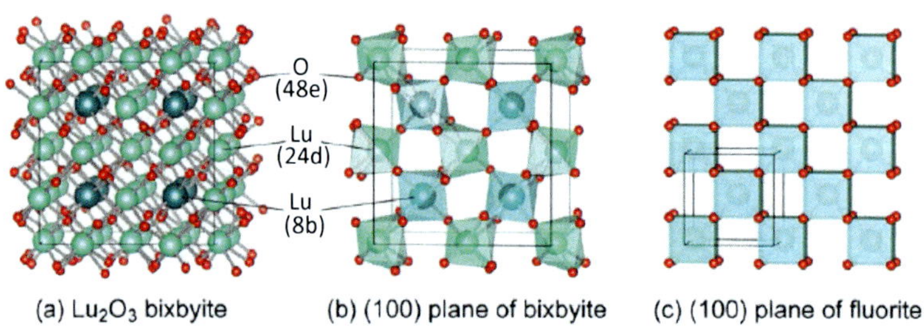

Figure 4.4 Schematic of crystal structures: Lu_2O_3 bixbyite (a), (100) planes in Lu_2O_3 bixbyite at x = 0.75 (b) and the simple fluorite at x = 0 (c) in <100> projection.

Among the rare earth sesquioxides, Y_2O_3, Lu_2O_3, and Sc_2O_3, the most promising gain medium for high-power and ultrashort pulsed lasers is Lu_2O_3 [72, 73]. The atomic mass of Lu^{3+} (174.97 g mol^{-1}) is approximately the same as that of Yb^{3+} (173.5 g mol^{-1}). Therefore, the thermal conductivity of $Yb{:}Lu_2O_3$ does not decrease with increasing Yb^{3+} content [74, 75], unlike the thermal conductivity of $Yb{:}Y_2O_3$, $Yb{:}Sc_2O_3$, and Yb:YAG (Fig. 4.5) [75–78].

Because Lu_2O_3 single crystals are difficult to grow as a result of their high melting points, transparent Lu_2O_3 ceramics have been prepared via several sintering process, i.e., pressureless and vacuum sintering techniques. Fully dense Lu_2O_3 ceramics were prepared using a homemade powder at a high sintering temperature and with a prolonged holding time. An et al. prepared transparent Lu_2O_3 ceramics by SPS using a common commercial powder. They proposed a two-step sintering process to improve the transmittance of the ceramics particularly in the visible range.

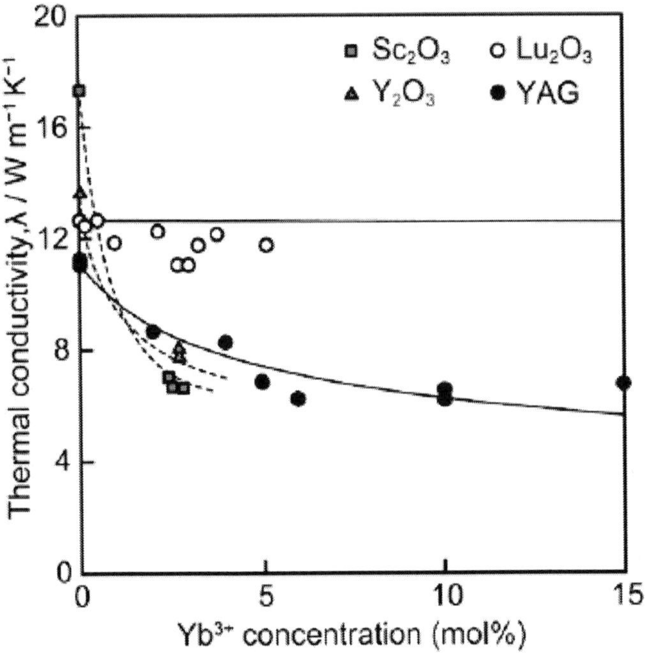

Figure 4.5 Effect of the Yb^{3+} dopant concentration on the thermal conductivity of Lu_2O_3, Y_2O_3, Sc_2O_3, and YAG ceramics. Solid lines indicate calculated values.

4.2.1 Effect of Temperature

Figure 4.6 shows the relative density, average grain size, and transmittance of Lu_2O_3 ceramics sintered at various temperatures [43]. The density of Lu_2O_3 ceramics increased significantly at 1300–1600 K without grain coarsening. The grain size of Lu_2O_3 ceramics was several micrometers. Significant grain coarsening occurred at temperatures greater than 1700 K. The grain coarsening at high temperatures caused the formation of residual pores at triple junctions of grains, which degraded the relative density and transmittance. The optimal sintering temperature should be approximately 1600 K, i.e., just below the temperature at which significant grain coarsening begins.

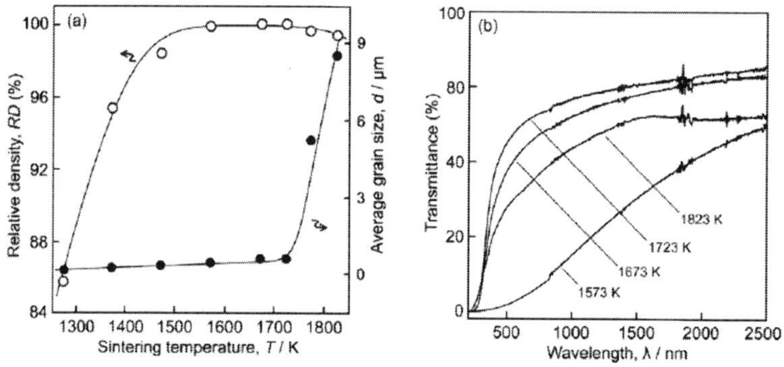

Figure 4.6 Effect of the sintering temperature on the relative density (open circles in (a)), average grain size (closed circles in (a)) and the transmittance of Lu_2O_3 ceramics (b).

4.2.2 Effect of Pressure and Holding Time

High pressures promote densification. An et al. [43] have reported that the densification of Lu_2O_3 ceramics was insufficient at low pressures, whereas a pressure greater than 60 MPa yielded fully dense ceramics (triangles in Fig. 4.7a). A high pressure is effective for consolidation without grain coarsening. The grain size was independent of pressure.

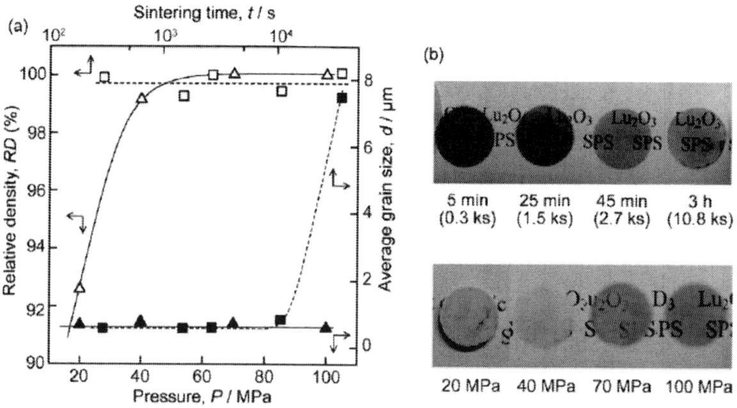

Figure 4.7 Effect of the applied pressure (triangles) and holding time (squares) on the relative density (open symbols) and average grain size (closed symbols). The images in (b) are photographs of Lu_2O_3 ceramics prepared at various sintering times (upper row) and pressures (lower row).

The relative density of Lu_2O_3 ceramics was observed to be independent of holding time (squares in Fig. 4.7a). Significant grain coarsening occurred when the holding time exceeded 10 hours. The holding time affected the microstructure observed at the later stage of sintering.

The transparency of Lu_2O_3 ceramics increased with increasing pressure and holding time (Fig. 4.7b). However, a prolonged holding time degraded the transparency due to grain coarsening, which was accompanied by pore agglomeration.

4.2.3 Two-Step Sintering

A preloading pressure can destroy agglomerates formed during the initial stage of sintering and suppress particle coarsening prior to the application of the final pressure. Although higher pressures commonly result in greater densification, high pressures at low temperatures might result in a high defect concentration, which would promote densification during subsequent high-temperature sintering. Although a high heating rate is conducive to fast densification, it may also result in pore trapping at high temperatures.

To remove defects from polycrystalline ceramics and improve their transparency, An et al. [8] proposed a two-step sintering process (Fig. 4.8a). The temperature was first increased to a moderate temperature under a low pressure and with a high heating rate during the first step. In the second step, the temperature was further increased to the final temperature at a low heating rate and the pressure was increased to the final pressure.

Figure 4.8b shows the shrinkage curves for the densification of Lu_2O_3 ceramics under various pressures. For high pressures applied during the first step, the initial packing density increased and densification commenced earlier. When the final pressure was applied during the second step, densification was significantly promoted. This change was more pronounced when the pressures applied during the first step were lower.

The heating rate used after the application of the final pressure affected the removal of pores (Fig. 4.9) [8]. Although fully dense Lu_2O_3 ceramics were obtained almost independent of the heating rate, smaller but more abundant pores were observed at the triple junctions of grain boundaries at higher heating rates (Fig. 4.9c). The residual pores became a scattering source for visible light and degraded the transparency of the ceramics.

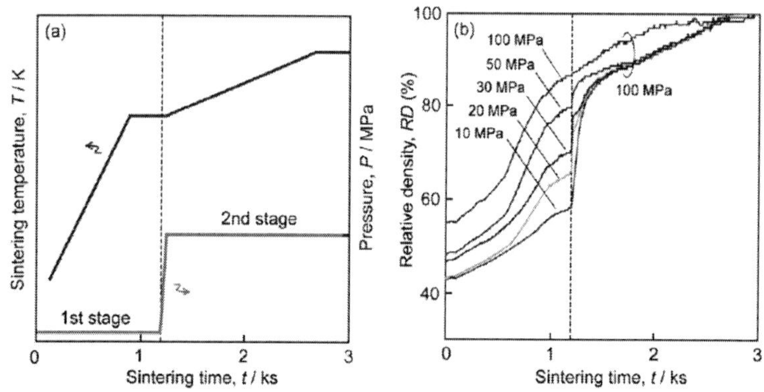

Figure 4.8 Schematic of a two-step sintering process (a) and the effect of the preloading pressure on the shrinkage curve of Lu_2O_3 ceramics (b). The sintering temperature was 1723 K.

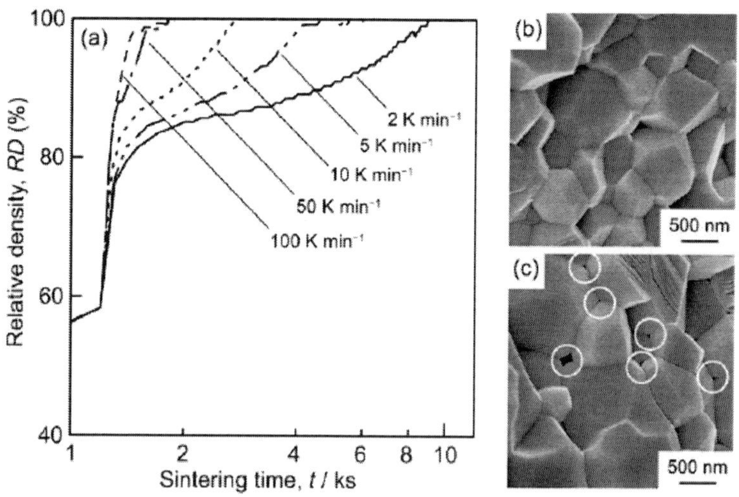

Figure 4.9 Effect of heating rate on the shrinkage curve of Lu_2O_3 ceramics (a), and the fracture surface of Lu_2O_3 ceramics prepared at a preloading pressure of 10 MPa, a sintering temperature of 1723 K, and a heating rate of 2 K min^{-1} (b) or 100 K min^{-1} (c). White circles indicate pores.

The in-line transmittance of the Lu_2O_3 ceramics exhibited a maximum when a moderate pressure was applied during the first step and a low heating rate was used in the second step (Fig. 4.10) [8]. Because the sintering time in the SPS method is short, the diffusion of pores through the grain boundaries is difficult. Therefore, the combination of pressure and heating rate in a two-step sintering method should be optimized to obtain highly transparent Lu_2O_3 ceramics.

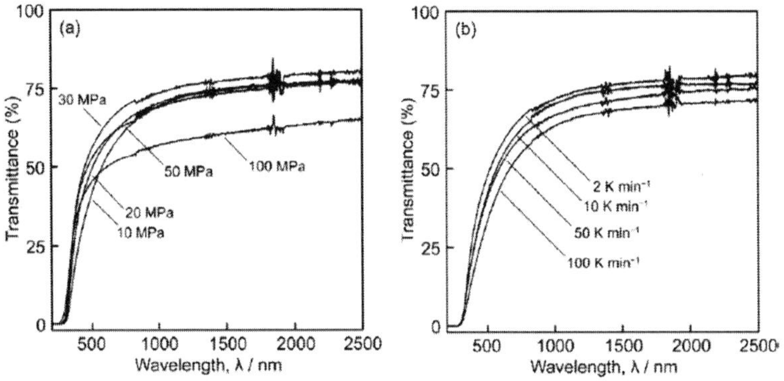

Figure 4.10 Effect of preloading pressure (a) and heating rate (b) on the transmittance of Lu_2O_3 ceramics.

4.2.4 Effect of LiF Addition

Lithium fluoride (LiF) has been used as a sintering aid to lower the sintering temperature of ceramics [79, 80]. It can be used in the SPS process to obtain highly transparent ceramics [81–85]. A small amount of LiF was observed to enhance the grain growth and diffusional creep by increasing the lattice and grain boundary diffusivities [86]. Moreover, LiF was observed to react with residual carbon on particle surfaces to form volatile carbon fluoride gas [84, 87].

The addition of a small amount of LiF to Lu_2O_3 and Nd:Lu_2O_3 ceramics promoted their densification and improved their transmittance in the visible range (Fig. 4.11a) [88]. However, the addition of a large amount of LiF caused cracking of the sintered ceramics. A similar behavior has been reported in the SPS of YAG ceramics [84]. The overall transmittance of Lu_2O_3 and Nd:Lu_2O_3 ceramics significantly increased with increasing LiF content (Fig. 4.11b,c).

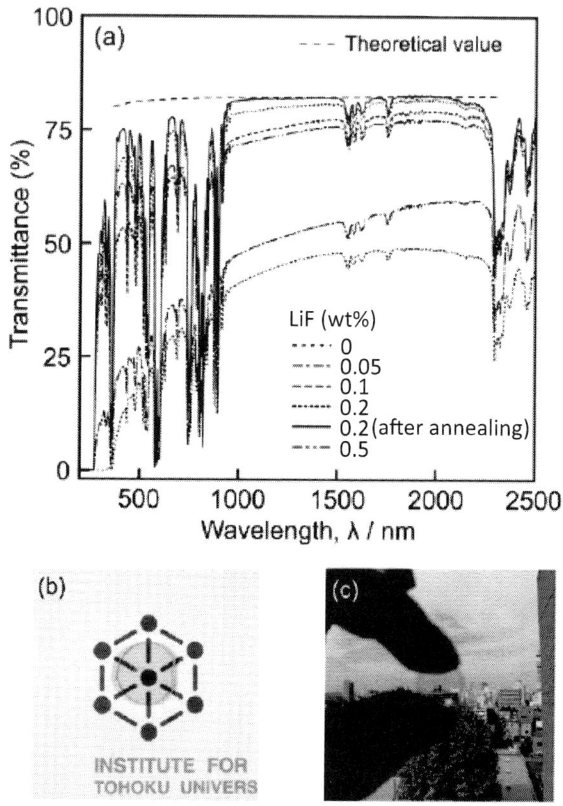

Figure 4.11 Effect of the amount of LiF on the transmittance of Nd:Lu_2O_3 ceramics (a) and photographs of transparent Lu_2O_3 ceramics prepared in the presence of LiF (b, c).

4.2.5 Comparison with the Other Sintering Techniques

Transparent Lu_2O_3 ceramics have been prepared by various sintering techniques using different powders, primarily homemade nanopowders. In addition, many researchers have been focusing on the preparation of rare earth–doped Lu_2O_3 as a host material for lasers.

Transparent nanocrystalline Eu:Lu_2O_3 [89] and Nd:Lu_2O_3 [90] ceramics were prepared by vacuum sintering. A two-step sintering method involving a combination of vacuum sintering and hot isostatic pressing was also used for the preparation of transparent microcrystalline Eu:Lu_2O_3 ceramics [91]. Nanocrystalline Eu:Lu_2O_3

ceramics were translucent, whereas microcrystalline $Eu:Lu_2O_3$ ceramics were highly transparent. Pressureless sintering under a H_2 atmosphere produced microcrystalline transparent $Eu:Lu_2O_3$ [92], $Nd:Lu_2O_3$ [93], and $Yb:Lu_2O_3$ [94] ceramics. Hot pressing produced transparent $Yb:Lu_2O_3$ [82] and $Ho:Lu_2O_3$ [95] ceramics with slightly smaller grain sizes than that of ceramics produced by pressureless sintering. Laser oscillations using $Yb:Lu_2O_3$ [82] and $Ho:Lu_2O_3$ [95] ceramics have been demonstrated. SPS produced transparent $Nd:Lu_2O_3$ ceramics with a transmittance of 81.8% (99.5% of the theoretical value) at 1080 nm after the ceramics were annealed in air [88], and their scintillation properties were studied as well [96].

Conventional sintering techniques (vacuum sintering and hot pressing) require a prolonged sintering time (on the order of several hours) to obtain highly transparent Lu_2O_3 ceramics. SPS can produce highly transparent $Nd:Lu_2O_3$ ceramics in a substantially shorter time.

4.3 Preparation of Transparent $Lu_2Hf_2O_7$ Ceramics

Dilutetium hafnate, $Lu_2Hf_2O_7$, has a defect-fluorite structure. Subramanian et al. have reviewed the relationship between the crystal structure of $A^{3+}{}_2B^{4+}{}_2O_7$ compounds and the ionic radius ratio between the cations (r_A/r_B) [97]. The combination of ions with similar ionic radii ($r_A/r_B < 1.46$) results in the disordering of the cations and anion deficiencies as a defect-fluorite structure (Fig. 4.12a, Fig. 4.15a), whereas smaller B^{4+} cations ($r_A/r_B = 1.46–1.80$) tend to stabilize the ordering of cations and anions in a pyrochlore structure (Fig. 4.12b, Fig. 4.15b) [69]. Because $Lu_2Hf_2O_7$ has a small r_{Lu}/r_{Hf} ratio of 1.38, $Lu_2Hf_2O_7$ exhibits a defect-fluorite structure with $Fm3m$ symmetry (a = 0.515 nm), as do the other rare earth hafnates $RE_2Hf_2O_7$ (RE = Y, Ho, Er, and Tm). Lu and Hf cations randomly occupy the 4a sites, and one-eighth of the 8c sites for O anions are statistically defective.

$Lu_2Hf_2O_7$ is a promising host material for a high stopping power scintillator for X- and γ-rays because of its combination of high atomic numbers (Z_{Lu} = 71 and Z_{Hf} = 72), high density (9.95 Mg m^{-3}), and high effective atomic number (Z_{eff} = 68.9) among rare earth hafnates ($RE_2Hf_2O_7$). The crystal structure of Lu_2HfO_7 has not been refined because of the difficulty associated with growing $Lu_2Hf_2O_7$ single crystals due to high melting point of this material of approximately 2700 K.

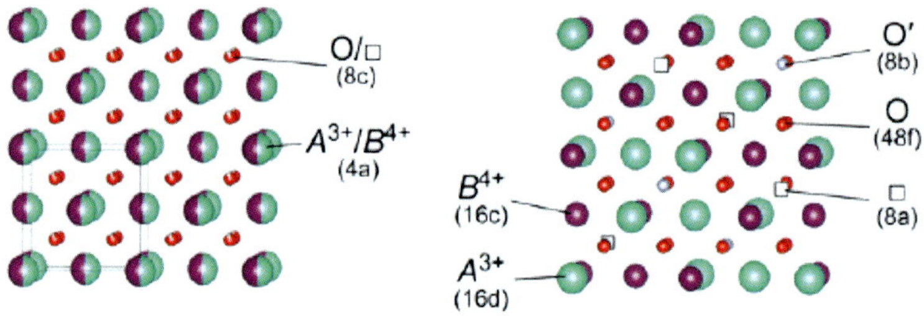

(a) Defect-fluorite (b) Pyrochlore (simplified)

Figure 4.12 Schematics of the crystal structures of defect fluorite (a) and simplified pyrochlore (b) in <100> projection.

Transparent $La_2Hf_2O_7$ [98], $Y_2Hf_2O_7$ [99], and $LaGdHf_2O_7$ [100] ceramics have been prepared by pressureless sintering. $Lu_2Hf_2O_7$ powders have also been synthesized [101–103]. Transparent $Lu_2Hf_2O_7$ ceramics were prepared by reactive SPS using Lu_2O_3 and HfO_2 powders [41]. $Lu_2Hf_2O_7$ ceramics does not have a pyrochlore superstructure; however, it can be indexed as a defect-fluorite structure. $Lu_2Hf_2O_7$ ceramics consists of submicron grains and exhibits high transparency. $Lu_2Hf_2O_7$ ceramics exhibits a dark-yellowish color before annealing but becomes yellow after being annealed in air. Although $Lu_2Hf_2O_7$ ceramics exhibited a slightly lower visible transmittance compared to $Y_2Hf_2O_7$ [99] and $LaGdHf_2O_7$ [100] ceramics, it exhibited a higher infrared transmittance [41].

4.4 Preparation of Transparent $Lu_3Al_5O_{12}$ Ceramics

Trilutetium dodecaoxides, $Lu_3M_5O_{12}$ (M = Al, Fe, Ga, and Ge), have a garnet structure ($a \approx 1.2$ nm). The garnet structure can be described as a general formula $Lu_3M_2M'_3O_{12}$, where Lu, M, M', and O ions locate at the 24c, 16a, 24d, and 96h sites, respectively. It consists of a framework of corner-sharing MO_6 octahedra and $M'O_4$ tetrahedra, and Lu occupies 8-coordinated sites within this framework (Fig. 4.13) [69].

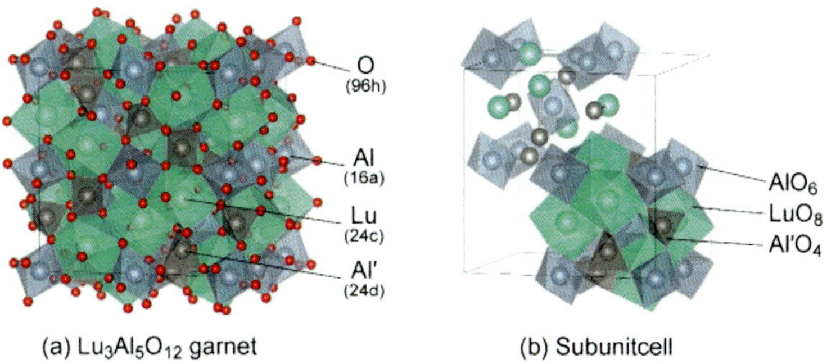

Figure 4.13 Schematic of the crystal structure of $Lu_3Al_5O_{12}$ garnet (a) and its ($^1/_2$, $^1/_2$, $^1/_2$) sub-unit cells (b) in <100> projection.

$Lu_3Al_5O_{12}$ (LuAG) is a frequently used host for fast-response scintillators and solid state lasers. In the former case the rare earth ions such as Ce^{3+} [104] or Pr^{3+} [105] provide suitable emission centers with decay times of several tens of nanoseconds. LuAG is more dense (6.7 Mg m^{-3}) and has a greater atomic number (Z_{eff} = 63) compared to those of YAG (4.55 Mg m^{-3} and 29, respectively), for the review of its scintillation characteristics, see [106]. In the case of laser host, namely the Yb doping has been applied to obtain femtosecond high-power infrared lasers [107].

Transparent LuAG ceramics have been prepared by vacuum sintering [108, 109] and pressureless sintering [110–113]. Although highly transparent LuAG ceramics have been prepared by these methods, high sintering temperatures, and prolonged sintering time resulted in a large grain size of several tens of micrometers, which in turn resulted in degraded mechanical properties.

4.4.1 Effect of Temperature

Fully dense transparent LuAG ceramics have been prepared by reactive SPS using Lu_2O_3 and α-Al_2O_3 powders [60]. At low sintering temperatures, the LuAG ceramics contained lutetium aluminum perovskite (LuAP), whereas single-phase LuAG ceramics formed at high sintering temperatures. The average grain size increased from the submicrometer level to tens of micrometers at high sintering temperatures (Fig. 4.14a). The transparency decreased because of significant grain coarsening that accompanied the formation of pores at the triple junctions of the grain boundaries.

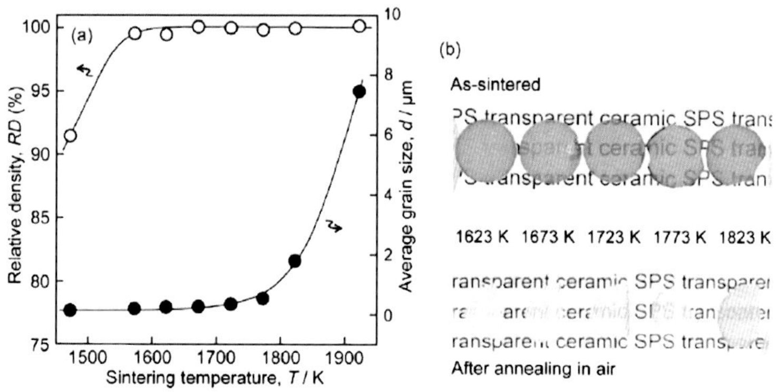

Figure 4.14 Effect of the sintering temperature on the relative density (open circles in (a)) and average grain size (closed circles in (a)). Image (b) shows photographs of transparent LuAG ceramics before (upper row) and after (lower row) being annealed in air.

As-sintered LuAG ceramics exhibit a yellow color. After being annealed in air, LuAG ceramics became colorless (Fig. 4.14b), and their transmittance increased with increasing sintering temperature. At high sintering temperatures, significant grain coarsening occurred, resulting in a diminished transmittance.

The Vickers hardness and indentation fracture toughness of transparent LuAG ceramics were 17 GPa and 1.6 MPa $m^{1/2}$, respectively. The hardness increased as the grain size decreased, which is consistent with the Hall–Petch relationship (Fig. 4.15). LuAG ceramics exhibited a combination of high transparency and good mechanical properties.

4.4.2 Reduction of Eu^{3+} to Eu^{2+} in LuAG Ceramics

The highly efficient and reasonably fast 5d–4f luminescence of Eu^{2+} cations makes them attractive for use in scintillators. However, Eu^{2+}-doped scintillators are limited mostly to halide materials. In the transparent Eu:LuAG ceramics prepared by SPS, Eu^{2+} 5d–4f luminescence was observed under ^{241}Am α-ray excitation [114]. Among transparent ceramics prepared by SPS, Eu^{3+} emission was dominant in Y_2O_3 and Lu_2O_3 ceramics, whereas Eu^{2+} emission was dominant in α-Al_2O_3, YAG, and LuAG ceramics. Al cation in host materials and a reducing atmosphere of SPS process would have caused the reduction of Eu^{3+} to Eu^{2+} in Eu:LuAG ceramics.

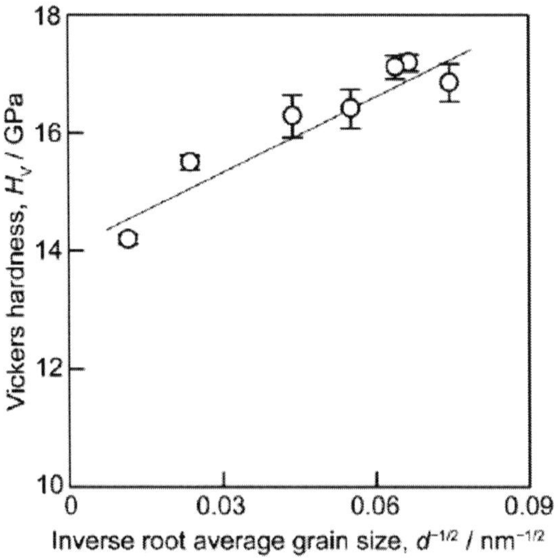

Figure 4.15 Effect of the average grain diameter on the Vickers hardness of transparent LuAG ceramics.

4.4.3 Comparison with the Other Sintering Techniques

Transparent Ce:LuAG and Pr:LuAG ceramics were prepared by vacuum sintering using nanopowders prepared by precipitation and combustion methods [109–113]. Pressureless sintering resulted in transparent LuAG ceramics with relatively large grains [110]. The sintering time for vacuum sintering and pressureless sintering was on the order of several days. In contrast, transparent LuAG ceramics were prepared by SPS within 1 hour.

4.5 Ceramic Scintillation Materials Prepared by SPS

Single crystals usually give rise to scintillation materials with the highest performance. However, in quite some inorganic materials with intrinsically promising scintillation performance the crystal growth is disabled, or becomes too complicated and expensive, due to high melting temperatures, phase transitions, non-equal volatility of constituents of the melt, and other reasons. That was why the scintillation ceramics become of interest already in the 1980s and found practical applications, e.g., in computed tomography [115, 116]. Completely different preparation routes of ceramic can avoid some specific defects present in crystals grown from the melt, e.g., the antisite defects in aluminum garnets [117]. On the other hand, segregation effects at grain boundaries and resulting traps and/or concentration quenching together with deteriorating effects of sintering aids can strongly degrade the ceramic scintillator's performance [118].

Scintillation ceramics described in [115] show rather slow response in the μs–ms time scale due to the employed emission centers from lanthanide group where emission transition is of 4f–4f character (Tb^{3+}, Eu^{3+}, Pr^{3+}). Fast scintillation ceramics based on Ce-doped and Pr-doped aluminum garnets were prepared by vacuum sintering combined with HIP or other classical ceramic preparation routes and their luminescence and scintillation characteristics were studied somewhat later in the 1990s [119] and in the first decade of 21st century [109–113, 120]; for review see [121]. 5d–4f allowed emission transition of these ions allow to obtain the dominant part of scintillation response of the order of several tens of nanoseconds, and such materials can directly compete with fast scintillators based on analogous single crystal compositions [105, 106].

4.5.1 Garnet-Based Ceramics

As mentioned above, fully dense and transparent undoped LuAG ceramic was prepared by SPS [60]. It was found that higher sintering temperatures around 1600–1700°C are more suitable for sample preparation and resulting properties, such as better densification, hardness, and higher transparency, were achieved. On the other hand, there is the question, how will the other defects and trapping phenomena related to the grain boundaries affect the material scintillation performance.

The Ce-doped LuAG ceramics was prepared by SPS and its luminescence and scintillation characteristics studied in Ref. [122].

Some of the samples were prepared from LuAG:Ce nanopowder synthesized by original radiation method and subsequent calcination [123]. The nanoparticle size was relatively uniform from 20 to 60 nm and the Ce concentration was around 1.5 mol. The nanopowder has very large surface, which leads to facilitation of the mass transfer and thermal surface processes. It makes the nanopowders excellent starting materials for SPS sintering [124, 125].

The sintered ceramic bodies are shown in the Fig. 4.16 together with their sintering temperatures. They are slightly translucent and of dark-green color. The high quality LuAG:Ce reference single-crystal is displayed for comparison.

Figure 4.16 LuAG:Ce ceramic bodies sintered at different temperatures from nanopowder.

The morphology of the surface of the sintered bodies was checked using backscattered electron (BSE) imaging. The BSE image of the LuAG:Ce sample sintered at 1700°C is shown in Fig. 4.17. There are large areas with many small pores and other large areas created by very large grains and some small pores. In the latter areas the grain growth progressed much, which can be caused by the above-mentioned enhancement of mass transfer and thermal surface processes due to large surface area of the nanopowders.

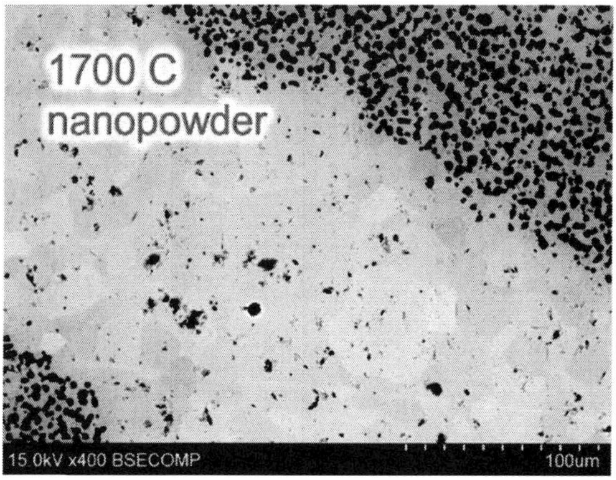

Figure 4.17 The BSE image of the LuAG:Ce sample sintered at 1700°C made from nanopowder.

The comparison of radioluminescence spectra for the LuAG:Ce ceramic samples sintered from nanopowder at 1700°C and that of single crystal is shown in Fig. 4.18 together in absolute comparison with BGO reference single-crystal scintillator. The LuAG:Ce spectra are dominated by the Ce^{3+} 5d–4f emission peaking around 510 nm. The intensities of this emission from ceramic and single crystal are comparable and almost 5 times as high as the emission peak of the BGO. The nanopowder sample shows some weak broadband defect emission within 260–400 nm, which appears much more intense in the single-crystal sample. Regarding the 400 nm band, it probably originates from the oxygen vacancies created during SPS sintering under reducing conditions (carbon die and vacuum), as it was shown that it is related to the F+ center in LuAG single crystal (one electron in oxygen vacancy) [126]. Complex emission peaking around 290 nm in single crystal is due to trapped exciton around the antisite defect [106], which points to much lower occurrence of such a defect in SPS ceramics which is expected.

The photoluminescence decay for the sample sintered from nanopowder at 1700°C is shown in Fig. 4.19a. The decay curve was fitted with sum of exponential using the deconvolution procedure with the instrumental response to obtain true decay times. The decay curve of the Ce^{3+} 5d–4f emission under the direct 4f–5d excitation at 452 nm could be approximated by a single-exponential function with the decay time of 54 ns, which is a value close to the values reported in the literature for LuAG:Ce ceramics and single crystals [106, 120]. Figure 4.19b shows the photoluminescence decay curve of the Ce^{3+} for the same sample under 193 nm excitation, which is close to the interband transition and might simulate the involvement of transport stage in the scintillation process. In this case the leading decay time is slightly prolonged to 64 ns and another slow component is present. It is apparently the sign of slower processes related to the energy exchange with the host and eventual interplay with defects and traps.

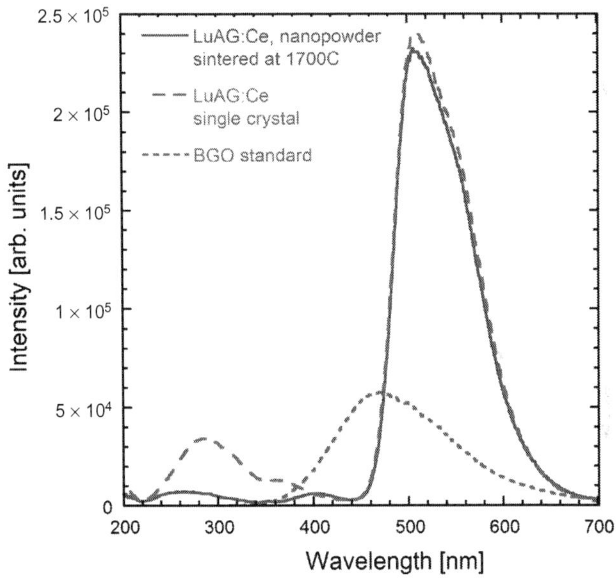

Figure 4.18 Radioluminescence spectra of LuAG:Ce SPS ceramics and single crystal with BGO standard in absolute comparison. Excitation X-ray tube (40 kV, 15 mA).

To summarize, LuAG:Ce ceramic samples were prepared by SPS method from nanopowder at different sintering temperatures (1500°C, 1600°C, and 1700°C). Very small pores were observed at all the sintering temperatures. Compact areas with large grains were observed at 1700°C, indicating the pronounced grain growth in the LuAG nanopowders at these temperatures. The overall scintillation efficiency for the sample sintered at 1700°C was as high as that for the single crystal. A much lower intensity of the defect luminescence compared to single crystal was found under X-ray excitation. The defect luminescence at 400 nm with decay time of a few nanoseconds was ascribed to F^+ centers (charged oxygen vacancies), which occur as a consequence of reductive effect of carbon die. The Ce^{3+} photoluminescence decay time under intracenter 4f–5d^1 excitation band at 452 nm shows the value of 54 ns close to that known from single crystals. However, under excitation in the absorption edge at 193 nm the decay slowed down with dominant decay time of 64 ns and longer submicrosecond decay component present.

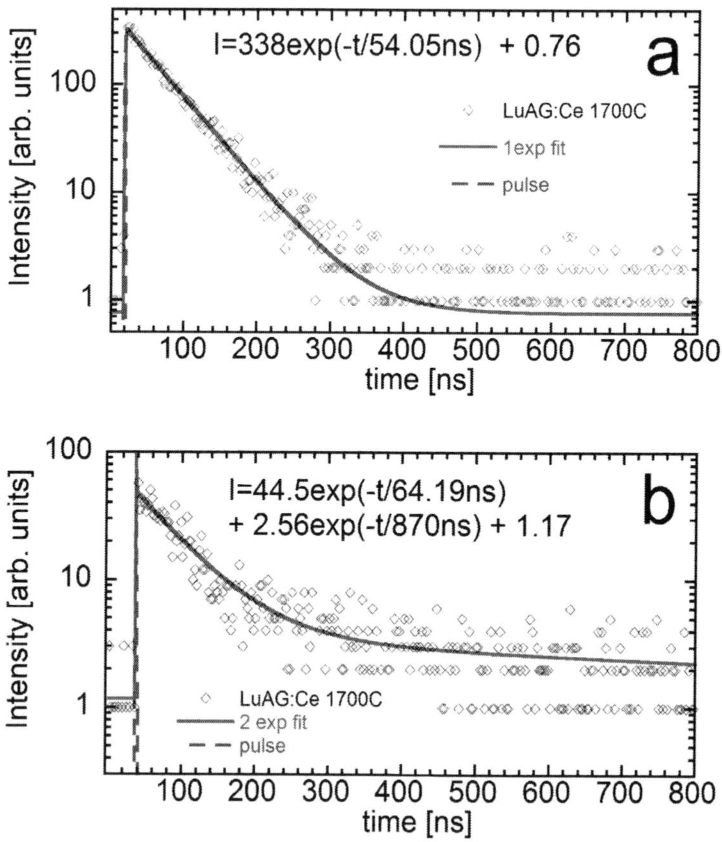

Figure 4.19 Photoluminescence decay of the Ce^{3+} emission (505 nm) under 452 nm excitation (4f–5d1 transition of Ce^{3+}) (a) and under 193 nm excitation (b) for the sample sintered from nanopowder at 1700°C.

4.5.2 $SrHfO_3$-Based Ceramics

The strontium hafnate $SrHfO_3$ (SHO)-based phosphors are attractive due to very high density (7.56 g/cm^3), low intrinsic radioactivity, and possibility of doping with various luminescence ions, such as Ce^{3+} [127] or Pb^{2+} [128]. Recently, an intense UV emission with submicrosecond decay time at room temperature was reported for nonstochiometric Sr-deficient material as well [129]. So far, most of the samples were studied in the form of powders, as the melting point of SHO is about 2700°C. Concerning the Ce-doped sample, the dominant scintillation emission is due to the Ce^{3+} 5d–4f transition at 400 nm. Its photoluminescence decay time value at room temperature was reported to be 15 ns [128] for powder sample prepared by acetate-citrate combustion method, while 36 ns was reported for that prepared by combustion method [127]. Very fast broadband luminescence was observed at 450 nm at low temperature and ascribed to the F^+ center [129].

Due to high melting point mentioned above the optical ceramic is the only way how to obtain bulk SHO-based optical materials. The study on the Ce-doped SHO optical ceramics together with $BaHfO_3$ one prepared by hot press or hot isostatic press was reported already in 2007 [130]. The light yield as high as 40,000 photons/MeV was found, which makes these materials competitive candidates for application in medical imaging. Therefore, SPS method has also been used to prepare SHO-based ceramics and their luminescence and scintillation properties were studied [131]. The sintered ceramic pellets are displayed in Fig. 4.20.

Figure 4.21 shows the transmission spectra of the samples. Due to the non-cubic structure of $SrHfO_3$ (orthorhombic phase Pbnm, ICDD Card No. 88-14) and porosity of the samples (see below), their transmittance is as low as several percentage points. The undoped SHO is transparent down to 250 nm, where sharp drop can be observed. This drop must be related to absorption of some defect centers, as the band-to-band transition is reported to be at shorter wavelengths around 200–220 nm [129]. The very sharp drop of transmittance at 540 nm in the SHO:Ce is the cause of the deep red color of the sample (Fig. 4.20) and must be related to a color center arising due to Coulombic mismatch between Ce^{3+} dopant and cation Sr^{2+} site where it is embedded. Improvement of transmittance at the short-wavelength region is apparent for the Al-codoped SHO:Ce due to improvement of charge balance in the lattice as Al^{3+} is expected to enter at Hf^{4+}cation site. The drop at 350 nm is ascribed to the Ce^{3+} 4f–5d transition [129, 130].

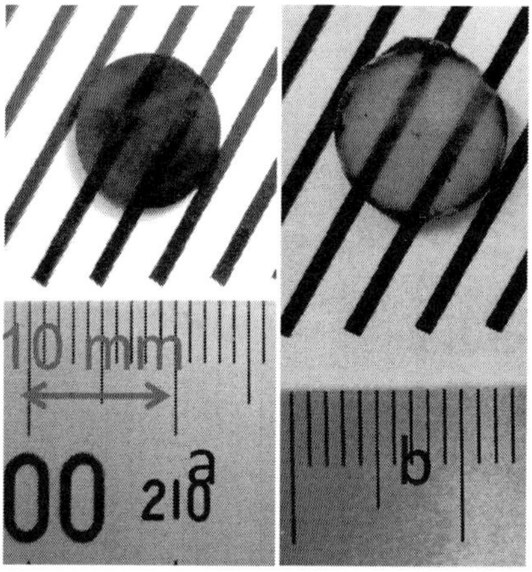

Figure 4.20 The SHO:Ce (a) and the Al-codoped SHO:Ce (b) ceramic bodies sintered by SPS method.

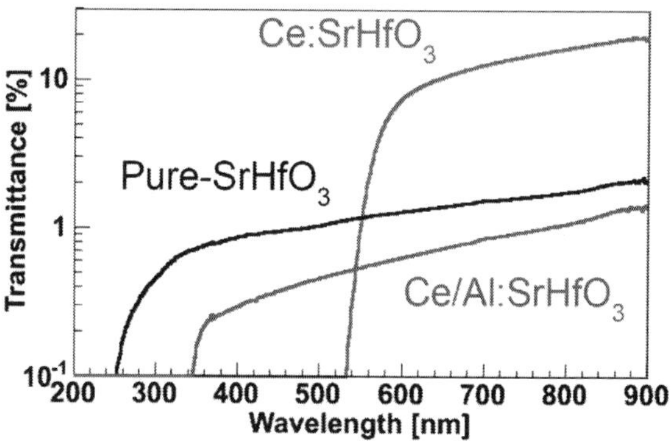

Figure 4.21 Transmission spectra of $SrHfO_3$-based SPS ceramics.

Figure 4.22 (upper part) shows the BSE image of the central part of the $Ce_{0.02}Sr_{0.98}HfO_3$ sample. The average grain size was roughly 30 μm. The Al-codoped SHO:Ce sample was chosen for the composition measurements. The composition was scanned using an electron probe microanalyzer on the diameter axis on a length of 10 mm, with a scanning interval of 0.1 mm as shown in Fig. 4.22 (lower part). The average of composition ratio for ($Sr_aCe_bAl_cHf_dO_e$) was (a,b,c,d,e) = (0.90 ± 0.02, 0.019 ± 0.005, 0.008 ± 0.006, 1 ± 0.02, and 2.62 ± 0.03), while the nominal composition was (0.96, 0.02, 0.02, 1, and 3). This result points to the escape of oxygen in the reducing conditions due to carbon and resulting release of CO_2 and creation oxygen vacancies and related complex defects.

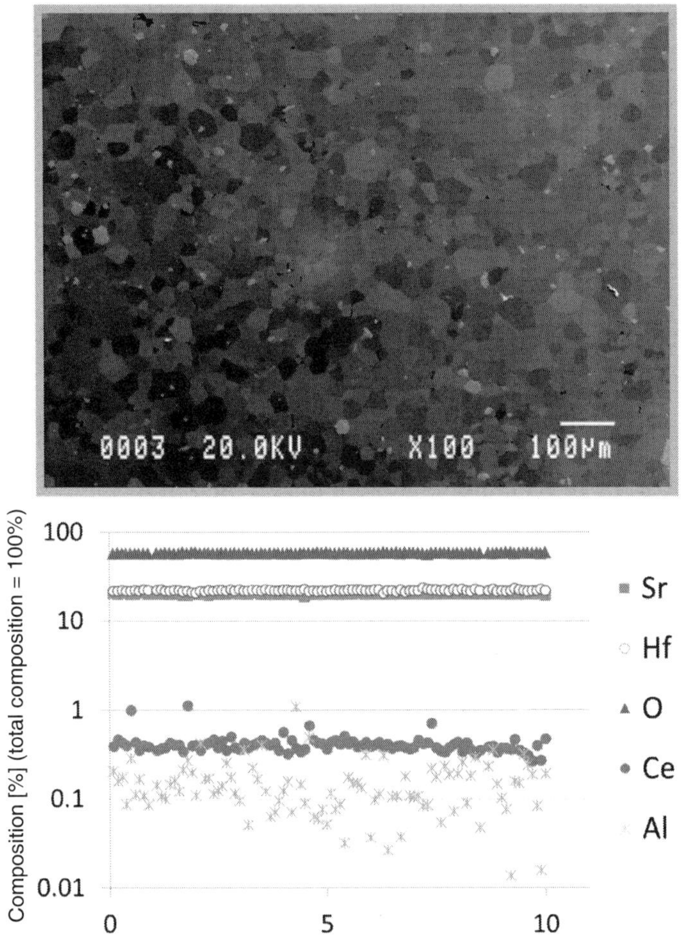

Figure 4.22 The BSE image of the surface of the SHO:Ce sample (upper part) and the composition analysis for the Al-codoped SHO:Ce sample performed by electron-probe microanalyzer (lower part).

The photoluminescence emission and excitation spectra are shown in Fig. 4.23a. The emission wavelength was near the maximum of the quantum efficiency of a PMT with bi-alkali photo cathode which makes the Al-codoped SHO:Ce suitable for coupling with such a type of PMT. The 395 nm emission can be ascribed to the 5d-4f emission of Ce^{3+} in agreement with existing literature [127, 129, 130]. The photoluminescence decay in Fig. 4.23(b) shows the decay time value of 18.2 ± 0.2 ns which is close to that reported in [129].

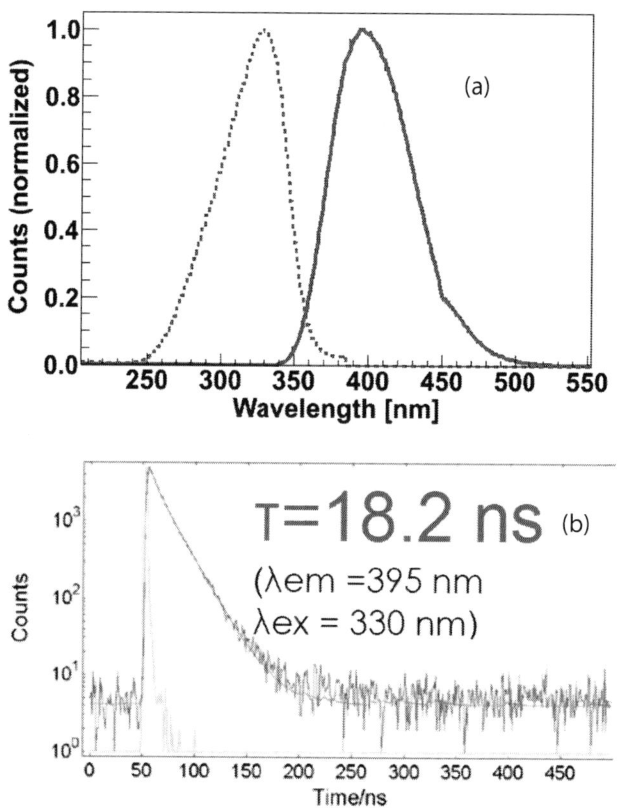

Figure 4.23 Photoluminescence characteristics of the Al-codoped SHO:Ce: (a) the excitation spectrum for 395 nm emission (dotted line) and emission spectrum for the 330 nm excitation (solid line); (b) decay kinetics of the 395 nm emission under 330 nm excitation.

Regarding scintillation characteristics the measured radioluminescence spectrum of SHO:Ce,Al sample under 5.5 MeV α-ray (^{241}Am) excitation is closely similar to the photoluminescence one in Fig. 4.23a. An attempt was made to measure light yield in a standard experiment which provides the photopeak position from an energy spectrum under excitation by ^{137}Cs radioisotope. Obtained value of 4000 ph/MeV is rather low compared to literature data in [130] and must be severely affected by low transmittance of the sample described above. Scintillation decay (see Fig. 4.24) is, however, close to the photoluminescence one, which points to fast and efficient energy transfer from the host towards Ce^{3+} centers.

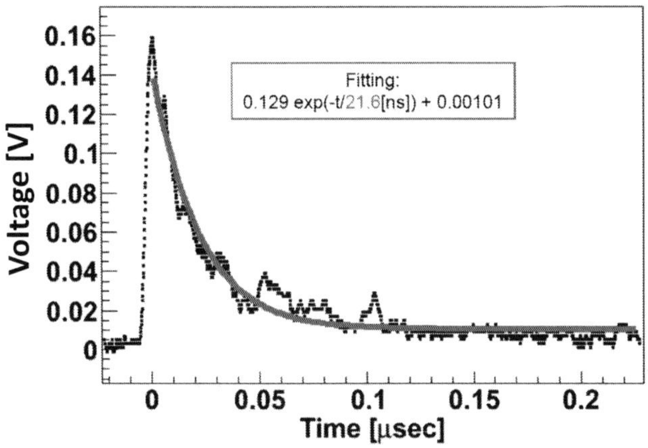

Figure 4.24 Scintillation decay curve for the Al-codoped SHO:Ce under ^{137}Cs 662 keV γ-ray excitation.

In conclusion, the $SrHfO_3$, $Ce_{0.02}Sr_{0.98}HfO_3$, and $Ce_{0.02}Al_{0.02}Sr_{0.96}HfO_3$ ceramic bodies were prepared by SPS. $Ce_{0.02}Sr_{0.98}HfO_3$ was not transparent in the wavelength region of the Ce^{3+} 5d–4f emission. On the other hand, the transparency region of the $Ce_{0.02}Al_{0.02}Sr_{0.96}HfO_3$ was wider, reaching up to 350 nm, which is the onset of the 4f–5d absorption of Ce^{3+}. The Ce^{3+} 5d–4f emission wavelength was 395 nm under 330 nm light and 5.5 MeV α-rays excitation. Despite the favorably fast scintillation response, the light yield value is low due to low sample transmittance. In order to improve it, the number of pores between the grain boundaries needs to be decreased, to prevent scattering of the scintillation light.

Additionally, the size of the particles of the starting SHO powder should be controlled approximately within 1–10 nm. Supposing the light output is 40,000 photons/MeV, the FWHM energy resolution is expected to be less than 6–7% at 662 keV, similarly as for Tl:NaI. Reasonably high energy resolution (less than 7%) and high-γ-ray stopping power are two of the main requirements for scintillators to be applied in Compton camera in order to obtain a good angular resolution of 4.6° at 662 keV (FWHM) [132]. If the transmittance of the Al-codoped SHO:Ce ceramic scintillators is improved, they could find application in the Compton camera due to the supposed good energy resolution, high effective atomic number, and negligible intrinsic background.

4.5.3 Lu_2O_3-Based Scintillation Ceramics

The high density of 9.4 g/cm^3 and atomic number of 67 makes the Lu_2O_3 an interesting host for scintillation materials. The scintillation properties of Ce-doped Lu_2O_3 single crystal fibers were investigated in the nineties [133]. It was found that the Lu_2O_3:Ce has low light output to find application in γ-ray detectors because the lowest $5d^1$ excited state of Ce^{3+} is located in the conduction band of Lu_2O_3, as is the case for the other materials [134]. Yen et al. [133] reported that the dominant absorption band at 400 nm based on 4f-$5d^1$ transition of Ce^{3+} in Lu_2O_3 corresponds to a true donor-like process that promotes electrons from Ce^{3+} to the conduction band and effectively quenches the Ce^{3+} 5d–4f luminescence.

However, due to very high melting point of Lu_2O_3, which is around 2500°C, and resulting difficult preparation of the single crystals, the ceramic counterparts again attract attention due to easier fabrication. Because of its very high thermal conductivity the Lu_2O_3 is attractive as a laser host for high-power lasers which has been already mentioned in Section 4.2.

The possibility of application of infrared scintillators for radiation dosimetry and online monitoring in medical imaging has been suggested recently [135]. The concept consists in implanting small infrared scintillating elements via catheter near the tumor under irradiation. The infrared photons can penetrate the human tissue within its transparency window around 1000 nm [136] and can be detected by infrared detectors such as charge-coupled device (CCD) or complementary metal oxide semiconductor detector outside the patient's body, as shown in the schematic in Fig. 4.25. The intensity of the light emitted by the scintillator would be proportional to the irradiation intensity. Moreover, the intensity of the scintillation light determines the current produced by a photodetector. Thus, the measured current can indicate the radiation dose received by the patient during therapy. This feedback allows immediate control of the irradiation intensity. The development of such a monitoring system has been started recently to prevent accidents related to under- or overdosage or beam misalignment [135].

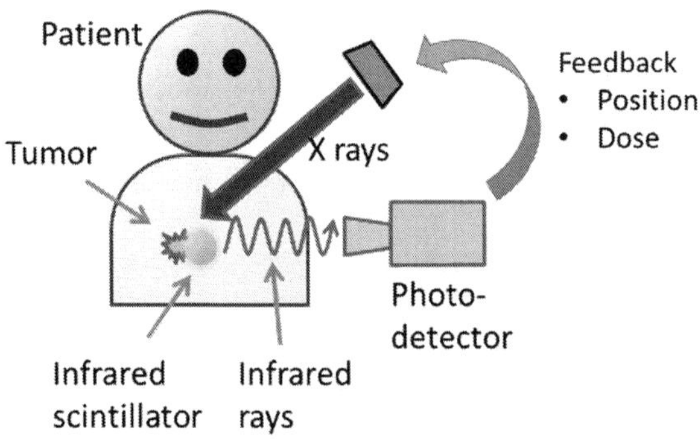

Figure 4.25 Schematic of the real-time radiation monitoring system employing an infrared scintillator.

The results summarized above stimulated the preparation and investigation of Nd:Lu_2O_3 ceramics to test its suitability also for the infrared scintillator application [96]. The 0.1, 0.5, 1.0, and 5.0 mol% Lu_2O_3:Nd ceramics were sintered by the SPS method. The examples of the sintered ceramic bodies are shown in Fig. 4.26.

The transmittance spectra of the sintered samples in near infrared spectra region are displayed in Fig. 4.27. In these spectra the 4f–4f absorption lines of the Nd^{3+} were identified. For the infrared scintillation materials intended to be used in the described application, the transparency on the infra-red region between 1000–1100 nm is of crucial importance. From Fig. 4.27 we can see that the transmittance of the studied samples exceeded 50%. Although this value can certainly be improved yet, the samples can be used to for further evaluation of scintillation properties as a first step in the development of these scintillator materials.

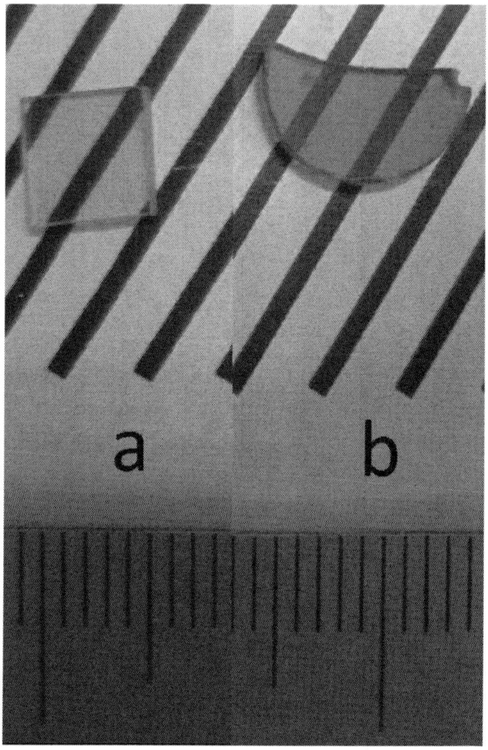

Figure 4.26 Examples of Lu_2O_3 sintered transparent ceramic bodies doped with 1 mol% Nd (a) and 5 mol% Nd (b).

The X-ray-excited radioluminescence of 0.5 and 1 mol% Nd:Lu_2O_3 ceramics in the infrared region is shown in Fig. 4.28. The observed peaks were assigned to the $^4F_{3/2}$-^4I_i (i = 11/2 and 9/2) transitions of the Nd^{3+} ions. They are placed exactly in the near-infrared region where the human tissue is transparent. Notably lower intensity of scintillation light in the 0.1 and 5.0 mol% Lu_2O_3:Nd (not shown) might be due to low Nd concentration in the 0.1% sample and an onset of concentration quenching effect in the 5% one, respectively. The Lu_2O_3:Nd samples showed the emission lines not only in the visible and infrared regions, but also in the UV region. Namely, presence of the Nd^{3+} 4f–4f emission at approximately 370 nm ascribed to $^4D_{3/2}$-^4I_i (i = 11/2 and 9/2) transitions suggests that the $^4D_{3/2}$ level is located still below the conduction band.

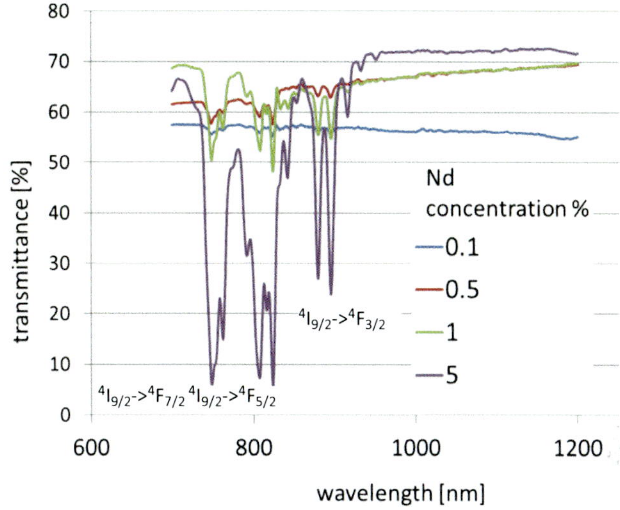

Figure 4.27 Transmittance spectra in the near-infrared region for the Nd:Lu_2O_3 ceramics with different Nd concentration. Absorption lines of the Nd^{3+} center are marked.

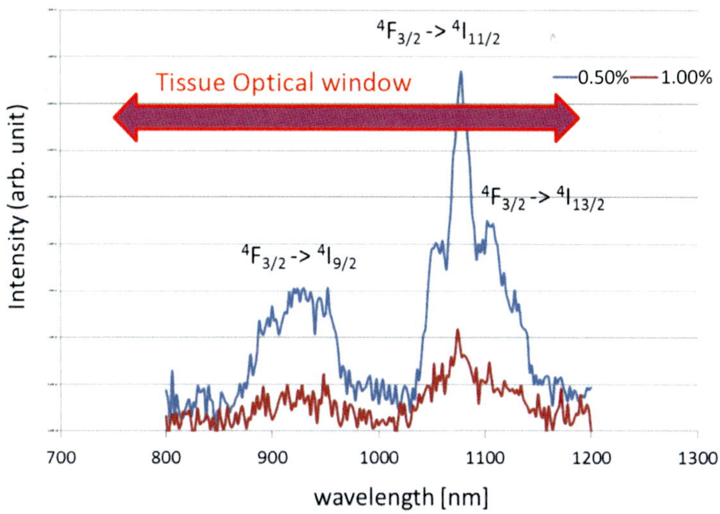

Figure 4.28 The α-ray excited radioluminescence spectra in the near-infrared region for the Lu_2O_3 ceramic samples with different Nd concentrations.

To summarize, Lu_2O_3:Nd ceramics with different Nd concentrations from 0.1 mol% up to 5 mol% were successfully prepared by SPS. Radioluminescence of these ceramics was studied and several scintillation emission peaks were observed at 300–1200 nm under X-ray excitation. The near infrared emission within 900–1100 nm was successfully detected and assigned to the $^4F_{3/2}$-^4I_i (i = 11/2 and 9/2) transitions of Nd^{3+} ions. Their spectral positions correspond to the transparency region of the human tissue and thus the Lu_2O_3:Nd ceramics can be proposed as a candidate for application as a radiation dosimeter in medical radiation therapy [96].

4.6 Summary

Transparent ceramics have attracted widespread attention as an alternative to single crystals because of their high performance and potential scalability. Although transparent ceramics have been prepared by traditional sintering methods, SPS is a candidate process for obtaining transparent ceramics from commercially available powders via a fast densification process. Given the speed of manufacturing process, samples can be prepared typically within one working day so that such a method is particularly useful for fast initial material screening and composition optimization. SPS proved to be promising for the R&D of practical and novel transparent high density scintillation oxide ceramics.

Acknowledgments

This work is partially supported by (i) the Adaptable & Seamless Technology Transfer Program through Target-Driven R&D (A-STEP), JST, (ii) the Japan Society for the Promotion of Science (JSPS) Grant-in-Aid for Exploratory Research (AY), (iii) Development of Systems and Technology for Advanced Measurement and Analysis, Japan Science and Technology Agency (JST), (iv) the funding program for next-generation world-leading researchers, JSPS, and (v) the Czech National Science Foundation No. 13-09876S project.

References

1. Handwerker, C. A., Cannon, R. M., French, R. H., and Coble, R. L. (1994). A retrospective, *J. Am. Ceram. Soc.*, **77**, pp. 293–297.
2. Apetz, R. and van Bruggen, M. P. B. (2003). Transparent alumina: a light-scattering model, *J. Am. Ceram. Soc.* **86**, pp. 480–486.
3. Ikesue, A. and Aung, Y. L. (2008). Ceramic laser materials, *Nat. Photonics.* **2**, pp. 721–727.
4. Kintaka, Y., Kuretake, S., Tanaka, N., Kageyama, K., and Takagi, H. (2010). Crystal structures and optical properties of transparent ceramics based on complex Perovskite $Ba(M^{4+}, B_1^{2+}, B_2^{5+})O_3$ (M^{4+} = Ti, Sn, Zr, Hf; B_1^{2+} = Mg, Zn; B_2^{5+} = Ta, Nb), *J. Am. Ceram. Soc.* **93**, pp. 1114–1119.
5. Omori, M. (2000). Sintering, consolidation, reaction and crystal growth by the spark plasma system (SPS), *Mater. Sci. Eng. A.* **287** pp. 183–188.
6. Munir, Z. A., Anselmi-Tamburini, U., and Ohyanagi, M. (2006) The effect of electric field and pressure on the synthesis and consolidation of materials: a review of the spark plasma sintering method, *J. Mater. Sci.* **41**, pp. 763–777.
7. Grasso, S., Sakka, Y., and Maizza, G. (2009). Electric current activated/assisted sintering (ECAS): a review of patents 1906–2008, *Sci. Technol. Adv. Mater.* **10**, pp. 053001.
8. An, L., Ito, A., and Goto, T. (2011) Two-step pressure sintering of transparent lutetium oxide by spark plasma sintering, *J. Eur. Ceram. Soc.* **31**, pp. 1597–1602.
9. Morita, K., Kim, B.-N., Yoshida, H., and Hiraga, K. (2009). Spark-plasma-sintering condition optimization for producing transparent $MgAl_2O_4$ Spinel polycrystal, *J. Am. Ceram. Soc.* **92**, pp. 1208–1216.
10. Kim, B.-N., Hiraga, K., Morita, K., and Yoshida, H. (2009). Effects of heating rate on microstructure and transparency of spark-plasma-sintered alumina, *J. Eur. Ceram. Soc.* **29**, pp. 323–327.
11. Hulbert, D. M., Anders, A., Dudina, D. V., Andersson, J., Jiang, D., Unuvar, C., Anselmi-Tamburini, U., Lavernia, E. J., and Mukherjee, A. K. (2008). The absence of plasma in "spark plasma sintering," *J. Appl. Phys.* **104**, pp. 033305.
12. Olevsky, E. A. and Froyen, L. (2009). Impact of thermal diffusion on densification during SPS, *J. Am. Ceram. Soc.* **92**, pp. S122–S132.
13. Munir, Z. A., Quach, D. V., and Ohyanagi, M. (2011). Electric current activation of sintering: a review of the pulsed electric current sintering process, *J. Am. Ceram. Soc.* **94**, pp. 1–19.

14. Wu, Y. J., Kimura, R., Uekawa, N., Kakegawa, K., and Sasaki, Y. (2002). Spark plasma sintering of transparent $PbZrO_3$-$PbTiO_3$-$Pb(Zn_{1/3}Nb_{2/3})O_3$ ceramics, *Jpn. J. Appl. Phys. Part 2 Lett.* **41**, pp. L219–L221.

15. Kim, B.-N., Hiraga, K., Morita, K., and Yoshida, H.(2007) Spark plasma sintering of transparent alumina, *Scr. Mater.* **57**, pp. 607–610

16. Jiang, D., Hulbert, D. M., Anselmi-Tamburini, U., Ng, T., Land, D., and Mukherjee, A. K. (2008) Optically transparent polycrystalline Al_2O_3 produced by spark plasma sintering, *J. Am. Ceram. Soc.* **91**, pp. 151–154.

17. Kim, B.-N., Hiraga, K., Morita, K., Yoshida, H., Miyazaki, T., and Kagawa, Y. (2009) Microstructure and optical properties of transparent alumina, *Acta Mater.* **57**, pp. 1319–1326.

18. Aman, Y., Garnier, V., and Djurado, E. (2009). Influence of green state processes on the sintering behaviour and the subsequent optical properties of spark plasma sintered alumina, *J. Eur. Ceram. Soc.* **29**, pp. 3363–3370.

19. Chakravarty, D. and Sundararajan, G. (2010). Effect of applied stress on IR transmission of spark plasma-sintered alumina, *J. Am. Ceram. Soc.* **93**, (2010) pp. 951–953.

20. Grasso, S., Kim, B.-N., Hu, C., Maizza, G., and Sakka, Y. (2010). Highly transparent pure alumina fabricated by high-pressure spark plasma sintering, *J. Am. Ceram. Soc.* **93**, pp. 2460–2462.

21 Jin, X., Gao, L., and Sun, J. (2010). Highly transparent alumina spark plasma sintered from common-grade commercial powder: The effect of powder treatment, *J. Am. Ceram. Soc.* **93**, pp. 1232–1236.

22. Stuer, M., Zhao, Z., Aschauer, U., and Bowen, P. (2010). Transparent polycrystalline alumina using spark plasma sintering: Effect of Mg, Y and La doping, *J. Eur. Ceram. Soc.* **30**, pp. 1335–1343.

23. Dang, K. Q., Takei, S., Kawahara, M., and Nanko, M. (2011). Pulsed electric current sintering of transparent Cr-doped Al_2O_3, *Ceram. Int.* **37**, pp. 957–963.

24. Aman, Y., Garnier, V., and Djurado, E. (2010). A screening design approach for the understanding of spark plasma sintering parameters: a case of translucent polycrystalline undoped alumina, *Int. J. Appl. Ceram. Technol.* **7**, pp. 574–586.

25. Roussel, N., Lallemant, L., Durand, B., Guillemet, S., Ching, J.-Y.C., Fantozzi, G., Garnier, V., and Bonnefont, G. (2011). Effects of the nature of the doping salt and of the thermal pre-treatment and sintering temperature on spark plasma sintering of transparent alumina, *Ceram. Int.* **37**, pp. 3565–3573.

26. Santanach, J. G., Weibel, A., Estournès, C., Yang, Q., Laurent, C., and Peigney, A. (2011). Spark plasma sintering of alumina: Study of parameters, formal sintering analysis and hypotheses on the mechanism(s) involved in densification and grain growth, *Acta Mater.* **59**, pp. 1400–1408.

27. Lallemant, L., Fantozzi, G., Garnier, V., and Bonnefont, G. (2012). Transparent polycrystalline alumina obtained by SPS: Green bodies processing effect, *J. Eur. Ceram. Soc.* **32**, pp. 2909–2915.

28. Makino, Y., Kawahara, M., Sakaguchi, M., Kaiyama, Y., Akamatsu, K., and Yasuno, T. (2012). Influence of raw powder on transparency of SPS-synthesized aluminas, *J. Jpn. Soc. Powder Metall.* **59**, pp. 538–543.

29. Makino, Y., Kawahara, M., Sakaguchi, M., Kaiyama, Y., Akamatsu, K., and Yasuno, T. (2012). Synthesis of nano-structured alumina with high transparency by SPS processing, *J. Jpn. Soc. Powder Metall.* **59** pp. 532–537.

30. Stuer, M., Zhao, Z., and Bowen, P. (2012). Freeze granulation: Powder processing for transparent alumina applications, *J. Eur. Ceram. Soc.* **32**, pp. 2899–2908.

31. Grasso, S., Yoshida, H., Porwal, H., Sakka, Y., and Reece, M. (2013). Highly transparent α-alumina obtained by low cost high pressure SPS, *Ceram. Int.* **39**, pp. 3243–3248.

32. Roussel, N., Lallemant, L., Chane-Ching, J.-Y., Guillemet-Fristch, S., Durand, B., Garnier, V., Bonnefont, G., Fantozzi, G., Bonneau, L., Trombert, S., and Garcia-Gutierrez, D. (2013). Highly dense, transparent α-Al_2O_3 ceramics from ultrafine nanoparticles via a standard SPS sintering, *J. Am. Ceram. Soc.* **96**, pp. 1039–1042.

33. Suárez, M., Fernández, A., Menéndez, J. L., Nygren, M., Torrecillas, R., and Zhao, Z. (2013). Effect of freeze-drying treatment on the optical properties of SPS-sintered alumina, *Ceram. Int.* **39**, pp. 6669–6672.

34. Nanko, M. and Dang, K.Q. (2014). Two-step pulsed electric current sintering of transparent Al_2O_3 ceramics, *Adv. Appl. Ceram.* **113**, pp. 80–84.

35. Chaim, R., Shen, Z., and Nygren, M. (2004). Transparent nanocrystalline MgO by rapid and low-temperature spark plasma sintering, *J. Mater. Res.* **19**, pp. 2527–2531.

36. Wakahara, S., Yanagida, T., Yokota, Y., Fujimoto, Y., Chani, V., Sugiyama, M., Futami, Y., and Yoshikawa, A. (2013). Phosphorescent luminescence of pure magnesium oxide transparent ceramics produced by spark plasma sintering, *Opt. Mater.* **35**, pp. 558–562.

37. Zhang, H., Kim, B.-N., Morita, K., Yoshida, H., Hiraga, K., and Sakka, Y. (2011) Fabrication of transparent yttria by high-pressure spark plasma sintering, *J. Am. Ceram. Soc.* **94**, pp. 3206–3210.

38. An, L., Ito, A., and T. Goto, (2012) Transparent yttria produced by spark plasma sintering at moderate temperature and pressure profiles, *J. Eur. Ceram. Soc.* **32**, pp. 1035–1040.

39. He, M., Li, J., Lin, H., Guo, G., and Liang, L. (2006). Fabrication of transparent polycrystalline yttria ceramics by combination of SPS and HIP, *J. Rare Earths.* **24**, pp. 222–224.

40. Fukabori, A., An, L., Ito, A., Chani, V., Kamada, K., Goto, T., and A. Yoshikawa, (2012). Scintillation characteristics of undoped single crystals and ceramics, *IEEE Trans. Nucl. Sci.* **59**, pp. 2594–2600.

41. An, L., Ito, A., and Goto, T., (2013). Fabrication of transparent $Lu_2Hf_2O_7$ by reactive spark plasma sintering, *Optical Materials* **35**, pp. 817–819.

42. Quesada, A., del Campo, A., and Fernández, V. (2014). Sintering behaviour and translucency of dense Eu_2O_3 ceramics, *J. Eur. Ceram. Soc.* **34**, pp. 1803–1808.

43. An, L., Ito, A., and Goto, T. (2011). Fabrication of transparent lutetium oxide by spark plasma sintering, *J. Am. Ceram. Soc.* **94**, pp. 695–698.

44. An, L., Ito, A., and Goto, T. (2011). Effects of ball milling and post-annealing on the transparency of spark plasma sintered Lu_2O_3, *Ceram. Int.* **37**, pp. 2263–2267.

45. Frage, N., Cohen, S., Meir, S., Kalabukhov, S.., and Dariel, M. P. (2007). Spark plasma sintering (SPS) of transparent magnesium-aluminate spinel, *J. Mater. Sci.* **42**, pp. 3273–3275.

46. Morita, K., Kim, B.-N., Hiraga, K., and Yoshida, H. (2008). Fabrication of transparent $MgAl_2O_4$ spinel polycrystal by spark plasma sintering processing, *Scr. Mater.* **58**, pp. 1114–1117.

47. Bernard-Granger, G., Benameur, N., Guizard, C., and Nygren, M. (2009). Influence of graphite contamination on the optical properties of transparent spinel obtained by spark plasma sintering, *Scr. Mater.* **60**, pp. 164–167.

48. Morita, K., Kim, B.-N., Hiraga, K., and Yoshida, H. (2009). Fabrication of high-strength transparent $MgAl_2O_4$ spinel polycrystals by optimizing spark-plasma-sintering conditions, *J. Mater. Res.* **24**, pp. 2863–2872.

49. Wang, C. and Zhao, Z. (2009). Transparent $MgAl_2O_4$ ceramic produced by spark plasma sintering, *Scr. Mater.* **61**, pp. 193–196.

50. Kim, B., Morita, K., Lim, J., Hiraga, K., and Yoshida, H. (2010). Effects of preheating of powder before spark plasma sintering of transparent $MgAl_2O_4$ spinel, *J. Am. Ceram. Soc.* **93**, pp. 2158–2160.

51. Fu, P., Lu, W., Lei, W., Xu, Y., Wang, X., and Wu, J. (2013). Transparent polycrystalline $MgAl_2O_4$ ceramic fabricated by spark plasma sintering: Microwave dielectric and optical properties, *Ceram. Int.* **39** pp. 2481–2487.

52. Khasanov, O., Dvilis, E., Khasanov, A., Polisadova, E., and Kachaev, A. (2013). Optical and mechanical properties of transparent polycrystalline $MgAl_2O_4$ spinel depending on SPS conditions, *Phys. Status Solidi C* **10**, pp. 918–920.

53. Rothman, A., Kalabukhov, S., Sverdlov, N., Dariel, M. P., and Frage, N. (2014). The effect of grain size on the mechanical and optical properties of spark plasma sintering-processed magnesium aluminate spinel $MgAl_2O_4$, *Int. J. Appl. Ceram. Technol.* **11**, pp. 146–153.

54. Bonnefont, G., Fantozzi, G., Trombert, S., and Bonneau, L. (2012). Fine-grained transparent $MgAl_2O_4$ spinel obtained by spark plasma sintering of commercially available nanopowders, *Ceram. Int.* **38**, pp. 131–140.

55. Yong, X., Ping, F., Baohua, Z., Juan, G., Lin, Z., and Xuehua, W. (2014). Optical properties of transparent $ZnAl_2O_4$ ceramics: A new transparent material prepared by spark plasma sintering, *Mater. Lett.* **123**, pp. 142–144.

56. Chaim, R., Kalina, M., and Shen, J. Z. (2007). Transparent yttrium aluminum garnet (YAG) ceramics by spark plasma sintering, *J. Eur. Ceram. Soc.* **27**, pp. 3331–3337.

57. Chaim, R., Marder-Jaeckel, R., and Shen, J. Z. (2006). Transparent YAG ceramics by surface softening of nanoparticles in spark plasma sintering, *Mater. Sci. Eng. A* **429**, pp. 74–78.

58. Spina, G., Bonnefont, G., Palmero, P., Fantozzi, G., Chevalier, J., and Montanaro, L. (2012). Transparent YAG obtained by spark plasma sintering of co-precipitated powder. Influence of dispersion route and sintering parameters on optical and microstructural characteristics, *J. Eur. Ceram. Soc.* **32**, pp. 2957–2964.

59. Palmero, P., Bonelli, B., Fantozzi, G., Spina, G., Bonnefont, G., Montanaro, L., and Chevalier, J. (2013). Surface and mechanical properties of transparent polycrystalline YAG fabricated by SPS, *Mater. Res. Bull.* **48**, pp. 2589–2597.

60. An, L., Ito, A., and Goto, T. (2012). Effect of sintering temperature on the transparency and mechanical properties of lutetium aluminum

garnet fabricated by spark plasma sintering, *J. Eur. Ceram. Soc.* **32**, pp. 3097–3102.

61. Koide, M., Takei, S., Sato, T., and Matusita, K. (2002). Preparation of silica glass by pulsed electric current method, *J. Ceram. Soc. Jpn.* **110**, pp. 867–869.
62. Zhang, J., Tu, R., and Goto, T.(2012). Fabrication of transparent SiO_2 glass by pressureless sintering and spark plasma sintering, *Ceram. Int.* **38**, pp. 2673–2678.
63. Wang, L., Jiang, W., Chen, L., and Shen, Z. (2009). Formation of a unique glass by spark plasma sintering of a zeolite, *J. Mater. Res.* **24**, pp. 3241–3245.
64. Delaizir, G., Dollé, M., Rozier, P., and Zhang, X.H. (2010). Spark plasma sintering: An easy way to make infrared transparent glass-ceramics, *J. Am. Ceram. Soc.* **93**, pp. 2495–2498.
65. Hubert, M., Delaizir, G., Monnier, J., Godart, C., Ma, H.-L., Zhang, X.-H., and Calvez, L. (2011). An innovative approach to develop highly performant chalcogenide glasses and glass-ceramics transparent in the infrared range, *Opt. Express* **19**, pp. 23513–23522.
66. Bertrand, A., Carreaud, J., Delaizir, G., Duclère, J.-R., Colas, M., Cornette, J., Vandenhende, M., Couderc, V., and Thomas, P. (2014). A comprehensive study of the carbon contamination in tellurite glasses and glass-ceramics sintered by spark plasma sintering (SPS), *J. Am. Ceram. Soc.* **97**, pp. 163–172.
67. Shamrai, G. V., Zagorodnyuk, A. V., Magunov, R.L., and Zhirnova, A. P. (1984). The Lu_2O_3–TiO_2 system, *Russ. J. Inorg. Chem.* **29** pp. 1811–1812.
68. Petrosyan, A. G., Popova, V. F., Gusarov, V. V., Shirinyan, G. O., Pedrini, C., and Lecoq, P. (2006). The Lu_2O_3–Al_2O_3 system: Relationships for equilibrium-phase and supercooled states, *J. Cryst. Growth* **293**, pp. 74–77.
69. Momma, K. and Izumi, F. (2011). VESTA 3 for three-dimensional visualization of crystal, volumetric and morphology data, *J. Appl. Crystallogr.* **44**, pp. 1272–1276.
70. Pauling, L. and Shappell, M. (1930). The crystal structure of bixbyite and the C-modification of the sesquioxides., *Z. Krist.* **75**, pp. 128–142.
71. Hoekstra, H. R. (1966). Phase relationships in the rare earth sesquioxides at high pressure, *Inorg. Chem.* **5**, pp. 754–757.

72. Peters, R., Kränkel, C., Petermann, K., and Huber, G. (2007). Broadly tunable high-power Yb:Lu_2O_3 thin disk laser with 80% slope efficiency, *Opt. Express* **15**, p. 7075.

73. Snetkov, I. L., Silin, D. E., Palashov, O. V., Khazanov, E. A., Yagi, H., Yanagitani, T., Yoneda, H., Shirakawa, A., Ueda, K., and Kaminskii, A. A. (2013). Thermo-optical constants of sesquioxide laser ceramics Yb^{3+}:Ln_2O_3 (Ln = Y, Lu, Sc), *Phys. Status Solidi C* **10**, pp. 907–913.

74. Klemens, P. G. (1960). Thermal resistance due to point defects at high temperatures, *Phys. Rev.* **119**, pp. 507–509.

75. Gaumé, R., Viana, B., Vivien, D., Roger, J.-P., and Fournier, D. (2003). A simple model for the prediction of thermal conductivity in pure and doped insulating crystals, *Appl. Phys. Lett.* **83**, pp. 1355–1357.

76. Griebner, U., Petrov, V., Petermann, K., and Peters, V. (2004). Passively mode-locked Yb:Lu_2O_3 laser, *Opt. Express* **12**, pp. 3125–3130.

77. Aggarwal, R. L., Ripin, D. J., Ochoa, J. R., and Fan, T. Y. (2005). Measurement of thermo-optic properties of $Y_3Al_5O_{12}$, $Lu_3Al_5O_{12}$, $YAlO_3$, $LiYF_4$, $LiLuF_4$, BaY_2F_8, $KGd(WO_4)_2$, and $KY(WO_4)_2$ laser crystals in the 80–300K temperature range, *J. Appl. Phys.* **98**, p. 103514.

78. Peters, R., Kränkel, C., Fredrich-Thornton, S.T., Beil, K. Petermann, K., Huber, G., Heckl, O. H., Baer, C. R. E., Saraceno, C. J., Südmeyer, T., and Keller, U. (2011). Thermal analysis and efficient high power continuous-wave and mode-locked thin disk laser operation of Yb-doped sesquioxides, *Appl. Phys. B* **102**, pp. 509–514.

79. Hart, P. E., Atkin, R. B., and Pask, J. A. (1970). Densification mechanisms in hot-pressing of magnesia with a fugitive liquid, *J. Am. Ceram. Soc.* **53**, pp. 83–86.

80. Reimanis, I., and Kleebe, H.-J. (2009). A review on the sintering and microstructure development of transparent spinel ($MgAl_2O_4$), *J. Am. Ceram. Soc.* **92**, pp. 1472–1480.

81. Sanghera, J., Frantz, J., Kim, W., Villalobos, G., Baker, C., Shaw, B., Sadowski, B., Hunt, M., Miklos, F., Lutz, A., and Aggarwal, I. (2011). 10% Yb^{3+}-Lu_2O_3 ceramic laser with 74% efficiency, *Opt. Lett.* **36**, pp. 576–578

82. Sanghera, J., Kim, W., Baker, C., Villalobos, G., Frantz, J., Shaw, B., Lutz, A., Sadowski, B., Miklos, R., Hunt, M., Kung, F., and Aggarwal, I. (2011). Laser oscillation in hot pressed 10% Yb^{3+}:Lu_2O_3 ceramic, *Opt. Mater.* **33**, pp. 670–674.

83. Meir, S., Kalabukhov, S., Froumin, N., Dariel, M. P., and Frage, N. (2009). Synthesis and densification of transparent magnesium aluminate spinel by SPS processing, *J. Am. Ceram. Soc.* **92**, pp. 358–364

84. Frage, N., Kalabukhov, S., Sverdlov, N., Ezersky, V., and Dariel, M. P. (2010). Densification of transparent yttrium aluminum garnet (YAG) by SPS processing, *J. Eur. Ceram. Soc.* **30**, pp. 3331–3337.

85. Marder, R., Chaim, R., Chevallier, G., and Estournès, C. (2011). Effect of 1 wt% LiF additive on the densification of nanocrystalline Y_2O_3 ceramics by spark plasma sintering, *J. Eur. Ceram. Soc.* **31**, pp. 1057–1066.

86. Hodge, J. D. and Gordon, R. S. (1978). Grain growth and creep in polycrystalline magnesium oxide fabricated with and without a LiF additive, *Ceram. Int.* **4**, pp. 17–20.

87. Merac, R. D. M., Reimanis, I. E., Smith, C., Kleebe, H.-J., and Müller, M. M. (2013). Effect of impurities and LiF additive in hot-pressed transparent magnesium aluminate spinel, *Int J Appl Ceram Technol.* **10**, pp. E33–E48.

88. An, L., Ito, A., and Goto, T. (2014). Effect of LiF addition on spark plasma sintering of transparent Nd-doped Lu_2O_3 bodies, *J. Asian Ceram. Soc.* **2**, pp. 154–157.

89. Zych, E., Hreniak, D., Strek, W., Kepinski, L., and Domagala, K. (2002). Sintering properties of urea-derived Lu_2O_3-based phosphors, *J. Alloys Compd.* **341**, pp. 391–394

90. Lu, J., Takaichi, K., Uematsu, T., Shirakawa, A., Musha, M., Ueda, K., Yagi, H., Yanagitani, T., and Kaminskii, A. A. (2002). Promising ceramic laser material: Highly transparent Nd^{3+}:Lu_2O_3 ceramic, *Appl. Phys. Lett.* **81**, pp. 4324–4326.

91. Seeley, Z. M., Kuntz, J. D., Cherepy, N. J., and Payne, S. A. (2011). Transparent Lu_2O_3:Eu ceramics by sinter and HIP optimization, *Opt. Mater.* **33**, pp. 1721–1726.

92. Chen, Q., Shi, Y., An, L., Chen, J., and J. Shi, (2006). Fabrication and photoluminescence characteristics of Eu^{3+}-doped Lu_2O_3 transparent ceramics, *J. Am. Ceram. Soc.* **89**, pp. 2038–2042.

93. Zhou, D., Shi, Y., Xie, J., Ren, Y., and Yun, P. (2009). Fabrication and luminescent properties of Nd^{3+}-doped Lu_2O_3 transparent ceramics by pressureless sintering, *J. Am. Ceram. Soc.* **92**, pp. 2182–2187.

94. Zhang, H., Yang, Q., Lu, S., and Shi, Z. (2012). Structural and spectroscopic characterization of Yb^{3+} doped Lu_2O_3 transparent ceramics, *Opt. Mater.* **34**, pp. 969–972.

95. Kim, W., Baker, C., Bowman, S., Florea, C., Villalobos, G., Shaw, B., Sadowski, B., Hunt, M., Aggarwal, I., and Sanghera, J. (2013). Laser oscillation from Ho^{3+} doped Lu_2O_3 ceramics, *Opt. Mater. Express* **3**, pp. 913–919.

96. Kurosawa, S., An, L., Yamaji, A., Suzuki, A., Yokota, Y., Shirasaki, K., Yamamura, T., Ito, A., Goto, T., Boulon, G., and Yoshikawa, A. (2014). Scintillation properties of Nd^{3+}-doped Lu_2O_3 ceramics in the visible and infrared regions, *IEEE Trans. Nucl. Sci.* **61**, pp. 316–319.

97. Subramanian, M. A., Aravamudan, G., and Subba Rao, G. V. (1983). Oxide pyrochlores: a review, *Prog. Solid State Chem.* **15**, pp. 55–143.

98. Ji, Y., Jiang, D., Fen, T., and Shi, J. (2005). Fabrication of transparent $La_2Hf_2O_7$ ceramics from combustion synthesized powders, *Mater. Res. Bull.* **40**, pp. 553–559.

99. Zou, X.-Q., Zhou, G.-H., Yi, H.-L., Yang, Y., and Wang, S.-W. (2011). Fabrication of transparent $Y_2Hf_2O_7$ ceramic from combustion synthesized powders, *J. Inorg. Mater.* **26**, pp. 929–932

100. Yi, H., Zou, X., Yang, Y., Zhou, G., and Wang, S. (2011). Fabrication of highly transmitting $LaGdHf_2O_7$ ceramics, *J. Am. Ceram. Soc.* **94**, pp. 4120–4122.

101. Borisevich, A., Korzhik, M., and Lecoq, P. (2003). Luminescence of Ce doped oxygen crystalline compounds based on Hf and Ba, *Nucl. Instrum. Meth. Phys. Res. A* **497**, pp. 206–209.

102. Mandal, B. P., Garg, N., Sharma, S. M., and Tyagi, A. K. (2006). Preparation, XRD and Raman spectroscopic studies on new compounds $RE_2Hf_2O_7$ (RE = Dy, Ho, Er, Tm, Lu, Y): pyrochlores or defect-fluorite? *J. Solid State Chem.* **179**, pp. 1990–1994.

103. Shlyakhtina, A. V., Belov, D. A., Stefanovich, S. Y., Kolbanev, I. V., Karyagina, O. K., Egorov, A. V., Savilov, S. V., and Shcherbakova, L. G. (2011). δ-Phase to defect fluorite (order–disorder) transition in the R_2O_3–MO_2 (R = Sc, Tm, Lu; M = Zr, Hf) systems, *Mater. Res. Bull.* **46**, pp. 512–517.

104. Nikl, M., Mihokova, E., Mares, J. A., Vedda, A., Martini, M., Nejezchleb, K., and Blazek, K. (2000). Traps and timing characteristics of LuAG:Ce^{3+} scintillator. *Phys. Stat. Sol. B* **181**, pp. R10–R12.

105. Nikl, M., Ogino, H., Krasnikov, A., Beitlerova, A., Yoshikawa, A., and Fukuda, T. (2005). Photo- and radioluminescence of Pr-doped $Lu_3Al_5O_{12}$ single crystal. *Phys. Stat. Sol. A* **202**, pp. R4–R6.

106. Nikl, M., Yoshikawa, A., Kamada, K., Nejezchleb, K., Stanek, C. R., Mares, J. A., and Blazek, K. (2013). Development of LuAG-based scintillator crystals: a review. *Prog. Cryst. Growth Charact. Mater.* **59**, pp. 47–72.

107. Boulon, G. (2008). Why so deep research on Yb3+-doped optical inorganic materials? *J. Alloys Compd.* **451**, pp. 1–11.

108. Yanagida, T., Yoshikawa, A., Ikesue, A., Kamada, K., and Yokota, Y. (2009). Basic properties of ceramic Pr:LuAG scintillators, *IEEE Trans. Nucl. Sci.* **56**, pp. 2955–2959.

109. Shi, Y., Nikl, M., Feng, X., Mares, J.A., Shen, Y., Beitlerova, A., Kucerkova, R., Pan, Y., and Liu, Q. (2011). Microstructure, optical, and scintillation characteristics of Pr^{3+} doped $Lu_3Al_5O_{12}$ optical ceramics, *J. Appl. Phys.* **109**, pp. 013522–013522-7.

110. Liao, Y., Jiang, D., and Shi, J. (2005). Transparent lutetium aluminum garnet sintered from carbonate coprecipitated powders, *Mater. Lett.* **59**, pp. 3724–3727.

111. Li, H.-L., Liu, X.-J., Xie, R.-J., Zhou, G.-H., Hirosaki, N., Pu, X.-P., and Huang, L.-P. (2008). Cerium-doped lutetium aluminum garnet phosphors and optically transparent ceramics prepared from powder precursors by a urea homogeneous precipitation method, *Jpn. J. Appl. Phys.* **47**, pp. 1657–1661.

112. Cherepy, N. J., Kuntz, J. D., Tillotson, T. M., Speaks, D. T., Payne, S. A., Chai, B. H. T., Porter-Chapman, Y., and Derenzo, S. E. (2007). Cerium-doped single crystal and transparent ceramic lutetium aluminum garnet scintillators, *Nucl. Instrum. Meth. Phys. Res. A* **579**, pp. 38–41.

113. Liu, X.-J., Li, H.-L., Xie, R.-J., Hirosaki, N., Xu, X., and Huang, L.-P. (2006). Cerium-doped lutetium aluminum garnet optically transparent ceramics fabricated by a sol-gel combustion process, *J. Mater. Res.* **21**, pp. 1519–1525.

114. Sugiyama, M., Yanagida, T., Fujimoto, Y., Yokota, Y., Ito, A., Nikl, M., Goto, T., and Yoshikawa, A. (2012). Basic study of Eu^{2+}-doped garnet ceramic scintillator produced by spark plasma sintering. *Optical Materials* **35**, pp. 222–226.

115. Greskovich, C. and Duclos, S. (1997). Ceramic scintillators. *Annu. Rev. Mater. Sci.* **27**, pp. 69–88.

116. van Eijk, C. W. E. (2002). Inorganic scintillators in medical imaging. *Phys Med Biol.* **47**, pp. R85–R106.

117. Mihóková, E., Nikl, M., A. Mareš, J., Beitlerová, A., Vedda, A, Nejezchleb, K., Blažek, K., and D'Ambrosio, C. (2007). Luminescence and scintillation properties of YAG:Ce single crystal and optical ceramics. *J. Lumin.* **126**, pp. 77–80.

118. Shen, Y, Shi, Y., Feng, X, Pan, Y., Li, J., Zeng, Y., Nikl, M., Krasnikov, A. Vedda, A., and Moretti, F. (2012). The harmful effects of sintering aids in Pr:LuAG optical ceramic scintillator, *J. Am. Ceram. Soc.* **95**, pp. 2130–2132.

119. Zych, E., Brecher, C., Wojtowitz, A. J., and Lingertat, H. (1997). Luminescence properties of Ce-activated YAG optical ceramic scintillator material. *J. Lumin.* **75**, pp. 193–203.

120. Nikl, M., Mares, J. A., Solovieva, N., Li, H., Liu, X., Huang, L., Fontana, I., Fasoli, M., Vedda, A., and D'Ambrosio, C. (2007). Scintillation characteristics of Lu3Al5O12:Ce optical ceramics. *J. Appl. Phys.* **101**, pp. 033515.

121. Nikl, M., Yanagida, T., Yagi, H., Yanagitani, T., Mihokova, E., and Yoshikawa, A. (2013). Optical ceramics for fast scintillator materials. In: *Recent Advances in Ceramic Materials Research*, ed. J. J. Roa Rovira and M. S. Rubi, Nova Science Publishers, pp. 127–176.

122. Pejchal, J., Yamaji, A., Kucerkova, R., Jary, V., Beitlerova, A., Barta, J., Cuba, V., Yokota, Y., Ito, A., Nikl, M., Yoshikawa, A., and Goto, T. (2013). Luminescence properties of LuAG:Ce ceramics and nano-ceramics prepared by spark plasma sintering method. Presented at STAC-7 conference, Yokohama, Japan.

123. Bárta, J., Čuba, V., Pospíšil, M., Jarý, V., and Nikl, M. (2012). Radiation-induced preparation of pure and Ce-doped lutetium aluminium garnet and its luminescent properties. *J. Mater. Chem.* **22**, pp. 16590–16597.

124. Whiffen, R. K., Antić, Ž., Milićević, B., Pošarac-Marković, M., Janaćković, D., Dramićanin, M. D., Brik, M. G., Steins, I., and Veljović, D. (2014). Polycrystalline $(Y_{0.7}Gd_{0.3})_2O_3{:}Eu^{3+}$ ceramics fabricated by spark plasma sintering: densification and microstructure development, *Ceram. Int.* **40**, pp. 8853–8862.

125. Groza, J. R. (1999). Nanosintering, *Nanostructured Mater.* **12**, pp. 987–992.

126. Babin, V., Laguta, V.V., Maaroos, A., Makhov, A., Nikl, M., and Zazubovich, S. (2011). Luminescence of F^+-type centers in undoped $Lu_3Al_5O_{12}$ single crystals. *Phys. Stat. Sol. B* **248**, pp. 239–242.

127. Rétot, H., Bessière, A., Kahn-Harari, A., and Viana, B. (2008). Synthesis and optical characterization of $SrHfO_3$:Ce and $SrZrO_3$:Ce nanoparticles, *Opt. Mater.* **30**, pp. 1109–1114.

128. Mihokova, E., Moretti, F., Chiodini, N., Lauria, A., Fasoli, M., Vedda, A., Nale, A., Nikl, M., and Boháček, P. (2010). Optical and structural properties of Pb and Ce doped $SrHfO_3$ powders, *IEEE Trans. Nucl. Sci.* **57**, pp. 1245–1250.

129. Nikl, M., Bohacek, P., Trunda, B., Jary, V., Fabeni, P., Studnicka, V., Kucerkova, R., and Beitlerova, A. (2011). $SrHfO_3$-based phosphors and scintillators. *Optical Materials* **34**, pp. 433–438.

130. Van Loef, E. V., Higgins, W. H., Glodo, J., Brecher, C., Lempicki, A., Venkataramani, V., Moses, W. W., Derenzo, S. E., and Shah, K. S. (2006). Scintillation properties of $SrHfO_3:Ce^{3+}$ and $BaHfO_3:Ce^{3+}$ ceramics, *IEEE Nucl. Sci. Symp. Conf. Rec.*, pp. 1538–1540.

131. Kurosawa, S., Pejchal, J., Wakahara, S., Yokota, Y., and Yoshikawa, A. Optical properties and radiation response of $Ce:SrHfO_3$ prepared by the spark plasma sintering method, *Radiat. Meas.* **56**, pp. 155–158.

132. Kurosawa, S., Kubo, H., Hattori, K., Ida, C., Iwaki, S., Kabuki, S., Miuchi, K., Nishimura, H., Okada, Y., Parker, J.D., Takada, A., Takahashi, M., Tanimori, T., Ueno, K., and Yanagida, Y. (2009). Performance of 8 × 8 pixel $LaBr_3$:Ce and Gd_2SiO_5:Ce scintillator arrays coupled to a 64-channel multi-anode PMT, *IEEE Trans. Nucl. Sci.* **56**, pp. 3779–3788.

133. Yen, W. M., Raukas, M., Basun, S. A., van Schaik, W., and Happek, U. (2006). Optical and photoconductive properties of cerium-doped crystalline solids, *J. Lumin.* **69**, pp. 287–294.

134. van der Kolk, E., de Haas, J. T. M., Bos, A. J. J., van Eijk, C. W. E., and Dorenbos, P. (2007). Luminescence quenching by photoionization and electron transport in a $LaAlO_3:Ce^{3+}$ crystal, *J. Appl. Phys.* **101**, pp. 083703.

135. Nakata, E., Hosokai, Y., Kanzaki, H., Shikama, T., Nagata, S., Matsuki, H., Ishii, K., Ogawa, Y., Ariga, H., and Yamada, S. (2008). Novel implantable wireless real time micro dosimeter system using radiation scintillator, presented at the *Radiological Society of North America*, Chicago, US.

136. Huang, Y.-Y., Chen, A. C.-H., Carroll, J. D., and Hamblin, M. R. (2009). Biphasic dose response in low level light therapy, *Dose-Response.* **7**, pp. 358–383.

Chapter 5

LPE-Grown Thin-Film Scintillators

Miroslav Kucera[a] **and Petr Prusa**[b,c]
[a]*Faculty of Mathematics and Physics, Charles University, Ke Karlovu 5, 12116 Prague, Czech Republic*
[b]*Faculty of Nuclear Sciences and Physical Engineering, Czech Technical University, Brehova 7, 11819 Prague, Czech Republic*
[c]*Institute of Physics, Czech Academy of Sciences, Cukrovarnicka 10, 16253 Prague, Czech Republic*
miroslav.kucera @mff.cuni.cz

In scintillators, the high-energy particles (electrons, protons, neutrons, alpha particles) or photons (X-rays or gamma rays) can excite electrons in atoms, which then return to the ground state and this return is accompanied by emission of visible photons, which can be easily detected (measured) by standard detectors such as photomultipliers, CCD, etc. The de-excitation can be prompt (in nanosecond time scale) or delayed (miliseconds up to hours)—both types are used in practice.

The single crystalline films with thickness of several micrometers can be used for high resolution electron, cathodoluminescent, or X-ray screens, in various applications for imaging of microscopic objects with submicron resolution. Such films can be grown by the

Nanocomposite, Ceramic, and Thin Film Scintillators
Edited by Martin Nikl

ISBN 978-981-4745-22-2 (Hardcover), 978-981-4745-23-9 (eBook)
www.panstanford.com

liquid phase epitaxy (LPE) from the flux. In this chapter the present status of epitaxial oxide films, particularly garnets, perovskites, and oxyorthosilicates, will be reviewed. We focus especially on the growth procedure, and their luminescent and scintillation properties in comparison with the properties of bulk single crystal.

5.1 Introduction

X-ray or electron microimaging with high spatial submicrometer resolution using a single crystalline thin-film scintillator have found number of applications [14, 15]. It is operating in many synchrotron hard X-ray light sources [19] and X-ray screens where luminescence image of a scintillation screen is coupled to a CCD digital camera [2, 24–26]. Transparent single crystalline films provide better spatial resolution compared to powder phosphor screens due to eliminating light scattered by phosphor particles [29]. On the other hand, their efficiency is somewhat reduced due to waveguide effects in high optical quality films. The films of high crystalline perfection are needed where number of impurity ions and defects, which would otherwise induce electron/hole traps in the energy gap, have to be radically reduced. Due to low thickness, the scintillating films cannot be used for gamma detection (large penetration depth). Their main application is thus in detection of energetic particles (e.g., electrons, cathodoluminescent screens) or soft X-rays. On the other hand, in imaging applications, the energy resolution of scintillators is often of minor importance.

Liquid phase epitaxy (LPE) is a versatile method suitable for the growth of single-crystal films on single crystal substrates from high-temperature solution. The method can be also used in material research for development of new scintillators and it is more flexible compared to growth of bulk single crystals by Czochralski or Bridgman methods.

In the next sections we focus on the LPE growth of single crystalline oxide scintillator films, especially aluminum garnets, perovskites, and orthosilicates doped with rare earth ions such as Ce^{3+}, Pr^{3+}, Tb^{3+}, etc. The growth process and factors influencing the scintillation efficiency such as melt composition, growth conditions (i.e., growth temperature, supercooling, growth rate, used substrates) are described. Analysis of various fluxes with respect of their influence on the LY, material performance, and defects in films is discussed. The characterization methods with special emphasis to examination of scintillation properties of thin films are described, as well. Finally, the optical and scintillation properties of selected garnet, perovskite, and orthosilicate epitaxial films are reviewed.

5.2 Liquid Phase Epitaxy

5.2.1 Principles of LPE of Oxide Scintillators

In the liquid phase epitaxy (LPE) technique a single-crystalline oxide film is grown from high-temperature solution, widely known as flux growth, on an oriented single crystalline substrate. Commonly used oxide fluxes are PbO, PbF_2, BaO, B_2O_3, Bi_2O_3, MoO_3, etc. and combination of them. The films are grown from supersaturated (supercooled) melt at temperatures much lower than the melting point of crystals. The supersaturation is usually obtained by cooling the melt below the liquidus (saturation) temperature. In the undercooled metastable region of the melt, the growth of crystalline films begins on the atomic flat surface of the substrate whereby the crystalline structure and orientation of the film is continuation of that of the substrate. The method is suitable also for crystals that melt incongruently, i.e., that decompose or have a phase transition below the melting point.

In the flux LPE technique a slow transport of the crystal species to the liquid–solid interface and low growth rate guarantee the growth conditions close to thermodynamic equilibrium. Lower growth temperature supports improved structural perfection and excellent stoichiometry and enables to use lower growth rates for improved thickness control and better surface morphology.

The LPE method was developed to high reproducibility in 1970s and applied for the growth of iron garnets [31–33] for magnetic bubble memories, or integrated optical devices—planar waveguide solid state lasers, e.g., YAG:Nd, Er, Yb, Ho, Tm [34–40], saturable absorbers [34], high-temperature superconductors [43], or compound semiconductors laser diodes, for review, e.g., Refs. [44, 45]. In other thin-film growth methods based on physical vapor depositions (MBE, MOCVD, etc.), the films are deposited at extremely high supersaturation far from equilibrium and they usually suffer from much lower crystallographic quality what brings problems in scintillation applications.

In this report we call attention to important aspects, which have to be considered in material research based on LPE. The oxides epitaxial layers are grown from the flux. Therefore, the flux ions are always present as impurities in the lattice as discussed below. This requires very careful examination of the physical properties of films, and scintillation properties combined with optical absorption and luminescence spectroscopies, PL decay kinetics, and composition analysis (EPMA, GDMS) are particularly useful.

It is worth noting that epitaxial films grown on a substrate are single phase, other phases are, in principal, suppressed and are not constituted during the epitaxial growth on the oriented crystalline substrate. This is essential advantage of the LPE process.

5.2.2 Isothermal LPE

In the epitaxial growth the melt is supercooled below saturation temperature. In this metastable temperature region, the nucleation of a crystalline material begins on a seed, i.e., on a substrate immersed into the melt. Since the deposition of 1 μm thick oxide film takes from tens of seconds up to a few minutes, the LPE growth can be conducted isothermally, which is easy to control. The isothermal LPE method provides high-quality films with uniform thickness, composition, and concentration of dopants.

The complex oxides (garnets, perovskites, etc.) are grown from the flux at temperatures ~1000 °C, i.e., considerably lower compared to single-crystal melting point (~2000°C), which is significant advantage. For typical practical supercoolings ~2–20 K, the melt is stable with respect to garnet nucleation and remain supercooled for periods as long as 48 hours [33].

The isothermal dipping technique is schematically shown in Fig. 5.1. The melt is first heated to temperatures 40–80 K above saturation and homogenized by stirring for several hours. Temperature is then decreased toward growth temperature and a polished substrate, mounted horizontally on a Pt holder, is dipped into the supersaturated melt. In order to prevent the depletion of garnet compounds at the growth boundary, the substrate is rotated in the melt and the direction of rotation is periodically reversed. In this way the melt is also homogenized during the LPE process especially with respect to temperature and concentration homogeneity, which results in uniform composition and thickness (~1–2%). Typical rotations are 50–200 rpm with reversion every 2–20 revolutions.

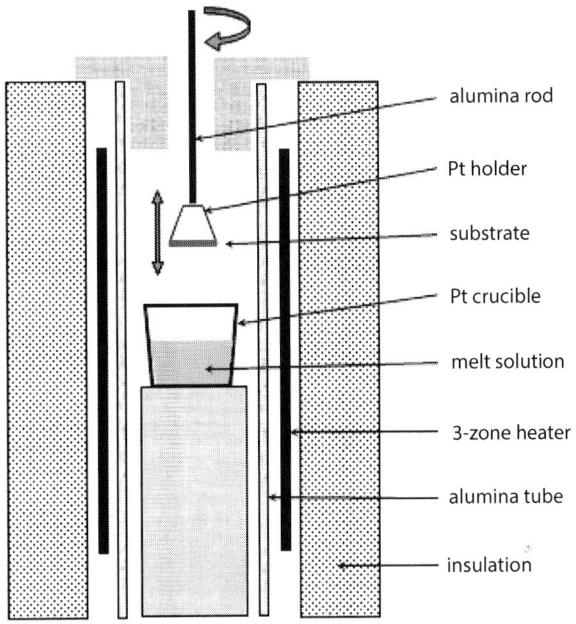

Figure 5.1 Schematic showing apparatus for isothermal LPE.

The saturation temperature, T_S, is determined from zero growth rate—below T_S the film growths on a substrate, while above T_S the substrate is dissolved in the melt. The growth rate depends on several factors: along with the flux type and melt composition, the most important factors are supersaturation, growth temperature, and substrate orientation. The growth rate is typically 1 μm/min but rates between 0.1 and 3 μm/min are ordinarily used. This enables to grow reproducibly films with thickness from submicrometers up to tens (or even hundreds) of micrometers. The best quality and uniform films are obtained at low growth rates, i.e., at low supersaturation.

There are several basic technological aspects which influence the emission efficiency and the LY of LPE film scintillators. Together with the flux composition it is primarily the supercooling, ΔT, and the growth temperature, T_G. Other important factors are growth rate (which is closely related to ΔT and T_G), crystal orientation and the lattice match with the substrate. These factors influence, e.g., the segregation coefficients of dopants and thus the concentration of both activators and impurities. They are discussed in the next sections.

Provided that oriented and perfectly polished substrates without structural defects (dislocation, inclusions, etc.) are available, perfectly uniform single-crystalline garnet films are fabricated using this technique from the $PbO\text{-}B_2O_3$ flux with low defect density and excellent surface morphology with nanometer surface roughness. Since the growth is near equilibrium, the film properties, such as concentration of dopants and impurities or surface morphology, principally depend on the growth conditions (melt composition, growth temperature, supercooling) and the growth process and temperature should be well controlled.

Advantages and properties of the LPE:

- The LPE technique provides single-crystalline films with structure quality comparable to that of used substrate; the films grown at low supersaturation have excellent stoichiometry with minimum defects.
- The quasi-equilibrium growth process of the LPE is favorable for very uniform distribution of activator ions. It ensures minimum concentration of intrinsic defects and high structural perfection (low dislocation density, low concentration of point defects) and high quality surface morphology.
- LPE is highly versatile method for material research—it is easy to change composition (or doping) of films by gradual modification of the melt, i.e., from a single melt a set of samples with different composition (doping) can be produced.
- Epitaxial growth of samples ensures that no second phases are present in the single-crystalline films.

Limitations and drawbacks of the growth from the flux:

- Impurity and contaminant ions from the flux (such as Pb^{2+}, B^{3+}, etc.) are inevitably present in the crystal and the growth of chemically pure samples is impossible. This is fundamental drawback of the flux growth especially in case when the flux impurities may have detrimental effect on physical properties of samples (e.g., quench emission of scintillation films).

- Most frequently used PbO flux partly attacks the Pt crucible, which results in contamination of the films also with Pt^{4+} impurity.
- LPE is unsuitable for growing very thin layers (<100 nm) with sharp interfaces, where accurate layer thickness uniformity and reproducibility are required.
- Availability of high-quality substrates with good surface quality and good lattice match to the film is essential.
- Film morphology is very sensitive to substrate properties (lattice misfit, misorientation), substrate surface quality and surface defects (dislocations, surface point defects, dust particles) and such defects are invariably replicated in the films.

5.2.3 Fluxes

The crystal-forming oxides are melted in the flux. Various fluxes are used for the growth of garnet, perovskite, or orhosilicate LPE films. Standard and commonly used flux is PbO-Ba_2O_3, alternate lead-free fluxes are BaO-B_2O_3-BaF_2, rarely MoO_3-LiO_2 or Bi_2O_3 fluxes (for growth of Bi-doped films) were also used. Aluminum garnets, perovskites, and orthosilicates melt congruently. However, when complex oxides are grown from the flux, the solute has usually different composition than the solution. One of the reasons is the necessity to stabilize a particular phase—e.g., while growing aluminum garnets it is necessary to select that region of the phase diagram in which garnet is the only stable phase. Therefore high excess of Al compared to Y is necessary in order to stabilize the garnet phase and to suppress the perovskite one. The second reason is different tendency of specific ions for incorporation into the crystal lattice. This tendency is described by the *segregation coefficient*, k, i.e., concentration ratio of a specific element in the melt and in the crystal, $k = c_{crystal}/c_{melt}$. Some large ions, e.g., Ce or Pr, have very low segregation in garnets or perovskites, $k \ll 1$. Basic properties of common fluxes are described in the following paragraphs.

5.2.3.1 PbO-B_2O_3 flux (here referred as PbO flux)

The PbO flux is the most common flux, considered as standard, for preparation of oxide films. Examples of garnet epitaxial films are shown in Fig. 5.2. Growth occurs under isothermal conditions. The advantages of this flux is reproducible growth, high structural quality films with uniform dopant distribution, smooth atomic-like surface can be achieved, reasonable growth temperatures 900–1050°C, low viscosity, and it is suitable for large-scale production. Addition of small amount of B_2O_3 into PbO largely improves the flux properties and especially formation of defects in films is considerably reduced [33, 46, 47]. The molar ratio PbO:B_2O_3 is usually between 10:1 and 20:1. The supercooled melt is stable with respect to garnet nucleation and highly diluted PbO-B_2O_3-garnet oxides melt can be supercooled up to 150 K without spontaneous nucleation of the garnet phase. In practice, however, films with the best properties (structure morphology, surface properties, low impurity concentration) are grown at low supercooloing (1–10 K) and at low or moderate growth rates < 1 μm/min. The growth rate is approximately proportional to the undercooling and rates from < 0.1 up to several μm/min can be achieved.

There are some severe disadvantages of the PbO flux, which have to be carefully considered in growing of scintillation films. The flux impurities, such as Pb^{2+} and Pt^{4+} ions from the flux (PbO flux attacks the Pt crucible), inevitably enter the crystal lattice. In scintillator oxides (garnets, orthosilicates, perovskites) especially Pb^{2+} ions quench the Ce^{3+} emission. Finally, the PbO flux suffers from higher evaporation and volatility above 1000°C. To overcome limitations of the PbO flux, especially to avoid Pb^{2+} impurity, alternative lead-free fluxes have also been employed.

Important parameter which determines the quality of films and concentration of dopants (and also impurities) is supersaturation, $\Delta T = T_S - T_G$, where T_G is the growth temperature. The best quality films grown from the PbO flux are obtained at low undercooling with well-controlled growth conditions. Under these conditions the concentration of impurities, including Pb^{2+} ions, in the garnet lattice is significantly reduced. Likewise higher T_G contributes to reduction of impurities; however, at temperatures above 1000°C, increased evaporation of the melt appears. Concentration of the doping ions, such as Ce or Pr, depends similarly on the undercooling due to their very low segregation coefficients. In optimized growth process at low supercoolings the Pb^{2+} impurity can be kept below 50 wt ppm in YAG or LuAG and the effect of Pb quenching on Ce emission can be reduced.

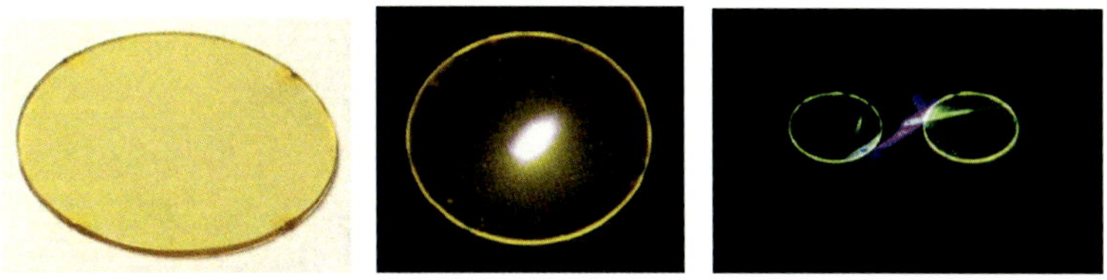

(a) (b) (c)

Figure 5.2 Garnet epitaxial films: (a) YAG:Ce film grown on YAG substrate; (b) emission from GdYAG:Ce film excited by a blue laser beam at λ = 450 nm; the white bright spot is due to yellow Ce^{3+} luminescence and scattered blue excitation light from rough surface; (c) green luminescence from LuAG:Ce films excited at 450 nm. The emission is observed from edges due to the waveguide effect. The thickness of films is 8–15 μm, diameter 20 mm.

5.2.3.2 BaO-B_2O_3-BaF_2 flux (here referred as BaO flux)

The BaO flux was developed for iron garnet film growth [48–50], and adapted for the growth of high-purity aluminum garnets [4, 5]. This flux is primarily suitable for the growth of high-purity samples with lowest-possible contamination, which depends only on purity of input raw materials. Large Ba^{2+} ions do not enter the garnet lattice (according to the GDMS analysis Ba content in garnet films is <5 ppm [12]). Although boron ions were detected in films 100–300 ppm, they have not any noticeable effect on the scintillation properties of garnet films [12]. The scintillation and emission properties of LPE films grown from the BaO flux are markedly better compared to PbO-grown films and comparable with high quality single crystal counterparts [51–54].

Evaporation of the BaO flux is low, the flux does not attack the Pt crucible, and high supercoolings up to 80 K were also demonstrated [4]. However, the basic disadvantage of the BaO flux is much higher viscosity compared to the PbO flux. The melt cannot be completely removed from the sample surface after the growth process and aftergrowth may occur. Likewise increased content of crystal forming oxides (which are close to the crystal stoichiometric ratio and form ~15– 25% of the melt) leads to higher viscosity. This all results in higher number of surface defects [21, 49] (especially surface pits, where some of them may come from imperfectly dissolved oxides) and worse surface quality compared to PbO-grown films, see Section 5.2.5.

5.2.3.3 Other fluxes

MoO_3-$LiMoO_4$ flux (gradient deposition technique)

This flux was rarely used for garnet growth [5, 55, 56]. This melt does not attack the Pt crucible and its low viscosity and low evaporation is an advantage. Garnet films with promising emission properties were reported [5]. However, the solubility of garnet oxides in this flux is very low, <2 %, and the growth in the temperature gradient is inevitable. Rather problematic growth reproducibility is fundamental disadvantage of this flux since the growth process is essentially dependent on the temperature profile and also on temperature history of the melt.

Bi_2O_3-B_2O_3 flux

This flux can be favorably used for the growth of Bi-doped (garnet) oxides [57–59]. Large Bi ions partly enter the garnet lattice, their maximum concentration is, however, limited and depends on the lattice constant of the host. This flux eliminates occurrence of Pb^{2+} ions in films; however, the films grown from Bi_2O_3 do not reach the quality of PbO grown films.

5.2.3.4 Ion segregation coefficients

In the growth of substituted or multicomponent oxide systems (e.g., garnets, silicates), the knowledge of segregation of ion species between the melt and the crystal is important. The effect of cation size of RE ions on the segregation coefficients in iron and gallium garnets was frequently studied, e.g., Refs. [60–62], and general conclusions are valid also for aluminum garnets. We report here results concerning doping of Ce, Pr, Gd, Tb, Ga, and Sc atoms into aluminum garnets, YAG, and LuAG.

The segregation depends on the radius of a doped ion, but also on growth conditions, such as the growth rate, supersaturation, growth temperature, used flux, and on lattice constant of grown crystal (i.e., on crystal composition). Segregation coefficients of several ions used in scintillation garnet systems grown by the LPE from PbO and BaO fluxes are surveyed in Table 5.1, data are from Refs. [4, 5, 12, 20, 48, 63]. The ionic radii of light RE ions in garnet crystal sites with CN = 8, Ce^{3+} (1.143 Å) and Pr^{3+} (1.126 Å) are much higher compared to those of the host ions Y^{3+} (1.019 Å) or Lu^{3+} (0.977 Å) and their segregation coefficients are very low. Therefore high excess of Ce or Pr oxides in the melt is inevitable. Furthermore, amount of CeO_2 or Pr_6O_{11} oxides in the melt is limited by their solubility in the flux: e.g., as for CeO_2 its maximum is about 0.2 g and 0.5 g in 100g of PbO [15] and BaO flux [4], respectively. Above these limits the oxides are not completely dissolved and form second phases on the melt surface with negative impact on the morphology of the epitaxial film. Second, the segregation depends significantly also on the growth rate especially for ions with low segregation such as Ce and Pr. In Fig. 5.3 concentration of cerium in Ce:LuAG films grown from PbO fluxes with various content of garnet forming oxides in the melt is shown as a function of the growth temperature. It is evident, that segregation coefficient of Ce can differ by a factor of 3 for a specific flux composition in dependence on the supersaturation (growth temperature). Similar development is observed also for films grown the BaO flux [4]. Since the growth rate is roughly proportional to supersaturation, we can infer general trends and common properties of both melts: the higher the growth rate (i.e., supersaturation), the weaker is segregation of ion species between liquid and solid, and vice versa. This indicates, that at high growth rates (high supersaturation) concentrations of Ce or Pr ions in the crystal are significantly increased. However, this tendency is observed also for other large ions including Pb^{2+}(1.29 Å) impurity, which quench the Ce^{3+}(5d–4f) emission.

While segregation of the RE ions does not differ much for both PbO and BaO fluxes, interestingly, it is very different for Ga ions, which substitutes for Al in octahedral and tetrahedral sites (see Table 5.1).

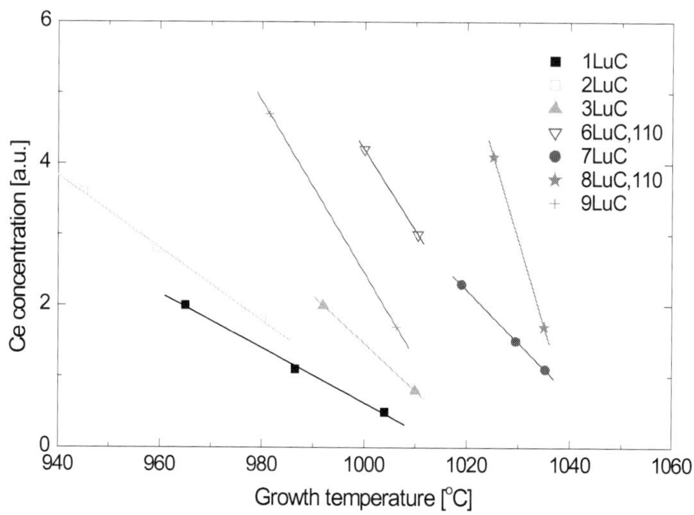

Figure 5.3 Concentration of cerium in LuAG:Ce films grown from various PbO-Ba_2O_3 fluxes as a function of the growth temperature. The films were grown on (111) oriented LuAG substrates except for series 6LuC and 8LuC which were grown on (110) oriented substrates. The content of garnet forming oxides in particular fluxes varied in the range from 4.1 to 5.6 mol. %, cerium content was in the range from 2.7 mol. % (1LuC) to 5.6 mol. % (9LuC) of garnet oxides. Reprinted from Ref. 4, Copyright 2010, with permission from Elsevier.

Table 5.1 Segregation coefficients for selected trivalent ions in LuAG and multicomponent GAGG garnet epitaxial films grown from PbO-B_2O_3 and BaO-B_2O_3-BaF_2 fluxes

	PbO flux		BaO flux	
	LuAG*	LGAGG**	LuAG	LGAGG
Y	0.9		1.1	
Ce	0.01–0.05	0.05–0.08	0.05–0.2	0.1–0.2
Pr	0.03		0.03–0.06	
Gd	0.65–0.75	0.8–0.9	0.5–0.7	0.8–0.95
Tb	0.5–0.7		0.7–0.9	
La	0.005			
Ga	0.35–0.5	0.55–0.65	1.3	1.3–1.5
Sc	1.2–1.8	2–4	1–1.6	

* LuAG = $Lu_3Al_5O_{12}$

** LGAGG = $(LuGd)_3(GaAl)_5O_{12}$

Furthermore, the lattice parameter plays also substantial role for ion segregation. LuAG and GAGG have different lattice constant, a = 11.91 and 12.3 Å, respectively. Higher k values for large RE ions (Ce, Pr and also Pb, Ga, Sc) in multicomponent GAGG films is mostly related to their larger lattice parameter and consequently more space available for large RE or Ga substitution.

5.2.4 Epitaxial Growth on Substrates

Growth of high quality single crystalline films by the LPE is limited by availability of suitable substrates. Since the LPE is a near-equilibrium process, it is very sensitive to growth parameters. Basic requirements on substrates are low lattice misfit at growth temperature, similar thermal expansion coefficients of the substrate and the film, low misorientation, and good chemical/thermal stability. In homoepitaxial growth, the substrate should have the same crystallographic structure and match the lattice constant with the film. The formation energy of dislocations in garnets is so high that mechanical stress induced by a difference in the lattice constants of film and substrate cannot be released through misfit dislocations. Instead, the film on the substrate is elastically strained. The lattice misfit (mismatch), i.e., difference between substrate and film lattice constants, $\Delta a = a_s - a_f^{\perp}$, can easily be determined by X-ray diffraction. At low misfit, $\Delta a/a_s < 0.2\%$, the film is grown under the tension or stress and in-plane lattice constant $a_f^{\parallel}$ of the film is the same as that of substrate, while out of plane constant $a_f^{\perp}$ differs. High quality garnet films with nanometer rms surface roughness (see Fig. 5.4) can only be grown on substrates at low lattice mismatch, <0.1 %. In heteroepitaxial growth, the growth of single crystal garnet (or silicates) films on garnet (silicate) substrate at mismatch as high as ~2 % was demonstrated [4, 63, 64]; however, at the expense of surface morphology and flatness. At high mismatch the rms surface roughness is as high as hundreds of nm and the film surface is matte.

The lattice misfit has a substantial effect on nucleation and epitaxial growth and consequently on structural perfection of an epitaxial layer. During the layer growth the misfit is first accommodated by a homogeneous strain, and after reaching a critical layer thickness, which depends on the degree of misfit and thickness of the film, the strain is released by cracking of the film-substrate structure. In garnets the misfit dislocations, which accommodate the strain are prevented—they are energetically unfavorable due to the high energy necessary for their formation. As a consequence, if the mismatch between film and substrate exceeds critical value (about ~2% in garnets or orthosilicates), nucleation does not occur even in strongly supersaturated melt. Critical mismatch depends to some extend on material to be grown, flux used, substrate parameters (orientation, composition).

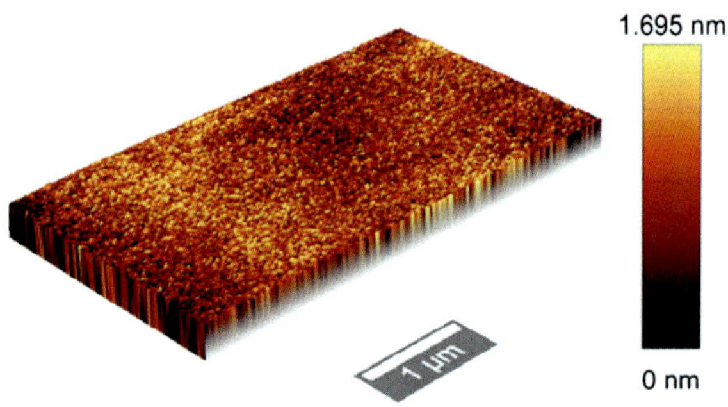

Figure 5.4 AFM topography of the surface morphology of $Yb_3(AlGa)_5O_{12}$:Er garnet epitaxial film grown homoepitaxially onto (111) oriented YAG substrate from the PbO flux. Reprinted from Ref. 16, Copyright 2015, with permission from Elsevier.

One of possible mechanisms of growth of films with large misfit is formation of a transient layer at the film–substrate interface [31]. It is formed both in homo- and heteroepitaxial growth. Its thickness is from tens of nanometers up to micrometers. The composition of the transient layer gradually varies, it contains usually surplus of flux or doping ions, e.g., Pb, Ga. At small misfit it balances the lattice constants of the epilayer and the substrate.

Homoepitaxially and heteroepitaxially grown films observed in optical microscope are shown in Fig. 5.5. The homoepitaxial films with lattice mismatch $\Delta a < 0.2$ % have smooth glossy surface, those grown heteroepitaxially with $\Delta a > 0.5$ % are matte with rough surface. The rms surface roughness measured by the Zygo interferometric profiler is ~1 nm and ~300 nm in homo- and heteroepitaxial films, respectively.

An important issue is the morphology of the film surface and film/substrate interface. Large lattice mismatch between the Ce:LuAG film and the YAG substrate at heteroepitaxial growth, Δa

= –0.09 Å, is the main problem in obtaining films of high surface perfection. The formation of a transient layer at the film-substrate interface is of real importance, especially for very thin films. In the transient layer the chemical composition changes and a crucial issue is its contamination by impurity ions from the flux and presence of other defects. For small film-substrate lattice mismatch, $\Delta a < 0.02$ Å, this transient layer is usually narrow from several to dozens of nanometers. For higher lattice mismatch the morphology of the transient layer is more complicated. The edges of samples cleaved normal to the film-substrate interface are shown in Fig. 5.6. Sharp film-substrate interface is observed at homoepitaxial growth of the Ce:LuAG/LuAG system (Fig. 5.6a), where the lattice constants of the film and substrate virtually merge. Because of nearly the same chemical composition, the optical contrast between film and substrate is very weak and the film-substrate interface is difficult to observe. On the other hand, at heteroepitaxial growth of Ce:LuAG/YAG system, where the lattice mismatch is high and the film-substrate interface is apparent, Fig. 5.6b. An oriented columnar-like structure of the film is observed in optical microscope. The morphology of the transient layer may be critical especially in micrometer thick films and a detailed study of its structure is called for.

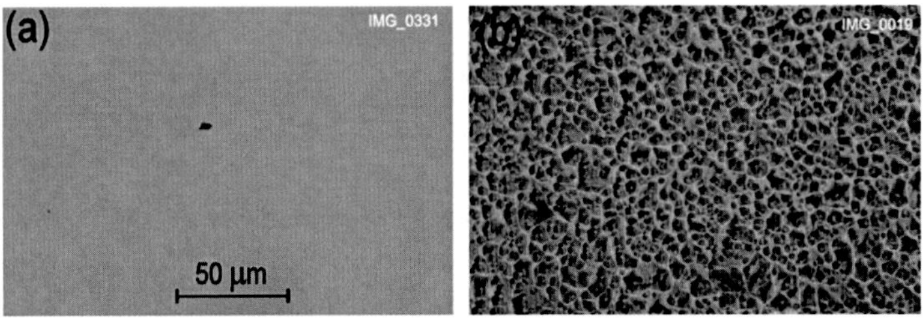

Figure 5.5 Optical microscopy of (111) oriented LuAG:Ce epitaxial films grown from PbO-flux: (a) high quality film grown on LuAG substrate, lattice misfit $\Delta a < 0.01$ Å (homoepitaxial growth), (b) LuAG:Ce film grown on YAG substrate (heteroepitaxial growth), lattice misfit $\Delta a = 0.09$ Å. Reprinted from Ref. 21, Copyright 2010, with permission from Elsevier.

The LPE growth process depends also on substrate crystallographic orientation. The misorientation of a substrate from optimal direction may lead to step-like surface of films. Only LPE at low supersaturation can yield atomically flat surfaces.

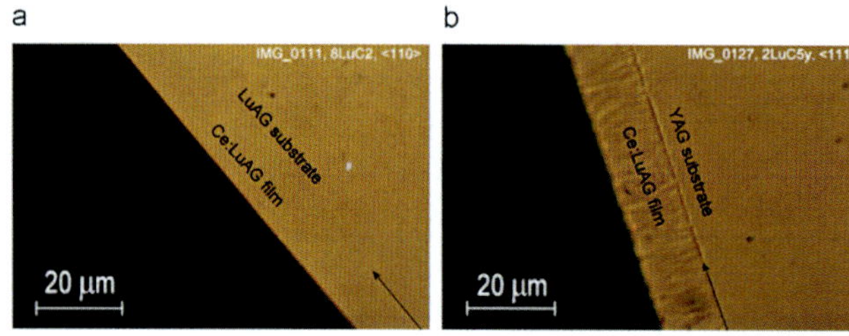

Figure 5.6 Cross section of cleaved samples observed in optical microscope; the arrows point to the film/substrate interface. (a) Homoepitaxially grown Ce:LuAG film on LuAG substrate, lattice constants of the film and substrate are almost the same, $\Delta a < 0.01$ A, and sharp film-substrate interface is observed. (b) Heteroepitaxially grown Ce:LuAG film on YAG substrate, the lattice mismatch is rather high, $\Delta a = 0.9$ A. Transient layer and columnar-like film growth is easily observed. Both films were grown from the PbO flux. Reprinted from Ref. 4, Copyright 2010, with permission from Elsevier.

Aluminum garnets are routinely grown on YAG substrates cut from Czochralski grown single crystals, which are available in diameters up to 100 mm. Other garnet substrates with different lattice parameters such as LuAG, GGG, rarely NdGG, YGG, etc., were also used, their structural parameters are shown in Table 5.2. Substrates with perovskite and orthosilicate crystal structure are of limited availability, the epitaxial films are usually grown on YAP or YSO, respectively.

The substrates reported in Table 5.2 suffer from one disadvantage: under high-energy excitation the host emission is induced in the UV spectral range. This may cause problems in some thin-film applications, since gamma or hard X-rays penetrates through the film into the substrate and the substrate emission is excited simultaneously with the emission from epitaxial film. Because of different timing characteristics of film and substrate emissions, it is be desirable to suppress the substrate emission. This can be carried out by intentional substitution of the host crystal by appropriate rare earth ions, e.g., Yb, Tm. At high RE concentrations, the RE emission also disappears due to concentration quenching [65]. As an example, crystals such as YbAG, YGG, GGG, YbSO have negligible emission. This approach was used for LSO:Tb scintillation films for microimaging using hard synchrotron radiation in ESRF [2].

Table 5.2 Parameters of typical garnet, perovskite, and orthosilicate substrates. CN, coordination number; Z_{eff}, effective atomic number; ρ [g/cm^3], density

Substrate	Crystal structure, space group	Lattice constant [Å]	Cation CN, (Y^{3+} symmetry)	Z_{eff}/ρ	References
Garnet					
YAG	cubic, Ia3d	a = 12.008	Y-8 (D_2), Al-6, 4	32/4.56	[31]
LuAG	"	a = 11.913	"	62/6.70	
GGG	"	a = 12.383	"	70/7.08	
Perovskite					
YAP	orthorhombic Pnma	a = 5.332 b = 7.356 c = 5.176	Y-12 (C_{1h}), Al-6	33/5.35	[66]
LuAP	"	a = 5.330 b = 7.294 c = 5.100	"	65/8.34	[66]
Orthosilicate					
YSO	monoclinic B2/b	a = 14.371 b = 10.388 c = 6.710 γ =122.17°	Y1-7, Y2-6 (C_1), Si-4	39/4.54	[67]
GSO	$P2_1/c$	a = 9.131 b =7.045 c = 6.749 γ =107.52°	Y1-9, Y2-7 (C_1)		[68]
LSO	B2/b	a = 14.277 b = 10.246 c = 6.640 γ =122.22°	Y1-7, Y2-6 (C_1)	66/7.4	[69]

5.2.5 Defects in LPE Films

Compared to Czochralski grown single crystals, the layers prepared by the liquid phase epitaxy (LPE) are grown from the flux at much lower temperatures and hence with lower structural disorder and lower number of intrinsic defects, e.g., antisite defects. On the other hand, various impurities can enter into the crystal lattice from the flux and some new defects specific to the LPE technology are induced during the growth. Furthermore, other factors inherent for epitaxial layers such as flux properties (composition, viscosity), growth conditions (growth temperature, supercooling, growth rate), substrate properties (film–substrate lattice match, defects in substrates) influence the quality of epitaxial films and subsequently their scintillation properties.

The crystallographic quality of substrates is of primary importance for growth of high-quality epitaxial films. All defects which intersect the substrate surface, e.g., dislocations, inclusions near the surface, scratches due to imperfect polishing, dust particles on the surface, etc., are replicated into the film and result in macroscopic defects. Although such defects usually do not affect notably the scintillation properties, however, they have to be eliminated in imaging screen application, where also high-quality surface morphology is of crucial importance.

The second type of defect are intrinsic ones on the microscopic (atomic) scale, such as impurity ions, various structural defects, vacancies, antisite defects, which may have, on the other hand, a dramatic impact on the scintillation properties since they may induce trap states in the forbidden gap.

Lattice (structural) defects

- Antisite defects (AD) Y_{Al}, Lu_{Al}, observed in oxides with two or more cation crystal sites (YAG, YAP), where Y or Lu ion is situated in octahedral (preferable Al) site instead of its preferred dodecahedral site. They form shallow trap states and induce slow component in the scintillation decay and seriously reduce the LY. AD defects are produced due to high crystal growth temperature in Czochralski or Bridgman techniques. In epitaxial films the AD are significantly reduced compared to SC due to lower growth temperature [70]; however, recent studies indicate non-negligible AD content even in films grown from the flux [71].
- Deep traps formed by oxygen vacancies, such as F+ or F color centers, drastically decrease the scintillation LY. These

vacancies form very effective electron traps. The vacancies are induced during the growth of single crystals in the reducing protective atmosphere [72] and they are eliminated by post growth heat treatment in oxidizing atmosphere. The color centers are effectively suppressed in epitaxial films due to the growth in oxidizing atmosphere and at significantly lower growth temperatures ~1000°C compared to single crystals.

Impurities

In the LPE techniques the films are grown from the flux and elimination of impurities is of significant importance.

- Isoelectronic trivalent impurity with closed electron shell, such as $B^{3+}(1s^2)$ $Sc^{3+}(3p^6)$ $La^{3+}(5p^6)$ do not show any radiative transitions in the visible and UV spectral regions. At low concentrations they do not influence notably the LY. However, as a point defect in the lattice they change the ion–ion distance and cause deformation of the crystalline field due to different ionic radii. Generally, the efficiency of scintillator materials is decreased by processes associated with electrons or holes trapped at point defects. When Sc- or La-doped YAG/LuAG is excited at the bandedge or by high-energy photons/particles, an intense broad emission band is induced near 275 nm in photoluminescence or radioluminescence spectra [73]. This emission, which has decay time of ~0.6–1 μs, originates from recombination of an exciton located near the Sc defect.
- Divalent or tetravalent impurities with closed electron shell np^6 (e.g., Mg^{2+}, Ca^{2+}, Zr^{4+}, Hf^{4+}): their effect on scintillation properties is divers depending on their concentration and type of the host lattice. At higher concentrations the non-isovalent ions always quench the 5d–4f luminescence of Ce^{3+} or Pr^{3+} ions since they are electron/hole capture centers. However, at low concentrations as trace impurities, they can sometimes even improve scintillation properties, e.g., co-doping with Ca^{2+} in LSO/YSO:Ce increases LY and shorten decays [74] and reduce afterglow. It has been proposed that doping YAP:Ce with Zr^{4+} can change the concentration of point defects [75], e.g., decrease concentration of oxygen vacancies [76].

Divalent ions Mg^{2+} or Ca^{2+} change valency of cerium to Ce^{4+}. Stabilization of a part of the cerium in the tetravalent charge state provides an additional fast radiative recombination pathway in the scintillation mechanism [77]. Discussion of intrinsic or extrinsic defects ("defect engineering") can be found in number of publications [70, 78–86].

- Transition metals (TM) usually quench RE emission, e.g., Fe^{3+} heavily quenches the PL, its emission is at 780 nm at RT.
- Pb, Pt are typical impurities in LPE films grown from the PbO flux. Pb^{2+} ion has large radius and its concentration is typically tens of ppm in YAG/LuAG [4, 63]. Nevertheless, presence of Pb^{2+} appeared problematic since they induce absorption in the deep UV range and quench effectively the Ce^{3+} emission [87, 88]. Furthermore, in garnets with larger lattice constant (GAGG) and especially in perovskites and othosilicates, where larger lattice space is available, concentration of Pb ions significantly increases.
- Energy levels of most of defects are situated in the energy gap. These trap states can capture electrons or holes, and represent thus a parallel channel for recombination (either radiative or nonradiative) with negative impact on scintillation efficiency (decreased LY, delayed decay or afterglow).

Macroscopic defects in epitaxial films

Good lattice match between the layer and the substrate is required in order to obtain good-quality single-crystalline layers. If the lattice mismatch exceeds about ± 0.02 Å, faceting or cracking appears in layers grown under pressure or under strain, respectively.

The epitaxial garnet films can be grown on substrates of various orientation, the best results are usually obtained for (111) oriented substrates. While a slight misorientation of (111) planes, even of several degrees, has a negligible effect on the surface quality, a misorientation of (110) plane of 0.5 deg resulted in faceted surface.

Many of surface defects of the LPE films are closely connected to the growth process and to the flux properties. The surface pits are frequent defects observed in LPE layers. Typical triangular and hexagonal pits observed in (111) surfaces are shown in Fig. 5.7a.

The orientation of pits is closely related to the crystal axis, the edges of triangles are in 110 directions. The origin of the pits is diverse—they may come from the point defects in the substrate (dislocation pits), from microscopic dust particles firmly attached on the surface during immersing the substrate into the flux, or directly from the flux (solvent particles).

In addition, the defects in the substrate that intersect the surface (such as dislocations, cation inclusions or mechanical defects) are often replicated into the film in the form of pits. Mechanical defects (scratches) in the substrate induced during the polishing are amplified in films and pit lines may appear (Fig. 5.7b).

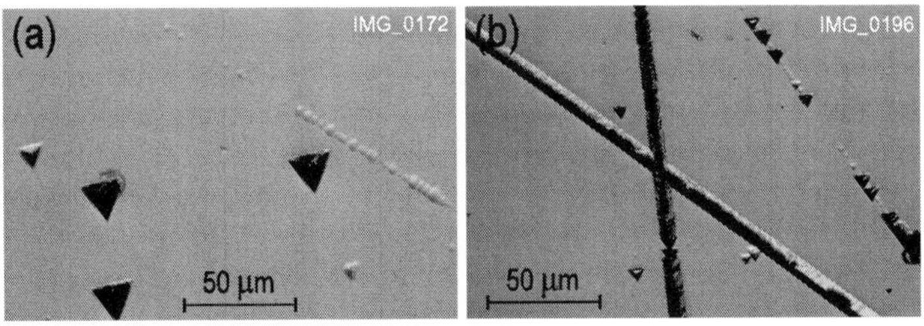

Figure 5.7 Triangular and hexagonal surface pits observed by optical microscopy in YAG:Ce films grown on (111) oriented YAG substrates (a), the pit lines in epitaxial film grown onto insufficiently polished YAG substrate (b); the films were grown from the MoO_3 flux. Reprinted from Ref. 21, Copyright 2010, with permission from Elsevier.

The pit defects are usually created at the very beginning of the growth process in the film/substrate interface due to the starting instability of the growth. However, a detailed inspection of Fig. 5.7 reveals also pits of smaller dimensions, which do not originate from the interface but were generated later during the layer growth. This is supported also by the fact that thick layers (dozens of micrometers) contain higher number of pits than the thin ones. Such pits come from "solvent particles," namely from some oxide components insufficiently dissolved in the flux. Particularly the layers grown from the BaO flux are susceptible to the pit creation. This flux suffers from high viscosity and high surface tension compared to the PbO or MoO_3 ones. This brings about a reduced flowing of the BaO flux and slows down dissolving of individual melt components. The surface pits cause scattering of emitted light and they should be eliminated in high resolution scintillation imaging screens.

Dislocation defects, shown in Fig. 5.8 (left), result in individual dislocation pits as those in Fig. 5.7a. Very rough surface in Fig. 5.8 (right), with large number of pits was grown from nonhomogenous flux with undissolved solvent particles.

Figure 5.8 Left: Dislocations in LuAG single-crystal substrate observed in polarizing microscope. Right: Large number surface pits in Ce:LuAG epitaxial film grown on (111) oriented substrate; the pits were induced during the growth due to undissolved solvent particles.

5.3 Characterization Methods: Experimental Techniques

Characterization methods for thin-film scintillator differ in many details from those commonly used for characterization of bulk materials. As an example, the LY or scintillation decays cannot be measured by standard γ-excitation, since the penetration depth of γ rays is much larger than the films thickness (mm x µm) and thus the effective signal comes mostly from the substrate. The films are usually excited either by soft X-rays (e.g., steady state RL measurements) or by alpha particles at 5.166 MeV (which have penetration depth about 10–15 µm in YAG/LuAG [41]) or by energetic electrons, penetration depth < 1 µm for 10 keV electrons [89].

5.3.1 XRD and Structural Properties of Films

The films grown by the LPE are single crystalline and single phase and their structural properties are comparable to those of substrate.

Combination of XRD, EPMA, and lattice constant calculations [90] provide valuable information not only on composition but also on occupation of individual crystallographic sites by particular ions.

XRD properties of several films are demonstrated in Fig. 5.9. The diffraction profile of the symmetric 444 diffraction of a Ce:YAG film grown on YAG substrate (homoepitaxy) from the PbO flux is shown in Fig. 5.9a. Lower intensity of substrate diffraction is due to radiation absorption in 11 μm thick film. Observed half-width of 444 diffraction peak in the epitaxial film, 12 ang. seconds, is comparable to that of the substrate (10 ang. seconds) giving evidence of high crystallographic quality of the film. From the position of the 444 diffraction peak in this sample, the lattice constant in the direction normal to the film surface was determined as $a_\perp = 12.0114(1)$ Å. The measured lattice constant of the YAG substrate was $a_S = 12.0093(1)$ Å. The higher lattice constant of the films compared to the substrate is in accord with the doping of large Ce^{3+} ions, ionic radius $r^{VIII} = 1.143$ Å, into the dodecahedral sites of YAG. On the contrary, the XRD study shows that the lattice constant of Ce-doped LuAG films is even lower than that of pure LuAG substrates. This discrepancy can be explained due to the antisite Lu_{Al} defects which occur in the Cz grown LuAG crystals, i.e., a part of small Lu^{3+} ions, $r^{VIII} = 0.977$ Å, is situated in the octahedral sites instead of regular Al^{3+} ions. These defects increase the lattice constant of a substrate compared to the stoichiometric crystal. In epitaxial LuAG films the concentration of these Lu_{Al} defects is presumed to be much smaller than in the substrates [7, 30] due to much lower growth temperature of films compared to that of the Czochralski or Bridgman grown crystals. As concerned the YAG substrates, the ionic radius of the Y^{3+} ions is much higher ($r^{VIII} = 1.016$ Å) compared to Lu^{3+} ions. Therefore the concentration of antisite Y_{Al} defects in YAG crystals is accordingly lower than that in LuAG crystals and the deviation from the stoichiometry of YAG is also significantly lower.

The asymmetrical reciprocal space map (RSM) around the asymmetric diffraction 664 (Fig. 5.9b) suggests that homoepitaxially grown samples, i.e. Ce:YAG/YAG and Ce:LuAG/LuAG, showed pseudomorphic film on substrate (the intensities of scattered X-rays in the reciprocal space exhibit the same Q_x coordinate in the RSM), i.e. the film lattice parameter in the plane is the same as that of substrate, $a_\parallel = a_S$. Since $a_\perp$ differs only slightly, $<2 \times 10^{-4}$, any misfit

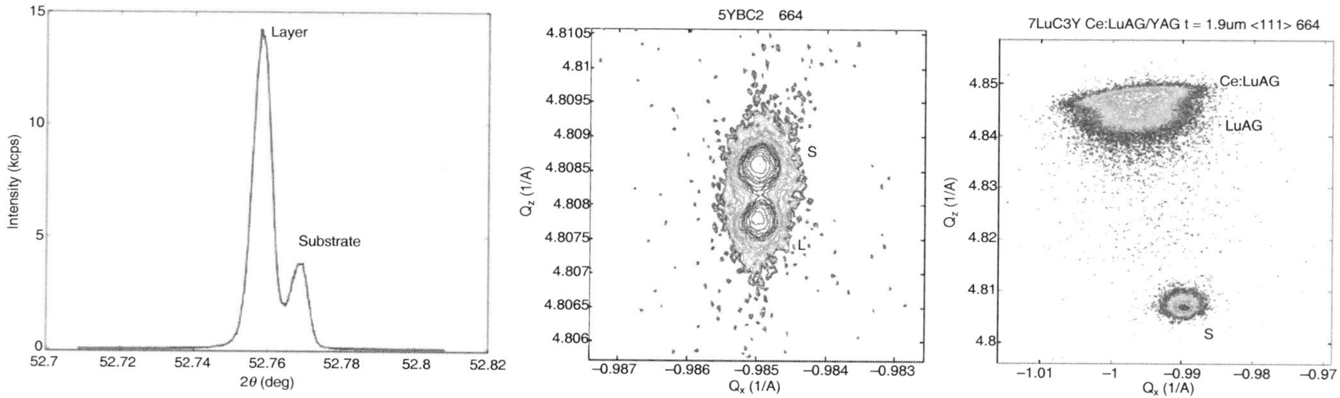

Figure 5.9 (a) 444 diffraction of 10.7 μm thick Ce:YAG film grown on (111) oriented YAG substrate from the PbO flux. (b) Asymmetrical reciprocal map of 664 diffraction of the same sample, z is direction normal to the film plane. A pseudomorphic film was grown on substrate, i.e. the in-plane lattice parameters of the film and substrate are the same; $a_\perp = 12.0114(1)$ Å, $a_\parallel = a_{\text{subs}} = 12.0093(1)$ Å. (c) Map of 664 diffraction of Ce:LuAG 13.4 μm thick film grown on (111) oriented YAG substrate from the PbO flux—a relaxed layer was grown: the lattice parameter of the film corresponds to that of the LuAG single crystal, $a_\perp = a_\parallel = 11.918(1)$ Å, $a_{\text{subs}} = 12.009(1)$ Å. Reprinted from Ref. 4, Copyright 2010, with permission from Elsevier.

dislocations are not expected in the film/substrate interface and a macroscopic elastic strain is present in the film.

In the hetereoepitaxy, i.e., Ce:LuAG films grown on the YAG substrates or vice versa, due to their different compositions, an emergence of the film-substrate mismatch is expected. The prepared films were admittedly crack-free but they were relaxed and exhibited different growth morphology than the homoepitaxial films. The lattice mismatch between the substrate and the film was as high as $\Delta a = a_s - a_f = -0.09$ Å ($\Delta a/a = 0.8$ %). This value is too large for the pseudomorphic growth. It is worth noting that for iron garnets the condition for the mismatch is rather severe and the mismatch should not exceed ±0.02 Å to avoid cracking or faceting of films. At excessive mismatch the films do not even nucleate [31]. In contrast, the crack-free LuAG epitaxial films on YAG substrates were successfully grown in spite of large mismatch. In this case the films were completely relaxed and both the normal and in plane lattice parameters of a film and a substrate strictly differ. The RSM around the asymmetric diffraction 664 for Ce:LuAG film grown on (111) oriented YAG substrate from the PbO flux is shown in Fig. 5.9c. The lattice parameter of the Ce:LuAG film corresponds to that of LuAG crystal, $a = 11.916$ Å, while the measured parameter of YAG substrate is $a_s = 12.009$ Å. Therefore the misfit dislocations that release the misfit strain are in the film/substrate interface. This fact enables to grow rather thick heteroepitaxial garnet films. The RSM of the film diffraction is broad compared to that of the substrate, Fig. 5.9c, which shows that macroscopic defects are present in films. Nevertheless, the XRD proved that the films are of single garnet phase and the films exhibit a distinct (111) orientation. The other parameters of heteroepitaxial films, e.g. growth rates, concentration of cerium or impurity ions, optical and emission properties, decay kinetics, are quite comparable with those for the homoepitaxial films. The major difference between these two types of films is surface roughness, film/substrate interface and different growth morphology.

5.3.2 Optical Methods: Absorption, Photoluminescence, PL Decay Kinetics

The optical methods—absorption, PL, and PL decay method—are in principle the same as those for bulk sample measurement.

We mention here only some problems specific to thin-film measurements.

Photoluminescence (PL) spectroscopy. Mutual comparison of the emission spectra in an absolute way (i.e., comparison of PL intensities) of thin-film samples is problematic due to various geometric factors influencing the PL intensity, which are difficult to take into consideration, as waveguide effect, quality of the edge profile, surface quality. Especially scattering centers on the surface—ideal optical waveguide with emission from the edge in high-quality films with perfect surface morphology vs. intense emission from the rough surface of heteroepitaxially grown films. Nevertheless, the PL spectra can be reliably compared in a relative way (normalized spectra). When the PL spectra are to be compared in an absolute way, great care should be devoted to the elimination of the geometric and surface factors (such as surface quality and scattering centers on the surface, edge profile, thickness of films, etc.). In the PL experiments the substrate can be also excited, since the films are semitransparent in the visible and UV spectral ranges up to 200 nm. If the substrate contains intentional RE or TM impurities or color centers it can contribute to the measured signal.

Thermo stimulated luminescence (TSL) is important technique providing information on shallow defects (e.g., AD). However, only limited number of experiments were published in thin films [13] due to very weak signal from the film (several micrometers thick). Attention should be paid to filter out the signal from the substrate, which can be orders of magnitude higher, the best way is to polish the substrate out.

Scintillation decay kinetics of emission centers—very important and reliable method providing information on physical processes in the scintillator (on emission centers, traps, energy transfer, possible quenching centers, etc.). In epitaxial films decays excited by alpha particles or by electrons (CL decay) are the only possible ways.

5.3.3 Scintillation Properties

Radioluminescence spectroscopy (RL) excited by X-rays or by alpha particles, cathodoluminescence (CL) is excited by an electron beam.

Both RL and CL originate from host-to-activator energy transfer mechanism and characterize the steady state scintillation properties. The RL is measured under X-ray excitation usually at lower excitation voltage of 10 kV (just to decrease the penetration depth) which would ensure that the film is primarily excited and contribution of the substrate to the RL emission is suppressed as much as possible. However, low excitation voltage is at the expense of decreased signal-to-noise ratio. Absorption efficiency as a function of X-ray energy, in % of absorbed radiation, for several garnets and oxide scintillators is displayed in Fig. 5.10. It is evident that thin films are not suitable for detection of high-energy X-rays.

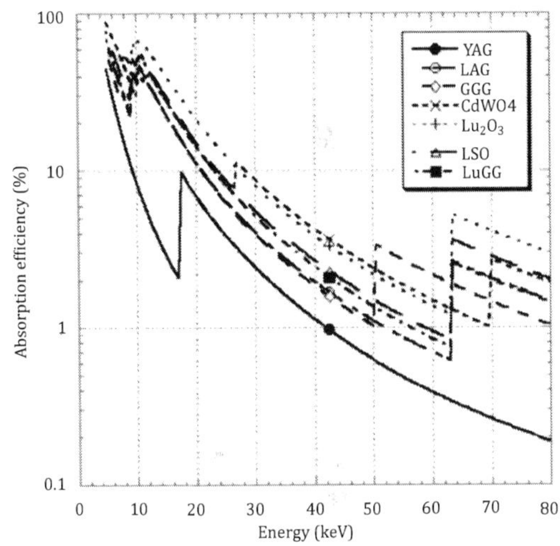

Figure 5.10 Absorption efficiency of oxide scintillators used for high-resolution X-ray imaging for 5 μm thick films. Reproduced from Ref. 19, with permission of the International Union of Crystallography.

The penetration depth of electrons [89] is significantly lower compared to α particles or X-rays and submicrons films can be reliably measured. Apparatus for cathodoluminescence characterization of materials is described in ref. [91].

5.3.4 LY Measurements of Scintillating Epitaxial Films

In this section we present analysis of light yield (LY) measurements of thin films using alpha particle excitation. The LY measurement of epitaxial films must face a problem of very low thickness of the films, from ~1 μm up to a few tens of micrometers [4, 5, 92, 93]. Probability of photon interaction in the film is very low, probability of interaction in the substrate is much higher. Unfortunately, substrate generates scintillation response also. In case of thin films, penetration range of secondary electrons must be considered. The range is at least several tens of μm in LuAG or YAG for photons over 100 keV, greater than film thickness.

Low-energy photon radiation is not of practical use either. It is absorbed in the film almost entirely, but low energy leads to low response, few tens of photoelectrons at best for relatively high light yield. Problems are caused by low quantum efficiency, low photon collection efficiency and low number of scintillation photons. It is very hard to measure such low response precisely. Beta radiation sources face the same problem and they are not even monoenergetic.

On the other hand, penetration range of alpha particles R_α emitted by common radionuclides is from 10 to 14 μm in LuAG/YAG [41, 51]. Energy over 5 MeV ensures sufficiently high response, R_α is usually lower than film thickness. Alpha particles may be considered monoenergetic. Measurement chamber must be either evacuated or source must be in direct contact with the film [41, 51]. Vacuum offers better reproducibility and safety, "contact" is simpler, cheaper and allows measurement of response of films with thickness $d < R_\alpha$, see below.

Unfortunately, low R_α excludes usage of reflector. Scintillation photon collection efficiency diminishes and is hard to quantify. Usage of reference bulk crystals with known light yield is recommended. Otherwise, photoelectron yield for given geometry would be known only.

Experimental setup for light yield (photoelectron yield) measurement is almost the same as for bulk crystals, the application of alpha particle sources is the only one significant difference. In the detection unit a hybrid photomultiplier (HPMT) [94, 95] coupled with preamplifier, amplifier, multichannel analyzer and PC is used. Calibration of setup is done by dark pulses spectrum measurement [96].

Measured spectra usually contain one broad peak with very long tail on low energetic side. The peak can be fitted by Gaussian. Thus, photoelectron yield and energy resolution are obtained [41].

If R_α is longer than film thickness, two peaks arise in the spectrum, see Fig. 5.11 [41]. Higher energy peak (HEP) is created by alpha particles depositing entire energy in the film, see particles C and D in Fig. 5.12. Lower energy peak (LEP) is created by particles only passing through the film, as A and B in Fig. 5.12. If number of detected particles is N_{DET}, number of pulses in HEP N_{HEP}, then relative peak area $N_{HEP}/N_{DET} = d/R_\alpha$ [41].

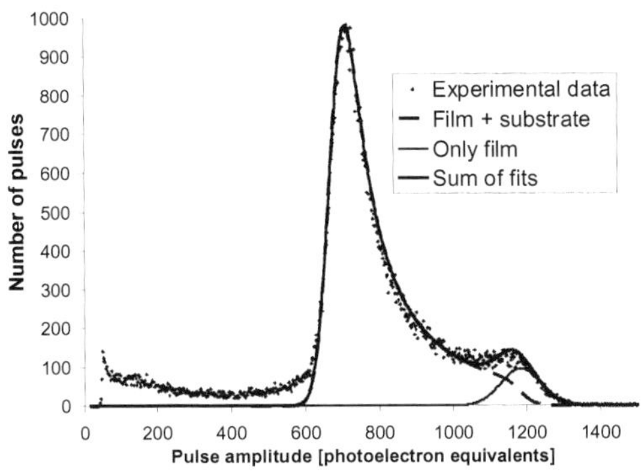

Figure 5.11 Spectrum of film with thickness below 2 μm fitted using equations (5.3) and $N_{HEP}/N_{DET} = d/R_\alpha$.

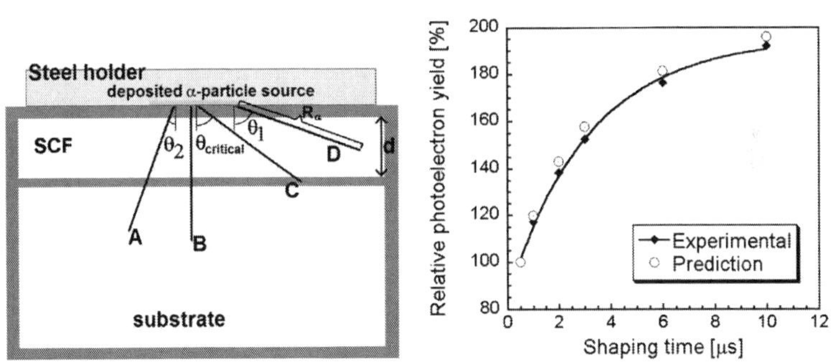

Figure 5.12 Left: Geometry of measurement. Particles C and D are depositing entire energy in the film, particles A and B are depositing part of the energy in the substrate. Right: Dependence of PhY(t_s)/PhY(t_r = 0.5 μs) on shaping time and the fit, see text for details. Results for LuAG:Ce bulk crystal.

Energy deposition of particles contributing in HEP is monoenergetic, response should be Gaussian. For LEP shape calculation, straight track of alpha particles, constant R_α and constant linear energy transfer (LET) were assumed, i.e., Bragg peak was omitted. Last assumption is acceptable, because response per track length is not strongly dependent on LET for high LET.

Let S_r be the scintillation response, film photoelectron yield is PhY_f, substrate photoelectron yield is PhY_s, E_s is energy deposited in the substrate, E_f is energy deposited in the film, E_α is energy of alpha particle, and θ is the angle in Fig. 5.12. $S_r = PhY_f E_f + PhY_s E_s$.

For $\theta < \theta_{critical}$:

$$S_r = PhY_f \frac{d}{R\cos\theta} E_\alpha + PhY_s \left(E_\alpha - \frac{d}{R\cos\theta} E_\alpha \right) \tag{5.1}$$

The goal is to find dN/dS_r, i.e., distribution of number of pulses in dependence on response (channel number). If dN is a number of particles emitted in the angle from θ to $\theta + d\theta$, $dN = A \cdot t(2\pi/4\pi) \sin(\theta)\, d\theta$, where A is the source activity and t is time of measurement.

Now $dN/d\theta = (dN/dS_r)\cdot(dS_r/d\theta)$; $dN/dS_r = (dN/d\theta)/(dS_r/d\theta)$. Let us calculate $dS_r/d\theta$ from Eq. 5.1. Now dN/dS_r is obtained:

$$\frac{dN}{dS_r} = \frac{1}{2(PhY_f - PhY_s)} \frac{R_\alpha}{d} \frac{\cos^2\theta}{E_\alpha} At \tag{5.2}$$

$\cos\theta$ can be expressed from Eq. 5.1 and substituted in Eq. 5.2. Finally,

$$\frac{dN}{dS_r} = \frac{d}{2R_\alpha} E_\alpha \frac{(PhY_f - PhY_s)}{(S_r - PhY_s E_\alpha)^2} At \tag{5.3}$$

Equation 5.3 is valid for S_r from $(PhY_f E_\alpha d/R_\alpha + PhY_s E_\alpha (1 - d/R_\alpha))$, i.e., perpendicular penetration, to $PhY_f E_\alpha$, i.e., entire energy deposition. Equation 5.3 does not take into account imperfect energy resolution, but using a simple convolution, real spectrum could be calculated, see Fig. 5.11. Small discrepancies of experiment and calculation should be contributed to initial assumptions.

Decay kinetics is important source for information about processes in the scintillator. If one is not able to measure decay curves conventionally, photoelectron yield measurement is able to give rough estimation of decay kinetics in the time interval of few μs [41, 51]. At first, dependence of photoelectron yield on amplifier shaping time is measured (see Fig. 5.12, right). This dependence can be fitted by equation $PhY(t_s)/PhY(t_r = 0.5\ \mu s) = m_1 - m_2 \cdot \exp(-m_3 t_s)$, t_r is reference shaping time, usually 0.5 μs. Percentage of "fast" photons (with decay times below hundreds of nanoseconds) $K_{f/e}$ can be calculated as $K_{f/e} = 100\ (1 - m_2/m_1)$ [41].

There is one more complex and precise method: the convolution of one photoelectron response with decay kinetics. If channels in decay curves measurement are short enough, i.e., one photoelectron response does not change significantly between channels i and i+1, pulse shape can be calculated as

$$R_k = \sum_{i=0}^{k} N_i \cdot RU_{k-i} \tag{5.4}$$

Where R_k is the response in time bin k, N_i is the number of photons emitted during time bin i, and $RU_{k\text{-}i}$ is one photoelectron response in k-th bin, which started in i-th bin. It means that entire response is a sum of all one-photoelectron responses. Relative photoelectron yield $PhY(t_s)/PhY(t_r) = \max(R_k(t_s))/\max(R_k(t_r))$, t_r is reference shaping time, usually 0.5 μs.

Calculation of decay curve from $PhY(t_s)/PhY(t_r)$ uses following method. At first, $N_i = f(t,p_i)$ is assumed. Parameters p_i are then changed iteratively, until the best agreement of experimental and calculated $PhY(t_s)/PhY(t_r)$ is found. Type of $f(t,p_i)$ must be chosen a priori. Single or double exponential, or exponential + power are used mostly.

The number of parameters p_i must be lower than number of available shaping times and the information about decay kinetics is reliable only in the time interval between 0.5 and 20 μs approximately.

Deconvolution and convolution have been tested on several samples, one example is in Fig. 5.12 (right) and Fig. 5.13. Also, $LiCaAlF_6$:Eu sample with single exponential decay was measured. Decay time measured by decay curve experiment was 1.68 μs and decay time calculated by deconvolution was 1.61 μs.

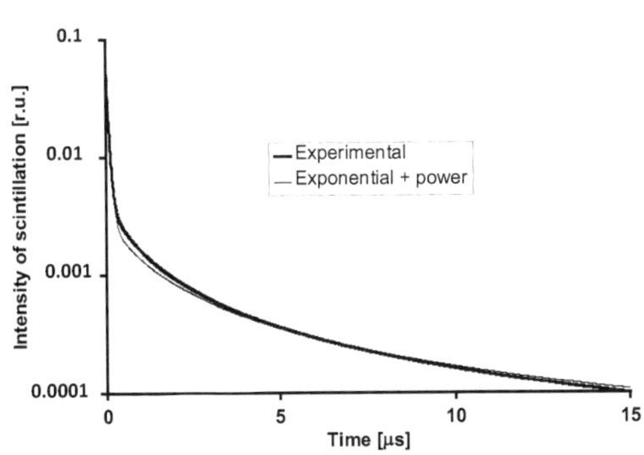

Figure 5.13 Comparison of measured decay curve with decay curve calculated using $PhY(t_s)/PhY(t_r = 0.5$ μs) measured dependence. Exponential + power decay function was assumed. Decay time of fast (exponential) component was assumed to be 60 ns.

5.4 Materials

Within the last two decades, a great number of oxide scintillators were prepared by the LPE; for review see [34, 97]. Main attention was paid to aluminum garnets [4, 5, 11, 30, 34, 52, 92], perovskites [3, 34, 98], and orthosilicates [2, 22, 34, 99–102]. Rare attempts were to grow other scintillators as ZnO [93, 103] or CdWO4 [104] were prepared by the LPE. The LPE was applied for production of waveguide solid state RE:YAG garnet lasers [34, 38, 39, 105]. A lot of oxide materials which are routinely prepared by other methods (Czochralski or Bridgman single crystals, polycrystals, solid state reaction or sol gel) were also grown by the LPE, for review until 1999 see ref. [34]. We shall not make here any exhausting survey but rather discuss some interesting material systems prepared by LPE. In Table 5.2 basic crystallographic properties of typical hosts are shown.

We shall focus here on scintillation properties of the most promising oxides especially Ce^{3+} and Pr^{3+} doped garnets, perovskites, and orthosilicates and compare their optical and scintillation properties with single crystal counterparts. These materials are non-hygroscopic, chemically stable, and have excellent optical and mechanical properties with large application potential. They have good light yield, fast scintillation response and adequate energy resolution—the properties of which have been incessantly improved. Luminescence and scintillation properties of above-mentioned SC materials have been recently described in several review papers [82, 106, 107] and references therein.

Ce^{3+} doped garnets, perovskites, and orthosilicates are considered as excellent and fast scintillator materials [82, 108, 109]. In the last decade Pr^{3+} doping has also been studied for even faster scintillators with comparable LY [110–113]. Doping with other RE ions, Tb, Eu, etc., provides very efficient PL and may be used for various phosphors in steady state applications, e.g., cathodoluminescent or X-ray screens [15, 29] where slow f–f emission in the millisecond time range is acceptable.

5.4.1 Garnets

Garnets have high-symmetry cubic structure with O_h^{10} space group, Table 5.2. The garnet structure is very flexible for cation substitution due to the three cation sites of tetra-, octa-, and dodecahedral symmetry.

Application of YAG:Ce in SEM detection was proposed in 1970s [114], its scintillation properties were reported 20 years ago [8]. It is excellent detector of electrons in SEM and for soft X-rays. Isostructural LuAG:Ce have higher density and higher effective atomic number, Z_{eff}, and associated higher stopping power of the lattice and is more efficient for hard X-ray and gamma-ray detection.

YAG:Ce films and Ga and RE (Gd, Tb, Eu, Tm) substituted films were first grown by Robertson in Philips Laboratories in the beginning of 1980s from the PbO flux [15, 29, 115, 116] for the cathodoluminescence application. LuAG:Ce film grown on YAG substrate from the PbO flux was reported in 2002 [92]. During last 15 years great number of substituted aluminum garnets has been grown and tested for scintillation applications. However, the scintillation efficiency, such as the LY, of epitaxial films grown from the PbO flux was lower as compared to single crystals grown from the melt (Czochralski or Bridgman). It was soon recognized that the impurity ions coming from the flux, especially Pb^{2+}, represent the main problem since they form trap states and parallel channels of nonradiative recombination [88, 117, 118]. Furthermore, effective nonradiative energy transfer from Pb^{2+} ions to Ce^{3+} ions results in appearance of slower components in the luminescence decay kinetics of Ce^{3+} centers [119]. Therefore, alternative lead-free fluxes have been tested and developed. Practical application of the BaO and MoO_3 fluxes for growth of aluminum garnet films was first reported in 2008 [4, 5]. Especially the BaO flux appeared very promising for growth of Ce and Pr doped epitaxial films with significantly improved scintillation properties [4, 11, 12, 51, 52, 120, 121]. Recently, it has been demonstrated that the scintillation properties (LY, suppressed slow components) of the epitaxial films of multicomponent GAGG:Ce garnets grown from the BaO flux are comparable to those of the best single crystals counterparts with the LY of ~50 phot/keV [11, 121], see below.

In YAG:Ce and especially in LuAG:Ce single crystals a substantial slow component in emission is observed which negatively affect their scintillation performance. This "slow light" is related to various types of shallow traps, which contribute to electron/hole trapping processes. The most important intrinsic defects, degrading the scintillation performance, are antisite defects (AD), which result in formation of shallow electron traps in the energy gap, and color centers related to oxygen vacancies [7, 13, 80]. Both AD and color centers are effectively suppressed during the LPE growth process due to the growth in oxidizing atmosphere and at significantly lower temperatures ~1000 °C.

TSL glow curves of single crystals and LPE films are compared in Fig. 5.14. In the LPE films the dominating TSL peaks at 120–170 K completely vanish. Furthermore, LPE samples show an absence of even deeper traps related to TSL peaks at 240 K and mainly at 280 K. Traps related to the latter peak may have detrimental influence to the overall stability of scintillator characteristics since the peak position is close to room temperature.

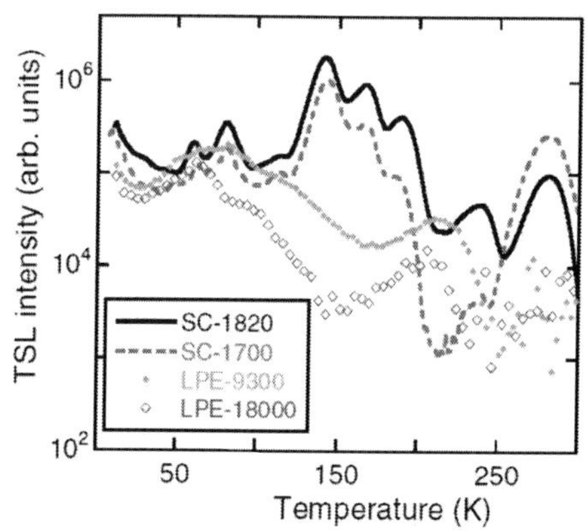

Figure 5.14 TSL glow curves of the LuAG:Ce single crystals, SC, and epitaxial films, LPE, grown from the PbO flux after X-ray irradiation with similar doses at 10 K. Reprinted from Ref. 13, Copyright 2005, with permission from John Wiley and Sons.

In Ce^{3+} and Pr^{3+} doped YAG and LuAG intense and fast luminescence originating from parity allowed 5d–4f transitions is observed in the transparency region of aluminum garnets. This shows on very good coupling of these ions to the lattice. Owing to high crystal field at c-sites in garnets the maximum of 5d broad band emission is located in the green-yellow range with maximum at 510–530 nm and at 308–320 nm in Ce^{3+} and Pr^{3+} ions, respectively. The scintillation decay is fast with decay times of ~60 ns and ~20 ns for Ce and Pr doping, respectively. The thermal quenching is well above the RT, for YAG/LuAG:Ce >700 K [122], for Pr:LuAG > 400 K and in Pr:YAG > 250 K [113].

However, because of their large ionic radii (Ce – 1.143 Å and Pr – 1.126 Å) only low content of Ce and Pr, rarely exceeding 1% in single crystals, is reached. Somewhat higher Ce concentration is obtained LPE films, especially those grown from the BaO flux [4, 5]. However, Ce and Pr concentration is constrained due to their very low segregation coefficients (Table 5.1), which would require high excess of Ce and Pr in the melt. The Ce content can be increased by growth at high supercooling or at lower growth temperatures (Fig. 5.3). However, this is not a very effective way—in films grown from PbO owing to the simultaneous increase of quenching Pb^{2+} ions (Fig. 5.15), and in films grown from BaO, the increase in the flux viscosity at temperatures below 1000°C and consequent worsened surface morphology is the main constraint.

The position of the 5d–4f emission band depends to some extent on the host composition—generally for large RE ions substituted in c-sites (Gd, Tb) the emission maximum of Ce^{3+} ions is red shifted towards 560 nm, for small RE ions (Lu) the blue shift is observed with maximum at 510 nm. The substitution of Ga ions causes blue shift of the emission and decrease of crystal field 5d splitting, presumably due to increased symmetry in c-sites of YGG compared to YAG; for details see Refs. [116, 123–126].

5.4.1.1 Ce^{3+}-doped YAG and LuAG

Comparison of scintillation properties of bulk Czochralski grown single crystals and epitaxial films of Ce-doped YAG and LuAG is shown in Table 5.3. The most obvious is the presence of strong slow component in scintillation and lower photoelectron yield in bulk crystals—e.g., in Ce:LuAG almost 70% of light of the Ce^{3+} emission center itself is related to the delayed radiation recombination [6]. This suggests a considerable influence of several traps of different origin and different energy depths, which participate in charge carrier trapping processes, cf. also Fig. 5.14. The antisite defects, i.e., Y or Lu situated at the octahedral Al site, $Lu(Y)_{Al}$, are often considered as an origin of the shallow traps in YAG or LuAG crystals [7]. These defects are strongly reduced in epitaxial films [13, 30]; however, they are not suppressed completely [71, 127]. The wide spectral band with slower decay situated near 300 nm in the bulk Cz-grown crystals was ascribed to the exciton localized near the $Lu(Y)_{Al}$ antisite defect [27].

Table 5.3 Comparison of scintillation properties of Czochralski single crystals and epitaxial Ce-doped YAG and LuAG films. τ_{PL} (530), photoluminescence decay time of Ce^{3+} emission; τ_{SC} (530), scintillation decay times of fast and slow components of Ce^{3+} emission for various excitation sources, CL – cathodoluminescence. T_G – growth temperature, N_{phe} –photoelectron yield excited by α particles at shaping time of 0.5 μs (increase of N_{phe} at 10 μs shaping time)

	YAG:Ce		LuAG:Ce		Refs.
	Single crystal	Epitaxial film	Single crystal	Epitaxial film	
τ_{PL} (530 nm) [ns]	62	60	54	54–56	[5, 6]
τ_{SC} (530 nm) [ns]:					
α excitation, 5.49MeV	68(35%) + 247		58(40%) + 1000		[7, 8]
γ excitation, 662 keV	96 + 230 + 1400	*	61 + 510 + 2400	*	[23]
X-ray excitation, 25 kV			58(30%) + 150 + 1600		[27]
synchrotron, 150 eV		81		10 + 70	[30]
CL, 10 keV	75			52	[10]
N_{phe} [phe/MeV] (N_{phe} increase)	320–360 (20%)	330–400 (15–30%)	380 (70%)	260–315 (50–70%)	[41] [42]
T_G [°C]	1970	950–1070	2020	950–1070	

* γ excitation cannot be used for thin epitaxial films because of low absorption.

The optical absorption spectra of Ce:LuAG films grown on LuAG and YAG substrates from the same PbO flux at various supercooling are displayed in Fig. 5.15. The first two low energy 4f–5d (Ce^{3+}) absorption bands are clearly resolved at 445 and 345 nm, but remaining three spectral bands situated below 280 nm are overlapped by Pb^{2+} absorption. The absorption peak near 262 nm and strongly increased absorption below 250 nm originate from allowed transitions in Pb^{2+} impurity ions [119, 128]. The peak at 262 nm is missing in films grown from BaO and MoO_3 fluxes [5]. It is worth noting that the content of both the Ce and impurity ions strongly depends on the growth conditions. It is obvious that concentrations of both Ce^{3+} and Pb^{2+} ions increase with supercooling, cf. peaks at 445 nm and 262 nm in Fig. 5.15. Undesirable consequence of Pb admixture is strongly increased absorption in the deep UV range which has negative impact on the emission efficiency of scintillator, especially those with luminescence in the UV (e.g., LuAG:Pr or perovskite and orthosilicate scintillators described below).

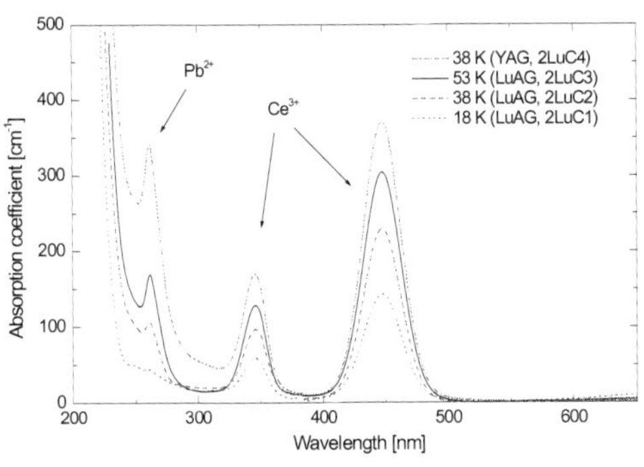

Figure 5.15 Optical absorption coefficient of epitaxial Ce:LuAG films grown from the same PbO flux under various undercoolings (substrates and undercooling are in the legend), saturation temperature T_S = 998 °C. Both Ce and Pb concentrations increase with increasing undercooling. Reprinted from Ref. 4, Copyright 2010, with permission from Elsevier.

The PL excitation and emission spectra of YAG:Ce and LuAG:Ce epitaxial films grown from both PbO and BaO fluxes along with YAG:Ce single crystal for comparison are displayed in Fig. 5.16. The spectra can be compared in an absolute way. The structure in the excitation spectra comes from the Ce^{3+} ions and it shows a close correspondence with the absorption curves. The broad emission band in the PL emission spectra has maximum around 520 nm. The maxima of the Ce peaks are mutually slightly shifted due to the different crystal fields of the host crystals. Generally, films grown from the BaO flux exhibit substantially higher intensity of PL emission compared to the PbO grown films. This can be partly explained by higher Ce content in the BaO grown films. Furthermore, the Pb^{2+} ions compete for the charge carrier capture with the Ce^{3+} ions and form additional nonradiative de-excitation channel which leads to quenching of the luminescence [88, 117, 119].

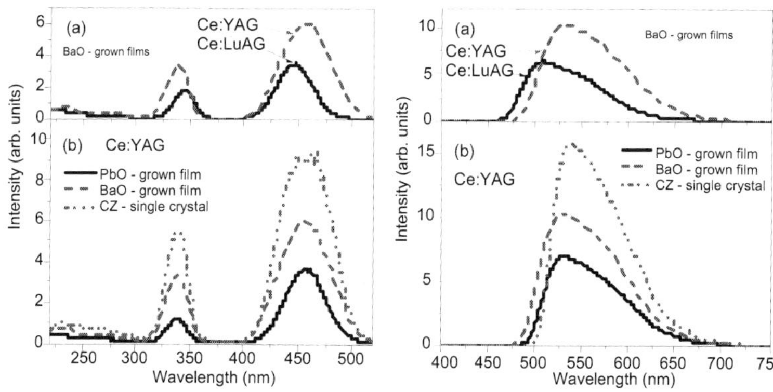

Figure 5.16 Left: excitation spectra (λ_{em} = 520 nm), right: emission spectra (λ_{ex} = 340 nm) of epitaxial YAG:Ce and LuAG:Ce films grown from PbO and BaO fluxes, also YAG:Ce single crystal is shown for comparison, [5].

The photoluminescence and scintillation decay curves of a LuAG:Ce film grown from the BaO flux are shown in Fig. 5.17. The PL Ce^{3+} related decays of films grown from the lead-free BaO flux exhibit one-exponential dependence over three orders of magnitude with the decay times t = 52–56 ns. These values are in close relation with those reported for high purity Cz single crystals. The inspection of Ce decay curves of films grown from the PbO flux shows some distortion [4, 5, 21] and additional fast component in the PL kinetics indicates nonradiative transfer of energy from the Ce^{3+} relaxed excited $5d_1$ state towards an impurity ion or nearby defect state [5, 6].

Spectrally unresolved cathodoluminescence decay is shown in Fig. 5.17. The decay is evidently multicomponent. The fundamental fast exponential component of 52 ns correlates with the PL decay of 5d–4f(Ce^{3+}) emission. At longer times, >300 ns, the decay can be approximated by power law function $\propto t^{-p}$, where p is a constant [121] and can be correlated with tunneling of a trapped electron to a nearby recombination center [129, 130]. De-excitation of an electron captured on a virtual trap can include also thermal ionization of shallow traps, which may occur simultaneously with tunneling. The possible pathways are discussed in more detail in [120, 121].

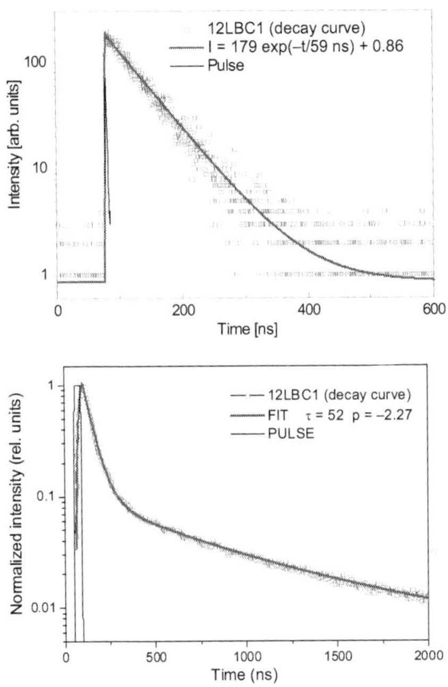

Figure 5.17 Decay kinetics of LuAG:Ce epitaxial film grown from the BaO flux; left: the PL decay, λ_{ex} = 340 nm, λ_{em} = 510 nm; right: spectrally unresolved e-beam excited scintillation decay. The solid lines are convolutions of excitation pulses with 1-exp decay (PL decay) and 1-exp decay with power law function (scintillation decay), for details see the text.

5.4.1.2 Ce^{3+}-doped multicomponent garnets

Recently, the "band-gap engineering" strategy was adopted to reduce the negative effects of shallow traps in bulk garnet crystals grown from the melt [123, 124, 131, 132]. Using the substitution of Ga and Gd ions into LuAG:Ce, the bottom of the conduction band is lowered, due to admixture of 4s(Ga^{3+}) states, and simultaneously the crystal field splitting of the excited 5d states is enlarged, due to Gd-induced decreased CF symmetry in c-states. This approach leads to strong suppression of the trapping effects, which degrade the scintillation response [133]. The scintillation properties of multicomponent $(Lu,Y,Gd)_3(Al,Ga)_5O_{12}$:Ce single crystals or ceramics are much better in comparison with LuAG:Ce or YAG:Ce. Very high light yield approaching 50 photons/keV, excellent energy resolution, and low afterglow were observed in bulk crystals [134–136]

The LPE technology was adopted also for production of multicomponent garnet films [63, 137, 138]. The first results in samples grown from the PbO flux were rather ambivalent. Some improvement of light yield and energy resolution was observed but any positive role of Ga admixture, observed in bulk crystals, was not found in films due to different origin of traps in epitaxial films which are connected likely to the Pb^{2+} impurities [63, 137].

Excellent scintillation properties were, however, very recently demonstrated in epitaxial garnet films $(Lu,Gd)_3(Al,Ga)_5O_{12}$:Ce grown from the BaO flux [11]. Compared to high quality bulk crystals with analogous composition, the films with optimized composition exhibit similar values of α-excited photoelectron yield and similar energy resolution. The obtained values of the LY were the highest in garnet epitaxial films exceeding at least two times the values previously reported for LuAG/YAG:Ce [41, 51, 53].

In Fig. 5.18 the Ce^{3+} -related emission and excitation PL spectra and X-ray excited RL spectra are shown demonstrating significant increase of the emission intensity in multicomponent GdGaLuAG:Ce films.

Photoluminescence 5d–4f(Ce^{3+}) decays and spectrally unresolved α-particle excited scintillation decays are shown in Fig. 5.19. The PL decay times of 5d(Ce) emission in GdYAG/GdLuAG:Ce are not affected by Gd substitution into these hosts. The PL decays are strictly one exponential with decay times 61 and 56 ns in Ce:GdLuAG and Ce:GdYAG, respectively, for Gd content up to 50% [63]. Due to Ga co-doping the decay time partially decreases and weak faster component ~20–30 ns appears in the PL decays indicating energy transfer from the 5d states due to beginning of the thermal ionization. Two-exponential decay was observed in all heavily Ga substituted samples. The main fast component of the scintillation decays under α-particle excitation was within 60–100 ns [11] corresponding to Ce^{3+} centers.

Another important progress of multicomponent garnets being considerable reduction of slow decay components in microsecond

– millisecond time ranges. Likewise the measured afterglow of the films is very low, reaching the value around 0.1% [11].

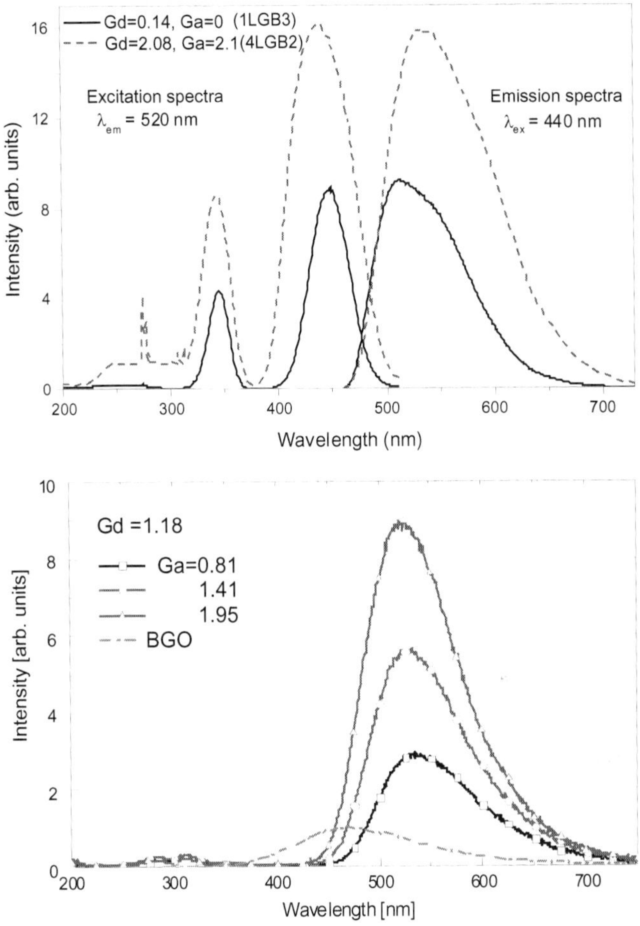

Figure 5.18 Emission and excitation spectra of epitaxial films (above) and X-ray excited RL spectra (below) for $Lu_xGd_{3-x}Ga_yAl_{5-y}O_{12}$:Ce samples with various content of Gd and Ga. Reprinted from Ref. 11, Copyright 2015, with permission from American Chemical Society.

The photoelectron yield measured under α-particle excitation of several epitaxial films exceeded 800 phels/MeV at 1 μs shaping times, the value quite comparable with multicomponent single crystals with gamma excited LY of 50 photons/keV, Fig. 5.20. These are the highest values ever reached in garnet epitaxial films, exceeding 2–3 times those measured in LuAG:Ce, YAG:Ce epitaxial films. The best scintillation performance was observed in films at higher Ga ($1.8 < y_{Ga} < 3$) and Gd ($x_{Gd} > 1.8$) concentrations, Fig. 5.20. Thermal quenching starts at Ga concentrations y > 3, and luminescence is completely quenched above y > 4 at RT. The optimal Gd content is 1.8 < x < 3 when slow f–f(Gd) emission at 312 nm is completely suppressed due to concentration quenching.

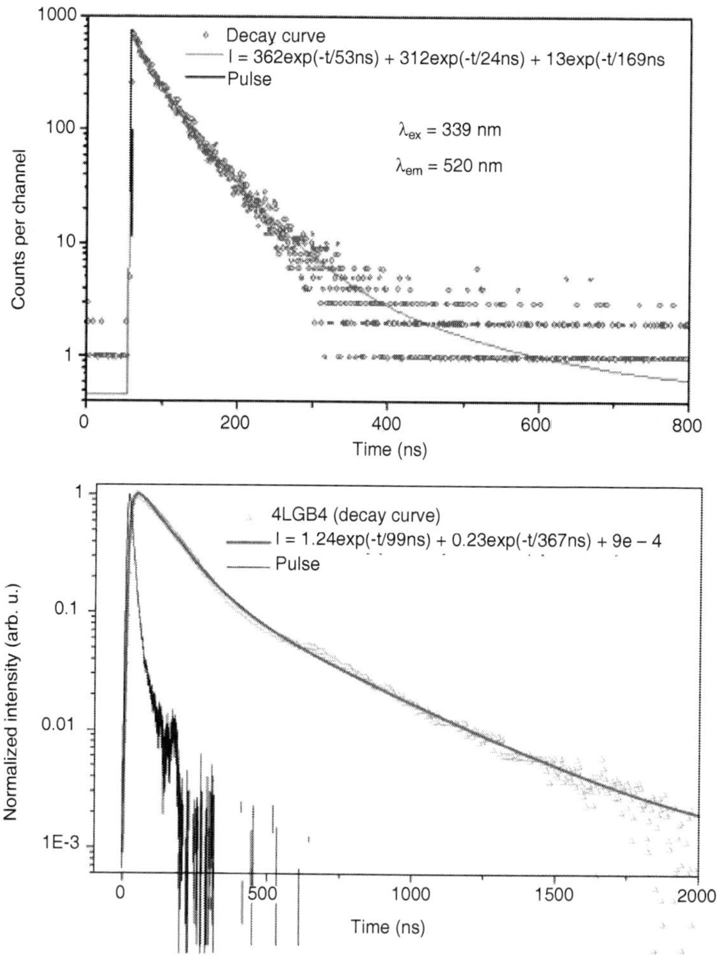

Figure 5.19 PL decay (above) and spectrally unresolved α-excited scintillation decay curves (below) of for $Lu_xGd_{3-x}Ga_yAl_{5-y}O_{12}$:Ce sample. Reprinted from Ref. 11, Copyright 2015, with permission from American Chemical Society.

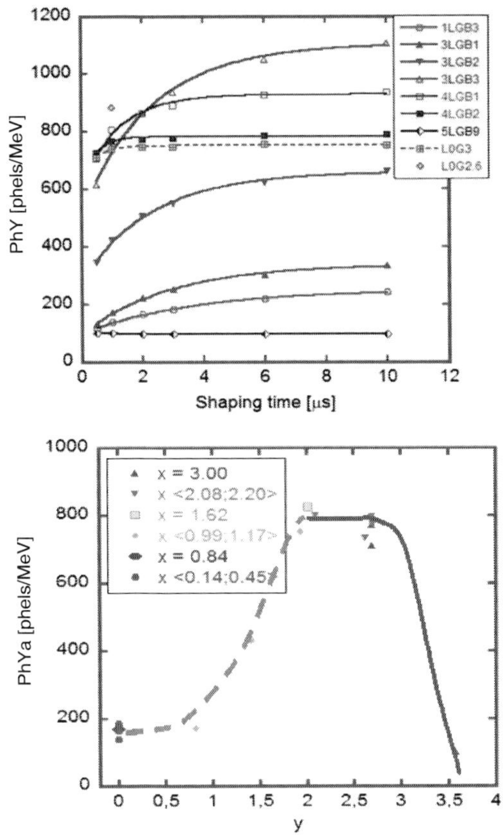

Figure 5.20 Dependence of the α-particle excited photoelectron yield on shaping time (above) and dependence of α-excited photoelectron yield on Ga content for some $Lu_xGd_{3-x}Ga_yAl_{5-y}O_{12}$:Ce films (below). Sample LOG3 in the left panel is reference GAGG:Ce single crystal wit LY of 50 phot/keV. Lines are only to guide the eye. Reprinted from Ref. 11, Copyright 2015, with permission from American Chemical Society.

In multicomponent garnets the role of Ga substitution is essential in suppression the shallow traps [133]. Most of excitation energy is, however, captured by heavy Gd ions. The nonradiative energy transfer from the Gd^{3+} sensitizer to Ce^{3+} activator ions was proved in (Ce, Gd) co-doped YAG and LuAG [139]. Such energy transfer is beneficial just for significant increase of scintillation efficiency in heavily Gd doped GAGG:Ce scintillators when the excitation energy migrating over the Gd sublattice is effectively transferred to Ce^{3+} activators [63, 120, 139]. The excitation energy migrates over the Gd^{3+} sublattice and (provided it is not lost at trap states originating, e.g., from structural defects or impurities) it is finally captured by Ce^{3+} ions and subsequently released by their characteristic green–yellow emission. The side effect of the fast migration is concentration quenching of slow f–f(Gd) emission at Gd content $x_{Gd} > 1.8$ at RT.

5.4.1.3 Pr^{3+}-doped garnets

Pr^{3+} doped LuAG and YAG with the dominant 5d–4f emission in the UV range between 300 and 400 nm belong to the family of fast (< 20 ns decay time), high performance (LY > 20 phot/MeV) complex oxide scintillators [110, 113, 140–145]. However, the epitaxial garnet YAG/LuAG:Pr films grown from the PbO flux [146, 147] suffer from serious negative influence of Pb^{2+}impurities, cf. previous section, and their scintillation performance was low. It was soon recognized that lead-free flux is necessary for the growth of epitaxial films. Here we review the scintillation properties and achievements of LuAG:Pr and LuAG:PrSc epitaxial films which were grown exclusively from the BaO flux [12, 52]. The reason for Sc,Pr codoping follows from expected LY increase due to energy transfer from Sc to Pr, and from the effort to suppress also the slow scintillation component in the decay kinetics.

The excitation and emission spectra of LuAG:Pr film are shown in Fig. 5.21a left. The broad emission double peak which dominates in the UV range with the main maximum at 308 nm originates from parity allowed interconfigurational 5d–4f transitions. Narrow 4f–4f transition observed near 480 and 615 nm are very weak. Almost all the emission in LuAG:Pr is located in the UV spectral range between 300 and 400 nm; integral intensity of the slow f–f emission is only 4%. The broad 5d(Pr^{3+}) bands in the excitation spectra at 240 and 282 nm are closely related to the absorption ones, Fig. 5.22 left. All these PL features of LuAG:Pr are virtually identical to those observed in single crystals.

Slow scintillation component in LuAG exists due to retrapping migrating electrons at shallow electron traps [143, 148]. The Pr^{3+}–Sc^{3+} codoping may speed up the scintillation response, because Sc^{3+} ions could simplify the energy migration towards Pr^{3+} centers due to good overlap of the Sc-related emission (Sc-trapped exciton) at around 275 nm with the 4f–5d$_1$ absorption band of Pr^{3+} center at 280 nm, Fig. 5.22.

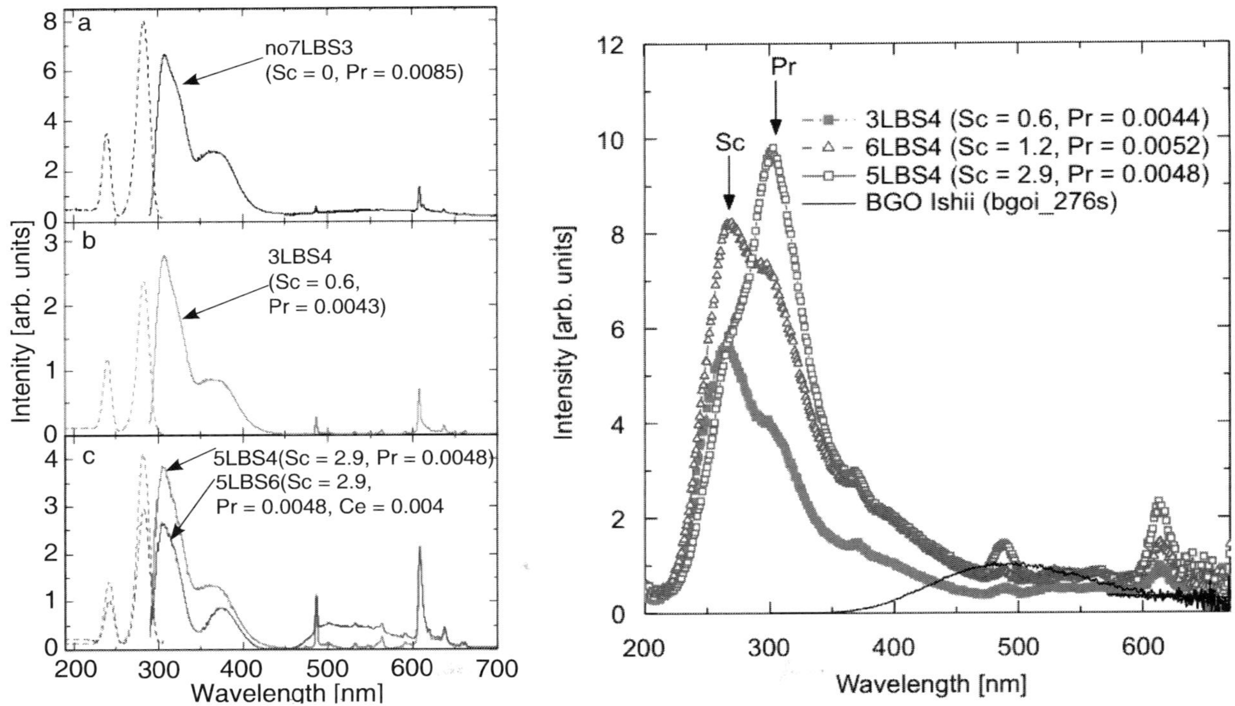

Figure 5.21 Left: PL excitation (λ_{em} = 308 nm) and emission (λ_{ex} = 282 nm) spectra of epitaxial LuAG:Pr (a) and co-doped LuAG:PrSc, LuAG:PrScCe (b, c) epitaxial films grown from BaO flux, Right: X-ray excited RL spectra of LuAG:PrSc films with various content of Sc. Reprinted from Ref. 12, Copyright 2011, with permission from Elsevier.

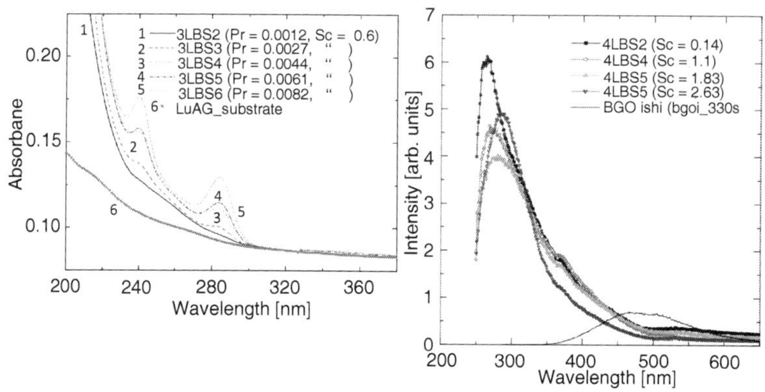

Figure 5.22 Left: Absorbance of ScPr co-doped LuAG epitaxial layers with 0.6 at% of Sc and Pr content specified in the legend grown on LuAG substrates; Right: X-ray excited radioluminescence spectra of LuAG:Sc epitaxial films. The RL of reference BGO crystal is also shown. Concentrations of Sc are shown in the legend. Reprinted from Ref. 12, Copyright 2011, with permission from Elsevier.

The PL spectra of ScPr co-doped LuAG films are shown in Fig. 5.21b left. Any Sc related emission is not observed. Additional Ce^{3+} co-doping, sample LuAG:ScPrCe in Fig. 5.21c left, however, partly quenches the UV emission of Pr^{3+} ions and Ce-Pr co-doping is useless for scintillation applications.

The X-ray excited RL spectra of co-doped LuAG:ScPr films are displayed in Fig. 5.21 right. The major Sc^{3+} and Pr^{3+}-related peaks are situated at 280 and 308 nm, respectively. The most remarkable feature of Sc co-doping is significant increase in intensity of the Pr^{3+} peak at 308 nm with increase in Sc content while the Pr content almost does not change, cf. Fig. 5.21 right. This notable dependency was explained by energy transfer from sensitizer Sc^{3+} to activator Pr^{3+} ions.

The Sc co-doping does not influence 5d–4f(Pr^{3+}) decay kinetics, Fig. 5.23 left. The dominating decay time, 18–20 ns, is in close relation with that reported for single crystals [113].

The photoelectron yield (PhY), of LuAG:PrSc films measured under alpha particle excitation for different shaping times is shown in Fig. 5.23 right. In films with optimized Sc content, $0.05 < x_{Sc} < 0.2$, the PhY exceeds the reference single crystal LuAG:Pr. Furthermore, the film is much better at shortest shaping times and its time dependence is also significantly improved.

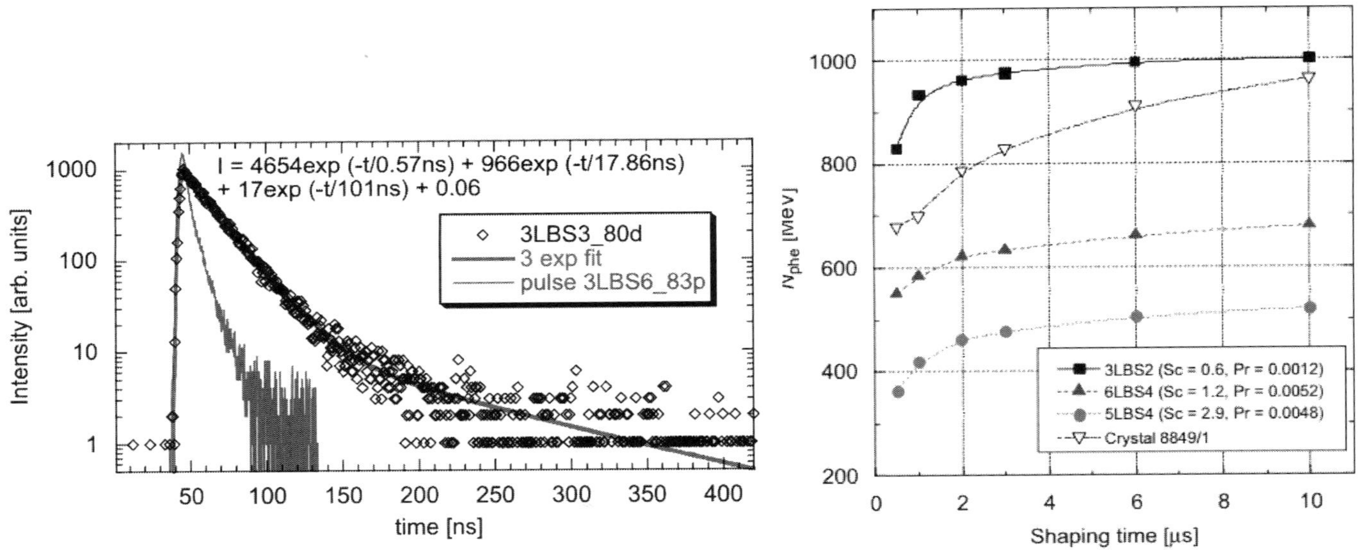

Figure 5.23 Left: Photoluminescence decay of PrSc-doped LuAG epitaxial film measured at Pr^{3+}(5d–4f) emission, λ_{em} = 320 nm and λ_{ex} = 282 nm; Right: Photoelectron yield *N*phe per 1 MeV at different shaping times under alpha particles (^{239}Pu) excitation of ScPr:LuAG epitaxial films, reference single crystal LuAG:Pr, sample 8849/1 is shown also. Reprinted from Ref. 12, Copyright 2011, with permission from Elsevier.

Co-doping of Sc^{3+} and Pr^{3+} thus appears as a suitable way of preparing high figure-of-merit scintillating films, even though optimum concentrations are rather critical.

Similar approach was also used for development of efficient Tb-Sc co-doped LuAG phosphors [20, 26]. Tb-doped garnets display intense f–f emission lines between 480 and 600 nm (at higher doping, 2–15 %) with dominant green emission at 543 nm, Fig. 5.24. Co-doped LuAG:TbSc shows improved PL and RL properties compared to LuAG:Tb. Furthermore, the Tb^{3+} doping quenches the host and Sc-related emissions situated in the UV range. Enhanced material performance can be understood as energy transfer from Sc sensitizer to Tb activator ions, due to the overlap of Sc-related emission and Tb(5d) absorption spectral bands.

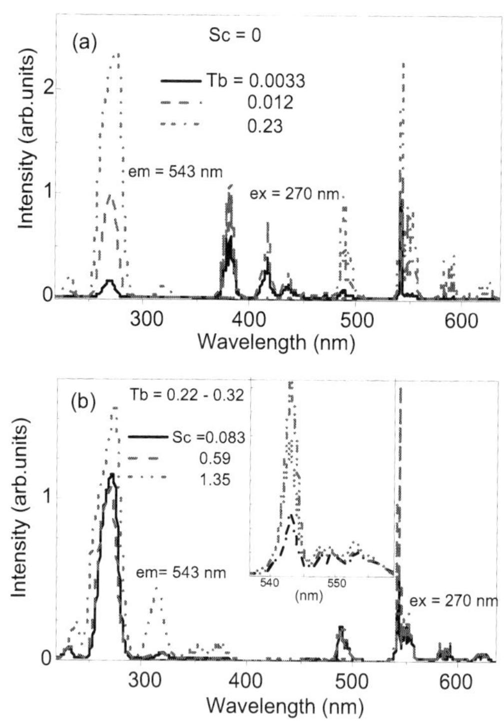

Figure 5.24 Photoluminescence excitation spectra at λ_{em} = 543 nm ($^5D_4$) and emission spectra excited at 5d(Tb^{3+}) band, λ_{ex} = 270 nm. (a) Tb:LuAG epitaxial films for several concentrations of Tb, (b) TbSc:LuAG films for various concentrations of Sc, see also Ref. [20].

5.4.2 Perovskites

Ce^{3+}- and Pr^{3+}-doped perovskites YAP/LuAP represent very fast scintillators with 5d–4f decay times of 18 ns [149] and 8 ns [1, 150, 151], respectively. YAP:Ce is excellent scintillator in SEM detector [152]. The LY of YAP:Ce about 20 000 ph/MeV was reported [96, 151, 153, 154], however, it is decreased in LuAP:Ce to half of that value due to shallow traps [155]. The LY of the best YAP:Pr is about 80% of that of YAP:Ce [1]. This suggests the presence of delayed radiative recombination processes. Most of the scintillation losses in these samples are due to shallow short-lived traps [156] and various charge carrier trapping centers, color centers, and yttrium antisite defects Y_{Al} directly proved by NMR [157], which play a major role in decreased LY.

Basic parameters of YAP/LuAP host systems are summarized in Table 5.2.

LuAP single crystal is difficult to grow from the melt due to instability of its perovskite phase (often twin formation and garnet phase appearance). The scintillation properties of more stable mixed (LuY)AP:Ce system is, however, significantly worsen compared to the end compounds due to increased content of slow components [155, 158].

On the other hand, both rare earth doped YAP, LuAP or TbAP epitaxial films were successfully grown onto YAP substrate from the PbO flux [34, 159]. The perovskite films are grown under similar conditions as garnets—the PbO/B_2O_3 ratio is chosen ~12 : 1, Y(Lu)/Al is close to stoichiometric ratio 1 : 1 in order to stabilize the perovskite phase and CeO_2 is in excess due to low Ce segregation coefficient, $k_{Ce} \sim 0.01$. At growth temperatures around 1000°C the growth rate is 0.5–1 μm/min [98].

Fundamental problem of perovskite films grown from the PbO flux is their contamination by Pb^{2+} ions due spacious crystal sites of 12 fold coordination, which is more favorable for positioning of large lead ions as compared to garnets. The Pb^{2+} ions quench the 5d–4f emission of Ce^{3+} and Pr^{3+} ions and significantly decrease the LY. Reported LY of the YAP:Ce epitaxial films grown from the PbO flux is < 10 % of the SC counterparts [3, 99, 147].

YAP and (YLu)AP epitaxial films grown from the lead-free BaO-B_2O_3-BaF_2 flux [3, 18] provide better emission and scintillation characteristics in spite of their worse structural and surface quality compared to PbO-grown films [3]. Although the LY of BaO-grown films is higher compared to those grown from the PbO flux, it is still far from the SC counterparts (4–15× lower) and growth from the BaO flux has potential for further development.

In the following paragraphs scintillation properties of most promising perovskites YAP and LuAP doped by Ce^{3+} and Pr^{3+} ions grown from the both fluxes are described.

5.4.2.1 Ce^{3+}-doped YAP/LuAP

The absorption spectra of the YAP:Ce and LuYAP:Ce films grown from the PbO and BaO are displayed in Fig. 5.25. In the UV range below 260 nm, the samples grown from BaO flux (curves 1 and 3) show much lower parasitic absorption than those grown from PbO flux. The strong absorption below 260 nm is connected primarily with Pb^{2+} impurity ions and originates from $^1S_0 \rightarrow {}^3P_1(Pb^{2+})$ transitions ions [128]. The split band centered around 290 nm is due to the $4f^1(^2F_{5/2}) \rightarrow 5d_1$ transitions of the Ce^{3+} ions [160].

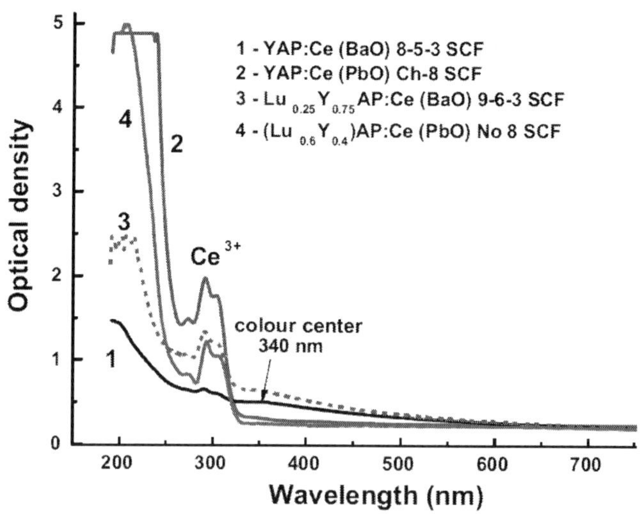

Figure 5.25 Absorption spectra of Ce^{3+}doped YAP:Ce and LuYAP:Ce films grown from BaO and PbO fluxes. The concentration of CeO_2 in the melt was 2 mol% (1,3) and 13–15 mol% (2,4). Reprinted from Ref. 18, Copyright 2013, with permission from Elsevier.

The CL spectra of YAP:Ce films grown from PbO and BaO fluxes are compared in Fig. 5.26. The emission bands centered at around

~375 nm originate from $5d_1$–$4f^1(^2F_{5/2,7/2})$ radiative transitions in the Ce^{3+} ions. However, in films grown from PbO this Ce^{3+}- related emission is overlapped with the emission of single and dimer Pb^{2+} centers [88]. This resulted in observed wider emission band of films grown from PbO, curve 1 and the inset in Fig. 5.26.

The effect of Pb^{2+} ions in the emission spectra is demonstrated in Fig. 5.27, where undoped YAP (BaO) and YAP:Pb (PbO) doped by intentional Pb impurities are compared. The low intensive luminescence in the UV and visible ranges of YAP (BaO), Fig. 5.27a, are most probably connected with the Ce^{3+} trace impurity and host defect emission centers, respectively [3]. The emission of YAP:Pb films is orders of magnitude higher, Fig. 5.27b, and is related to the Pb impurity and to various contributions in UV and vis ranges; for details see, e.g., Refs. [88, 117, 118, 157, 161].

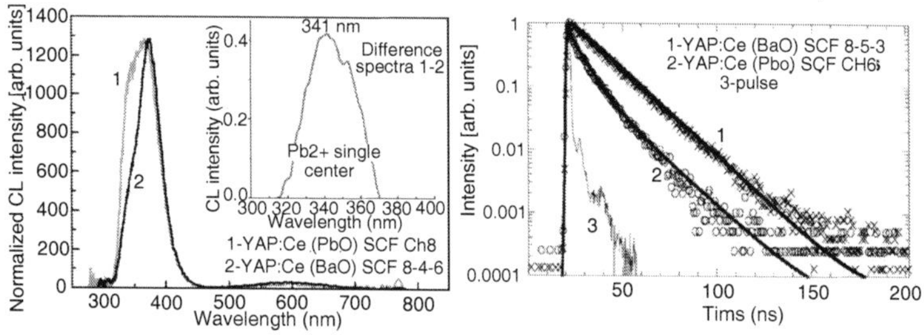

Figure 5.26 Left: Normalized CL spectra of YAP:Ce epitaxial films at 300 K grown from the PbO (1) and BaO (2) fluxes; inset: difference between the spectra; Right: the PL decay kinetics of Ce^{3+} luminescence in YAP:Ce grown from the BaO (1) and PbO (2) fluxes under excitation in the Ce^{3+} absorption band at 300 nm. The decay times from the exponential fits: τ = 15.7 ns and (6.5 + 16.1) ns for respective BaO and PbO grown samples. Reprinted from Ref. 3, Copyright 2009, with permission from John Wiley and Sons.

Typical PL decay kinetics of YAP:Ce films grown from BaO and PbO are shown in Fig. 5.26 right. The single-exponential decay up to four orders of magnitude is obtained for BaO-grown YAP:Ce films and the average decay time for several samples is 16±1 ns. Such values are close to typical decay time (17–18 ns) SC cozunterparts [107].

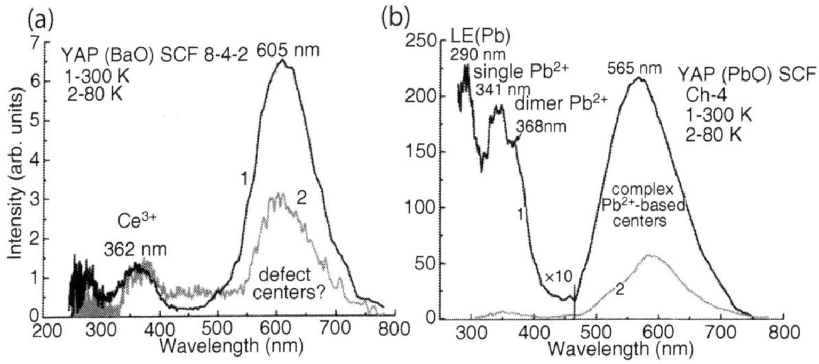

Figure 5.27 The CL spectra of undoped YAP grown from BaO (a) and PbO (b) fluxes at 80 and 300 K. Spectra in both panels can be mutually compared in an absolute way. Reprinted from Ref. 3, Copyright 2009, with permission from John Wiley and Sons.

Visible acceleration of the PL decay kinetics is observed in YAP:Ce films grown from PbO, curve 2 in Fig. 5.26 right. The decay curve can be quantitatively approximated by the superposition of two components with decay times of 6.5 and 16.1 ns. This can be explained by energy transfer from Ce^{3+} ions to the Pb^{2+}-based centers (Pb^{2+} dimer or more complex Pb^{2+} centers) [119]. The excitation of these Pb centers is possible due to the overlap of their excitation bands with the 350–370 nm Ce^{3+} emission band in YAP:Ce. Integration of decays shows that in films grown from PbO the losses due to energy transfer away from Ce^{3+} $5d_1$ relaxed excited state are more than 50% [3]. Such huge losses partly explain low scintillation efficiency of PbO-grown YAP:Ce films.

Photoelectron yield (LY) of YAP:Ce films measured under excitation by alpha-particles ^{241}Am (5.48 MeV) is definitely significantly reduced compared to SC. The reported LY of YAP:Ce grown from PbO is only 2–10 % of the value of SC counterpart [3]. The LY of films grown from BaO is notably higher but still only about 12–24 % [3, 18]. It shows the large potential of the BaO-based flux for improving scintillation properties of aluminum perovskites.

5.4.2.2 Pr^{3+}, Tb^{3+}-doped perovskites

The Pr^{3+} ions in YAP/LuAP show intensive and fast emission in the UV spectral range, Fig. 5.28 left. The wide double band with maxima at about 242 nm and 280 nm is due to allowed Pr radiative transition from its $5d_1$ level to $4f_2$ (^3H_J,^3F_J) levels. Corresponding PL decay is very fast, ~8ns, the shortest in the family of complex oxides. The structure between 480 and 650 nm comes from f–f(Pr^{3+}) transitions. A striking difference between spectra of SC and films is evident—the UV emission dominates in the SC, Fig. 5.28 left, while it is drastically suppressed in films grown from the PbO flux, Fig. 5.28 right, at the expense of slow f–f emission in the visible range. Therefore the LY of YAP/LuAP:Pr films is enormously low and reaches only about ~5 % of reference YAP:Ce SC [146]. The main reason for very low LY of Pr^{3+}-doped films is the strong quenching influence of Pb^{2+} impurity dopant. Namely, strong Pb^{2+}-related absorption below 260 nm is overlapped with 5d–4f(Pr^{3+}) emission. This leads to energy transfer from Pr^{3+} ions to Pb^{2+} centers, which is the main reason for the reduced LY in perovskites grown from the PbO flux [117]. Another reason for reduced LY of Pr^{3+}-doped epitaxial films of perovskites is the presence of the f–f luminescence of Pr^{3+} ions in the visible range, Fig. 5.28, which plays a role of competing channel for dissipation of the excitation energy [146].

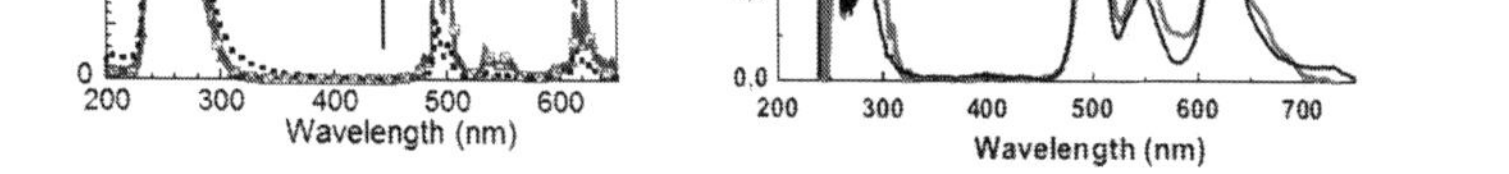

Figure 5.28 Left: X-ray excited RL spectra of Pr-doped YAP single crystals at RT, see also Ref. 1. Right: The CL spectra and decay kinetics of Pr^{3+} luminescence in YAP:Pr (1) and LuAP:Pr (2) epitaxial films grown from the PbO flux. Reprinted from Ref. 146, Copyright 2010, with permission from Elsevier.

On the other hand, the sharp f–f emission in trivalent rare earth ions is not markedly influenced by Pb^{2+} impurity. The problem of low LY of YAP:Ce or YAP:Pr films can be partly eliminated by co-doping with Tb^{3+} ions. Tb has large segregation coefficient (about 0.7–1.0) as compared to Ce^{3+} ions (~0.006–0.01) [98]. In Fig. 5.29 the RL spectra of LuAG:Ce and LuAG:Ce,Tb films co-doped with different content of Tb are displayed. The intensity of the Ce^{3+}-related luminescence in Ce,Tb co-doped films is strongly decreased. This observation confirms effective energy transfer $Ce^{3+} \rightarrow Tb^{3+}$ which is due to the overlap of the Ce(5d–4f) emission with transitions in Tb^{3+} between 380 and 420 nm [17]. However, due to this energy transfer, the light output of X-ray excited luminescence of from Tb^{3+} ions is about 2 times larger in LuAP:Ce,Tb as compared to LuAP:Tb counterpart and 30–40% larger than that of a bulk YAG:Ce standard sample [162]. The as-grown LuAP:Tb and LuAP:Tb,Ce films were tested for high-resolution X-ray imaging using a spatial resolution target at the ESRF, and displayed promising result [162].

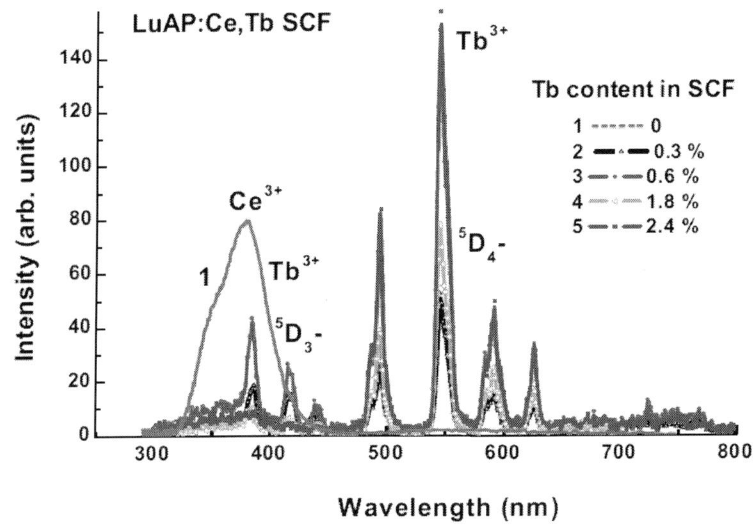

Figure 5.29 The X-ray excited spectra of LuAP:Ce and LuAP:CeTb epitaxial films at 300 K grown from PbO flux. Reprinted from Ref. 17, Copyright 2013, with permission from Elsevier.

5.4.2.3 Concluding remarks

The results reported above demonstrate detrimental influence of Pb^{2+} contamination and lead-induced centers on the scintillation characteristics of Ce^{3+} and Pr^{3+} doped aluminum perovskites. Adoption of a lead-free BaO flux is essential for production fast, high performance Ce, Pr doped YAP/LuAP thin-film scintillators.

5.4.3 Orthosilicates

Ce^{3+}-doped oxyorthosilicates RE_2SiO_5, where RE = Y, Lu, Gd, or combination of these ions, are promising scintillating oxide materials for gamma ray and high energy particle applications. Orthosilicates crystalize in monoclinic structure [163]. Crystals with small rare earth ions, e.g., Y_2SiO_5–YSO, Lu_2SiO_5–LSO, have space group C2/c (structure X1) and those with large RE ions, e.g., Gd_2SiO_5 – GSO, have space group $P2_1/c$ (structure X2). In both these structures, there are two crystallographically unequal rare earth sites, RE1 and RE2, with coordination numbers 7 and 6 (X1 structure) and 9 and 7 (X2 structure), respectively, for more details see also [69, 163–166]. The dopants and impurities with large ionic radii (Ce, Pr, Pb, etc.) prefer to occupy larger RE1 sites with higher coordination numbers of 7 and 9, respectively. Not all oxygen ions are bound to both yttrium and silicon and possible defects, especially the oxygen vacancies and related color centers F^+ or F, are more easily created in oxygen sites bound only to the yttrium ions.

The epitaxial films have been grown so far in several laboratories exclusively from the PbO-Ba_2O_3 flux [2, 34, 102, 167]. The reported quality of the YSO/LSO:RE films can be very good under optimal growth conditions (at low undercooling and suitable appropriately (010) oriented substrate). The melt used for the epitaxial growth has composition PbO-Ba_2O_3 with molar ratio 1:10 – 1:20 and solute Y_2O_3 + SiO_2 with SiO_2 in excess for stabilization of the orthosilicate phase [34]. Due to very low segregation coefficient of Ce, k_{Ce} ~0.01, large excess of CeO_2 in the melt is necessary, however, samples with higher Ce content is difficult to obtain due to limited solubility of CeO_2 in the melt. The growth temperatures are around 1000 °C.

5.4.3.1 Ce^{3+}-doped YSO, LSO

YSO:Ce has good LY and fast scintillation decay of 30–40 ns and is used for electron detection in SEM. Practical application of YSO/LSO:Ce crystals is limited owing to fairly high afterglow [168], likely due to electron trapping centers related to oxygen vacancies [169, 170]. This disadvantage was eliminated in mixed LYSO:Ce system and further improved by aliovalent co-doping by divalent ions LYSO:Ce,Ca,Mg [74, 171]. This optimized material found application in medical positron emission tomography PET due to its excellent LY, fast response, high stopping power and low afterglow. On the other hand, in epitaxial films the oxygen vacancies and related deep traps are substantially reduced due to much lower growth temperature and afterglow is suppressed in films.

Large Ce^{3+} ions are located predominantly in bigger coordination 7, site Ce1, in LSO/YSO (X2 structure), however, a part of Ce ions is in site Ce2 with coordination number 6. Ce1 and Ce2 have different absorption and emission spectral properties and kinetics of luminescence.

The principal disadvantage of YSO/LSO films grown from the PbO flux is quite high Pb^{2+} content due to spacious crystal sites more favorable for positioning of large lead ions into the orthosilicate lattice. According to the EPMA analysis, content of Ce and Pb in LPE films is comparable [102, 172, 173]. The divalent Pb^{2+} ions have, however, detrimental effect on the PL properties of YSO/LSO:Ce systems. In some films any Ce^{3+} centers were not detected in the absorption or PL excitation spectra and related PL emission was not observed by direct excitation into ground 4f(Ce^{3+}) state at 3.45 eV (360 nm) [174]. The emission from these samples with maxima in blue ~2.8 eV and UV ~3.5 eV ranges were identified as Pb^{2+} origin [87, 173, 175]. However, under excitation by ionizing radiation (X-ray, e-beam) and/or photoexcitation in the absorption region above 4.2 eV, the typical 5d–4f(Ce) luminescence is excited even in films where Ce^{3+} centers are absent [174].

These contradictory results indicate that Ce^{3+} content was negligible in studied LPE films grown from PbO. This might be result of large amount of Pb^{2+} ions in YSO:Ce films resulting in changing cerium valency into tetravalent state, Ce^{4+}, due to charge compensation. Small tetravalent Ce^{4+} ions also contribute to effective lattice volume compensation [176].

The absorption spectra of several YSO:Ce epitaxial films and reference Czochralski grown SC are displayed in Fig. 5.30 left. The maxima at 360, 300, and 265 nm of YSO:Ce crystal come from 4f–5d transitions in Ce^{3+}. In YSO:Ce films only low energy Ce^{3+} related absorption band at 360 nm is discernible in the spectra and below 350 nm strong broad band absorption is observed in all films. This broad band absorption originates from charge transfer transitions from valence oxygen 2p(O^{2-}) band to empty metal 4f(Ce^{4+}) states. The spectral peak located at 240–260 nm, which overlaps with charge transfer 2p–4f transitions, is observed in all LPE films and originates from Pb^{2+} centers ($^1S_0$ - $^3P_1$) [87, 128, 175].

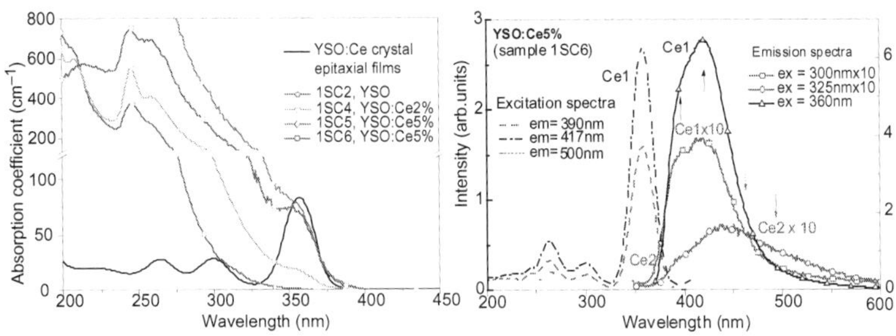

Figure 5.30 Left: Comparison of absorption coefficient spectra of YSO:Ce single crystal and epitaxial films grown from the PbO flux. Ce concentrations refer to the melt, note change of scale. Right: PL excitation and emission spectra of Ce1 and Ce2 centers in YSO:Ce epitaxial film, see also Ref. [9].

Nevertheless, in films grown using optimized LPE process the Ce^{3+} related luminescence is clearly observed, Fig. 5.30 right, in spite of significant amounts of Pb^{2+} and Ce^{4+} ions. Both the excitation and emission spectra of these films are in close correlation with those measured in YSO:Ce crystal [177–179] and the excitation spectra of films (Fig. 5.30 right) well correlate with the absorption spectrum of the reference SC (Fig. 5.30 left). Arrows in Fig. 5.30 right indicate Ce1 and Ce2 centers: emission doublet at 397 and 427 nm is ascribed to Ce1 center with 7-fold coordination and the wavelength tail (doublet at 460–500 nm) to Ce2 center with 6-fold coordination [180].

As mentioned above the luminescence can be excited by high energy excitation also in YSO:Ce films where content of trivalent Ce^{3+} is substantially reduced. This is due to the recombination of electrons coming from the conduction band with the Ce^{4+} centers and the two-photon excitation process, as proposed in Ref. [174]. The RL and CL spectra of LPE films and reference SC are shown in Fig. 5.31. The long-wavelength wings of the emission are notably red-shifted in comparison with the YSO:Ce single crystal. This is most likely caused by relative higher content of Ce^{3+} ions in Ce2 positions in films than in SC [22, 102]. The steady state emission efficiency is fairly high in LPE films grown under optimized conditions regardless of reduced number of Ce^{3+} centers—integrated RL intensity is ~10× higher as compared to reference BGO crystal, see Fig. 5.31a. However, reported LY measured by alpha particle excitation was only about 25–40% value of LYSO:Ce SC [22], in optimized films reaches up to 70% [9]. Lower LY of LPE films is a consequence of quenching effect of Pb^{2+} ions on Ce^{3+} emission. This is demonstrated in Fig. 5.32 where comparison of PL decays of Ce1 centers is compared for LSO:Ce LPE film with sizable Pb impurity content and a SC counterpart. Significant nonlinearity of the initial part of the decay curve of LPE film indicates considerable energy transfer away from the Ce1 emission centers.

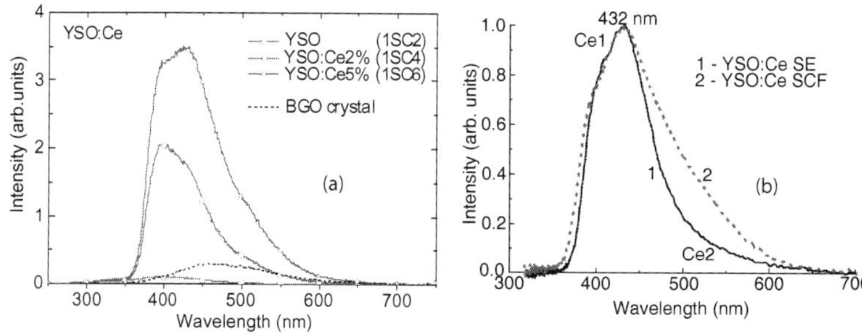

Figure 5.31 (a) X-ray excited RL spectra of YSO:Ce films with various Ce content [9]. BGO reference crystal is shown for comparison. (b) Normalized CL spectra of YSO:Ce film and single crystal SC. Reprinted from Ref. 22, Copyright 2013, with permission from Elsevier.

Important advantage of orthosilicates as compared to garnets is absence of shallow electron traps and consequential absence of slow components in the scintillation decay. Indeed, in pulsed height spectra the LY of YSO:Ce films is saturated within 0.5 μs [9].

Afterglow is substantially reduced (in ~2 orders of magnitude) in mixed LYSO:Ce and LGSO:Ce [181] systems usually co-doped by divalent Ca^{2+} or Mg^{2+} ions. Such co-doping in concentrations of tens of ppm has only marginal effect on the LY and decay kinetics.

The LY of LGSO:Ce films grown from PbO flux becomes significantly lower at increasing Gd in comparison with that of LSO:Ce films presumably due to the intensified quenching influence of Pb^{2+} impurity ions and increasing occupancy of Ce2 sites in the LGSO host [102].

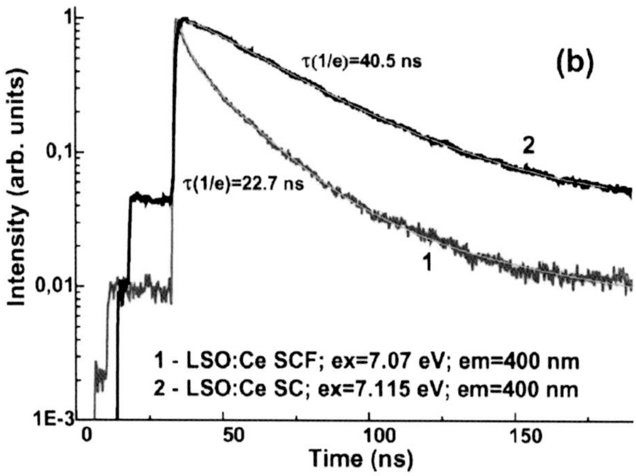

Figure 5.32 Comparison of LSO:Ce LPE film and single crystal (SC) decays (Ce1 centers) under excitation at the edge, 7.07–7.15 eV, multiexponential approximation of corresponding decay curves: 3.7+19+107 ns (LSO:Ce SCF) and 31.6+225 ns (LSO:Ce SC). Reprinted from Ref. 22, Copyright 2013, with permission from Elsevier.

5.4.3.2 Other dopants

Intense green emission is observed in Tb^{3+} doped YSO and LSO films grown from PbO flux [2, 25], Fig. 5.33. The lead Pb^{2+} impurity ions coming from the flux do not quench noticeably the 4f–4f (Tb^{3+}) emission (in contrast to Ce^{3+} emission in LSO/YSO:Ce mentioned above) and the emission is stable well above the RT, Fig. 5.33 left.

LSO:Tb films were used in high resolution X-ray microimaging detection screens in low and high dose beam lines of ESRF synchrotron [25, 100, 101, 182]. The thickness of films was between 2 and 30 μm and optimum efficiency was obtained for films with Tb concentration of 8–15% in the melt. In order to eliminate parasitic emission from standard LSO substrates, the best performance films were grown onto non-emitting YbSO substrates with suppressed afterglow [2, 183].

A bi-layer scintillator screen has been produced by the LPE process—Tb:LSO + LYSO:Ce double layer, 7 μm thick—with improved performance, ~1μm spatial resolution [25]. Double layer system enables combination most suitable activator ions for proposed application and prevents mutual quenching of dopants.

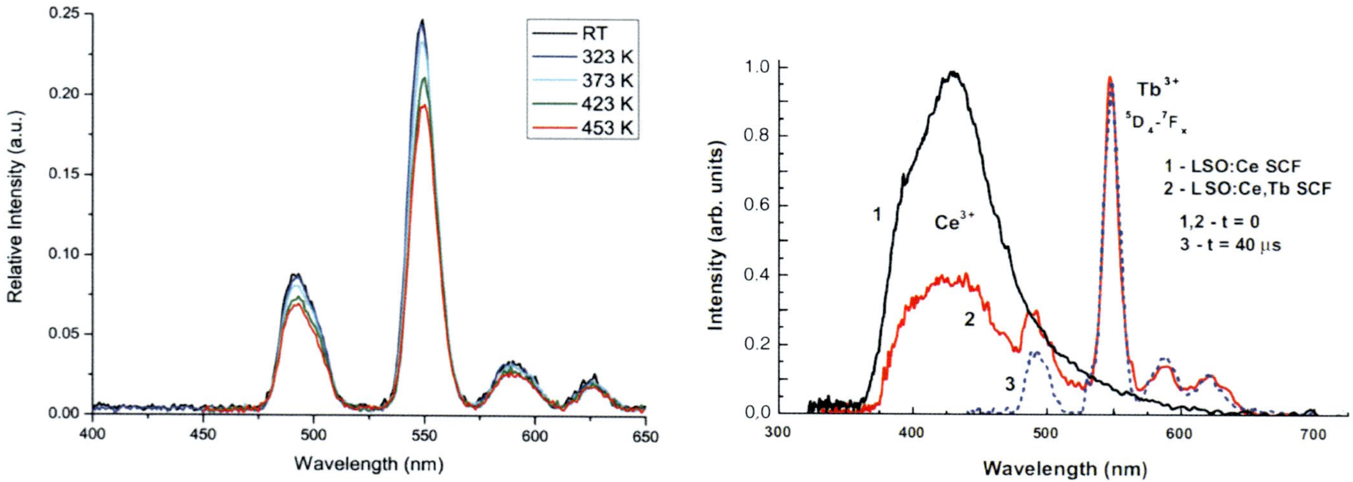

Figure 5.33 Left: X-ray excited RL spectrum of LSO:Tb at various temperatures and Tb content of 8%. Reproduced from Ref. 2, with permission of the International Union of Crystallography. Right: Time resolved normalized CL spectra of LSO:Ce (1) and LSO:Ce,Tb (2,3) films at delays (t = 0 (1,2) and 40 μs (3). Reprinted from Ref. 17, Copyright 2013, with permission from Elsevier.

Co-doped LSO:CeTb films were grown with the aim to increase the LY and/or for application in detection of complex events with different timing characteristics (ns × ms), Fig. 5.33b. Effective Ce → Tb energy transfer was observed, however, the LY under alpha excitation was only a fraction of LYSO:Ce or YAG:Ce reference single crystals [17, 99].

5.4.3.3 Concluding remarks

The results reported so far for the LPE grown orthosilicate demonstrate detrimental influence of Pb^{2+} contamination and lead-induced centers on the scintillation characteristics (LY and decay kinetics) of Ce^{3+} doped orthosilicates. Development of a lead-free flux for the growth of orthosilicate epitaxial films will be crucial for production fast, high performance Ce or Pr doped LYSO scintillators.

5.5 Applications of the LPE Films

5.5.1 Electron Detection in SEM

Epitaxial films of Ce-doped multicomponent garnet scintillator are especially perspective in the SEM detection units due to their high LY and low afterglow. Furthermore, the penetration depth of energetic electrons is typically only several micrometers and all the energy is deposited in the film. The practical impact of the scintillation properties on the quality of a SEM image is demonstrated in Fig. 5.34. When the YAG:Ce scintillator in the detector was replaced by epitaxial film of the multicomponent garnet GAGG:Ce scintillator with x_{Ga} = 2.7, the SEM image quality was obviously improved. This is consequence of higher electron-photon conversion efficiency of the GAGG:Ce scintillator and significantly reduced afterglow in single crystalline films as compared to the YAG:Ce single crystal. Such scintillator is suitable also for backscattered electron detectors where the high light yield of the scintillator is crucial.

5.5.2 X-ray Microimaging Screens

The efficiency of high-resolution pixel detectors for (hard) X-rays is important factor for X-ray imaging applications [2, 24–26, 100, 101]. The detector able to provide images with submicrometer spatial resolution used at ESRF synchrotron facility consists of a scintillator screen (epitaxial film with thickness typically 5–20 μm), optical microscope and a digital camera, Fig. 5.35. The scintillator converts the X-rays into a visible light image which is detected by the camera. Due to the low absorption of X-rays in the scintillator, Fig. 5.10, films of high density and high effective atomic number are preferred. Image in Fig. 5.35 was obtained by means of 10 μm thick LSO:Tb scintillator film grown onto YbSO substrate. Improved spatial resolution < 1 μm was demonstrated using a double layer LSO:Tb/LYSO:Ce scintillator grown by LPE [25].

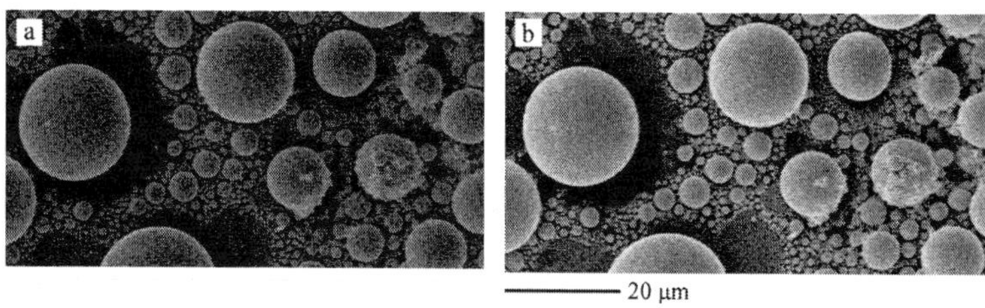

Figure 5.34 A SEM image of tin balls observed using the secondary electron detector with (a) commercial Czochralski grown YAG:Ce single crystal, and (b) GAGG:Ce multicomponent garnet epitaxial film grown from BaO flux. Images were recorded under the same conditions with e-beam energy of 10 keV and current of 57 pA. High scan speed with the dwell time (time per pixel) of 100 ns was used and the images were obtained by averaging of 3 consecutive scans. Reprinted from Ref. 10, Copyright 2016, with permission from Elsevier.

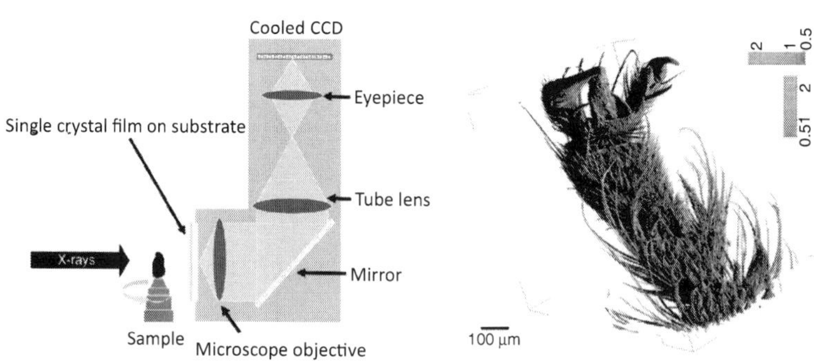

Figure 5.35 X-ray microtomography Left: Principle of an indirect high-resolution X-ray imaging system used in hard X-ray imaging in ESRF. Right: Rendering of a volume image acquired by synchrotron microtomography of a section of a honey bee tarsus. Reproduced from Ref. 2, with permission of the International Union of Crystallography.

5.5.3 Other Applications of Epitaxial Films: Waveguide Planar Lasers, Warm LEDs

Planar waveguides and channel planar lasers epitaxial films were demonstrated in number of rare earth doped garnets: Nd:YAG [105, 184], Tm:YAG [39, 105], Dy:YAG [185], Yb:YAG [186] Pr:YAG [187–189] and YbAG:Er [16, 190]. These waveguide garnet systems were grown from PbO flux, which provides the best surface morphology necessary for intended applications. Minor Pb contamination does not mean serious problem as in scintillators, since the devices are usually optically pumped in f states.

Phosphors for warm white LEDs are another prospective application. The emission wavelength can be properly red shifted by Gd substitution and required roughness of the surface is reached during the epitaxial growth, see Fig. 5.2b.

Acknowledgments

The authors are grateful to GACR, project No. P204/12/0805, for financial support of this work.

References

1. Nikl, M., Mares, J.A. et al. (2010), *IEEE Transactions on Nuclear Science,* **57**, pp. 1168–1174.
2. Douissard, P.A., Cecilia, A. et al. (2010), *Journal of Synchrotron Radiation,* **17**, pp. 571–583.
3. Zorenko, Y., Nikl, M. et al. (2009), *Physica Status Solidi a-Applications and Materials Science,* **206**, pp. 2586–2592.
4. Kucera, M., Nitsch, K. et al. (2010), *Journal of Crystal Growth,* **312**, pp. 1538–1545.
5. Kucera, M., Nitsch, K. et al. (2008), *IEEE Transactions on Nuclear Science,* **55**, pp. 1201–1205.
6. Nikl, M. (2005), *Physica Status Solidi a-Applied Research,* **202**, pp. 201–206.
7. Nikl, M., Vedda, A. et al. (2007), *Physical Review B,* **76**, pp. 195121.
8. Moszynski, M., Ludziejewski, T. et al. (1994), *Nuclear Instruments & Methods in Physics Research A,* **345**, pp. 461–467.

9. Kucera, M., Hanus, M. et al. (2015), unpublished results.
10. Bok, J., Lalinsky, O. et al. (2016), *Ultramicroscopy*, **163**, 1–5.
11. Prusa, P., Kucera, M. et al. (2015), *Crystal Growth & Design*, **15**, pp. 3715–3723.
12. Kučera, M., Nikl, M. et al. (2011), *Journal of Crystal Growth*, **318**, pp. 813–819.
13. Nikl, M., Mihokova, E. et al. (2005), *Physica Status Solidi B-Basic Solid State Physics*, **242**, pp. R119–R121.
14. Koch, A., Raven, C. et al. (1998), *Journal of the Optical Society of America a-Optics Image Science and Vision*, **15**, pp. 1940–1951.
15. Robertson, J.M., Vantol, M.W. et al. (1980), *Philips Journal of Research*, **35**, pp. 354–371.
16. Hlásek, T., Rubešová, K. et al. (2015), *Optical Materials*, **49**, pp. 46–50.
17. Zorenko, Y., Gorbenko, V. et al. (2013), *Radiation Measurements*, **56**, pp. 415–419.
18. Zorenko, Y., Gorbenko, V. et al. (2013), *Radiation Measurements*, **56**, pp. 159–162.
19. Martin, T. and Koch, A. (2006), *Journal of Synchrotron Radiation*, **13**, pp. 180–194.
20. Kucera, M., Nikl, M. et al. (2012), *IEEE Transactions on Nuclear Science*, **59**, pp. 2275–2280.
21. Kucera, M., Nitsch, K. et al. (2010), *Radiation Measurements*, **45**, pp. 449–452.
22. Zorenko, Y., Gorbenko, V. et al. (2013), *Radiation Measurements*, **56**, pp. 84–89.
23. Chewpraditkul, W., Swiderski, L. et al. (2009), *Physica Status Solidi a-Applications and Materials Science*, **206**, pp. 2599–2605.
24. Horodysky, P., Tous, J. et al. (2010), *Radiation Measurements*, **45**, pp. 628–630.
25. Martin, T., Douissard, P.A. et al. (2009), *IEEE Transactions on Nuclear Science*, **56**, pp. 1412–1418.
26. Tous, J., Blazek, K. et al. (2012), *Radiation Measurements*, **47**, pp. 311–314.
27. Nikl, M., Mares, J.A. et al. (2004), *Physica Status Solidi a-Applied Research*, **201**, pp. R41–R44.
28. Bruza, P., Fidler, V. et al. (2011), *Journal of Instrumentation*, **6**, pp. P09007–P09007.

29. Robertson, J.M. and Vantol, M.W. (1984), *Thin Solid Films,* **114**, pp. 221–240.
30. Zorenko, Y., Gorbenko, V. et al. (2005), *Journal of Luminescence,* **114**, pp. 85–94.
31. Winkler, G., *Magnetic garnets.* 1981, Braunschweig: Vieweg & Sons.
32. Robertson, J.M. (1978), *Journal of Crystal Growth,* **45**, pp. 233–242.
33. Blank, S.L. and Nielsen, J.W. (1972), *Journal of Crystal Growth,* **17**, pp. 302–311.
34. Ferrand, B., Chambaz, B. et al. (1999), *Optical Materials,* **11**, pp. 101–114.
35. Ferrand, B., Pelenc, D. et al. (1993), *Journal of Crystal Growth,* **128**, pp. 966–969.
36. Hanna, D.C., Large, A.C. et al. (1993), *Applied Physics Letters,* **63**, pp. 7–9.
37. Chartier, I., Ferrand, B. et al. (1992), *Optics Letters,* **17**, pp. 810–812.
38. Pelenc, D., Chambaz, B. et al. (1995), *Optics Communications,* **115**, pp. 491–497.
39. Rameix, A., Borel, C. et al. (1997), *Optics Communications,* **142**, pp. 239–243.
40. Sugimoto, N., Ohishi, Y. et al. (1995), *Applied Physics Letters,* **67**, pp. 582–584.
41. Prusa, P., Nikl, M. et al. (2009), *Physica Status Solidi a-Applications and Materials Science,* **206**, pp. 1494–1500.
42. Prusa, P., Cechak, T. et al. (2008), *Applied Physics Letters,* **92**.
43. Klemenz Rivenbark, C.F. (2010), Liquid phase epitaxy of advanced materials, in *Springer handbook of crystal growth*, Dhanarah, G. et al., Editors, Springer: Berlin. p. 1041–1068.
44. Capper, P. and Mauk, M. (2007), *Liquid Phase Epitaxy of Electronic, Optical and Optoelectronic Devices*, Chichester: J. Wiley.
45. Elwell, D. and Scheel, H.J. (2011), *Crystal Growth From High Temperature Solution,* 2 ed., London: Academic Press.
46. Tolksdorf, W. and Klages, C.-P. (1984), *Thin Solid Films,* **114**, pp. 33–43.
47. Giess, E.A. (1975), *Journal of Crystal Growth,* **31**, pp. 358–365.
48. Kucera, M., Nitsch, K. et al. (2003), *Physica Status Solidi a-Applied Research,* **198**, pp. 407–414.
49. Hiskes, R. (1974), *Journal of Crystal Growth,* **27**, pp. 287–298.

50. Suemune, Y. and Inoue, N. (1974), *Journal of Crystal Growth,* **24**, pp. 646–650.
51. Prusa, P., Mares, J.A. et al. (2010), *Optical Materials,* **32**, pp. 1360–1363.
52. Prusa, P., Kucera, M. et al. (2011), *Journal of Crystal Growth,* **318**, pp. 545–548.
53. Mares, J.A., Beitlerova, A. et al. (2007), *Radiation Measurements,* **42**, pp. 533–536.
54. Mares, J.A., Prusa, P. et al. (2010), *Radiation Measurements,* **45**, pp. 369–371.
55. Bonner, W.A., Lecraw, R.C. et al. (1978), *Journal of Applied Physics,* **49**, pp. 1871–1872.
56. Tamanoi, K., Nomoto, T. et al. (1991), *Japanese Journal of Applied Physics Part 1-Regular Papers Short Notes & Review Papers,* **30**, pp. 3516–3517.
57. Zorenko, Y., Gorbenko, V. et al. (2013), *Journal of Luminescence,* **141**, pp. 137–143.
58. Zorenko, Y., Gorbenko, V. et al. (2013), *Journal of Luminescence,* **134**, pp. 539–543.
59. Gorbenko, V., Krasnikov, A. et al. (2013), *Journal of Luminescence,* **134**, pp. 469–476.
60. Chani, V., Shimamura, K. et al. (1997), *Materials Science and Engineering,* **R20**, pp. 281–338.
61. Chani, V.I., Boulon, G. et al. (2010), *Japanese Journal of Applied Physics,* **49**, pp. 075601.
62. Chani, V. (2003), *Journal of Ceramic Processing Research,* **4**, pp. 67–70.
63. Kucera, M., Hanus, M. et al. (2014), *IEEE Transactions on Nuclear Science,* **61**, pp. 282–289.
64. Zorenko, Y., Gorbenko, V. et al. (2013), *Radiation Measurements,* **56**, pp. 150–154.
65. Robbins, D.J., Cockayne, B. et al. (1979), *Physical Review B,* **19**, pp. 1254–1269.
66. Petrosyan, A.G., Shirinyan, G.O. et al. (1999), *Journal of Crystal Growth,* **198**, pp. 492–496.
67. Leonyuk, N.I., Belokoneva, E.L. et al. (1999), *Journal of Crystal Growth,* **205**, pp. 361–367.
68. Cong, H.J., Zhang, H.J. et al. (2009), *Journal of Applied Crystallography,* **42**, pp. 284–294.
69. Gustafsson, T., Klintenberg, M. et al. (2001), *Acta Crystallographica Section C-Crystal Structure Communications,* **57**, pp. 668–669.

70. Zorenko, Y., Voloshinovskii, A. et al. (2007), *Physica Status Solidi (b)*, **244**, pp. 2180–2189.

71. Przybylińska, H., Wittlin, A. et al. (2014), *Optical Materials*, **36**, pp. 1515–1519.

72. Kvapil, J., Manek, B. et al. (1981), *Journal of Crystal Growth*, **52**, pp. 542–545.

73. Ryskin, N.N., Dorenbos, P. et al. (1994), *Journal of Physics-Condensed Matter*, **6**, pp. 10423–10434.

74. Koschan, M., Yang, K. et al. (2012), *Journal of Crystal Growth*, **352**, pp. 133–136.

75. Nikl, M., Mares, J.A. et al. (2002), *Nuclear Instruments & Methods in Physics Research A*, **486**, pp. 250–253.

76. Stanek, C.R., Levy, M.R. et al. (2005), *Physica Status Solidi (b)*, **242**, pp. R113–R115.

77. Nikl, M., Kamada, K. et al. (2014), *Crystal Growth & Design*, **14**, pp. 4827–4833.

78. Stanek, C.R., McClellan, K.J. et al. (2006), *Physica Status Solidi (b)*, **243**, pp. R75–R77.

79. Stanek, C.R., McClellan, K.J. et al. (2006), *Journal of Applied Physics*, **99**, pp. 113518.

80. Stanek, C.R., McClellan, K.J. et al. (2007), *Nuclear Instruments and Methods in Physics Research A*, **579**, pp. 27–30.

81. Stanek, C.R., Levy, M.R. et al. (2008), *Nuclear Instruments and Methods in Physics Research B*, **266**, pp. 2657–2664.

82. Nikl, M., Yoshikawa, A. et al. (2013), *Progress in Crystal Growth and Characterization of Materials* **59**, pp. 47–72.

83. Stanek, C.R., Jiang, C. et al. (2013), *Physica Status Solidi B-Basic Solid State Physics*, **250**, pp. 244–248.

84. Zorenko, Y., Voloshinovskii, A. et al. (2004), *Radiation Measurements*, **38**, pp. 677–680.

85. Ashurov, M.K., Rakov, A.F. et al. (2001), *Solid State Communications*, **120**, pp. 491–494.

86. Kuklja, M.M. (2000), *Journal of Physics-Condensed Matter*, **12**, pp. 2953–2967.

87. Babin, V., Gorbenko, V. et al. (2013), *Radiation Measurements*, **56**, pp. 124–128.

88. Babin, V., Bichevin, V. et al. (2009), *Physica Status Solidi B-Basic Solid State Physics*, **246**, pp. 1318–1326.

89. Schauer, P. and Bok, J. (2013), *Nuclear Instruments & Methods in Physics Research B,* **308**, pp. 68–73.
90. Strocka, B., Holst, P. et al. (1978), *Philips Journal of Research,* **33**, pp. 186–202.
91. Bok, J. and Schauer, P. (2014), *Measurement Science & Technology,* **25**, pp. 075601.
92. Zorenko, Y., Gorbenko, V. et al. (2002), *Nuclear Instruments & Methods in Physics Research B,* **486**, pp. 309–314.
93. Yoshikawa, A., Yanagida, T. et al. (2012), *Ieee Transactions on Nuclear Science,* **59**, pp. 2286–2289.
94. D'Ambrosio, C. and Leutz, H. (2003), *Nuclear Instruments & Methods in Physics Research A,* **501**, pp. 463–498.
95. D'Ambrosio, C., Leutz, H. et al. (2003), *Nuclear Instruments & Methods in Physics Research A,* **497**, pp. 186–197.
96. Mares, J.A., Nikl, M. et al. (2003), *Nuclear Instruments & Methods in Physics Research A,* **498**, pp. 312–327.
97. Zorenko, Y., Gorbenko, V. et al. (2010), *Radiation Measurements,* **45**, pp. 444–448.
98. Zorenko, Y.V. and Gorbenko, V.I. (2009), *Physics of the Solid State,* **51**, pp. 1800–1808.
99. Zorenko, Y., Gorbenko, V. et al. (2012), *IEEE Transactions on Nuclear Science,* **59**, pp. 2260–2268.
100. Cecilia, A., Rack, A. et al. (2011), *Nuclear Instruments and Methods in Physics Research A,* **648, Supplement 1**, pp. S321–S323.
101. Cecilia, A., Rack, A. et al. (2011), *Nuclear Instruments and Methods in Physics Research A,* **633, Supplement 1**, pp. S292–S293.
102. Zorenko, Y., Gorbenko, V. et al. (2011), *Journal of Crystal Growth,* **337**, pp. 72–80.
103. Ehrentraut, D., Sato, H. et al. (2006), *Journal of Crystal Growth,* **287**, pp. 367–371.
104. Zorenko, Y., Gorbenko, V. et al. (2012), *Ieee Transactions on Nuclear Science,* **59**, pp. 2281–2285.
105. Szachowicz, M., Moretti, P. et al. (2007), *Applied Physics Letters,* **90**.
106. Nikl, M., Vedda, A. et al., (2010), Single-crystal scintillation materials, in *Springer Handbook of Crystal Growth,* Dhanarah, G. et al., Editors., pp. 1663–1700, Springer: Berlin.
107. Nikl, M. and Yoshikawa, A. (2015), *Advanced Optical Materials,* **3**, pp. 463–481.

108. Nikl, M. (2000), *Physica Status Solidi a-Applied Research,* **178**, pp. 595–620.

109. Nikl, M. (2006), *Measurement Science & Technology,* **17**, pp. R37–R54.

110. Nikl, M., Ogino, H. et al. (2005), *Physica Status Solidi a-Applied Research,* **202**, pp. R4–R6.

111. Nikl, M., Ogino, H. et al. (2005), *Chemical Physics Letters,* **410**, pp. 218–221.

112. Ogino, H., Yoshikawa, A. et al. (2006), *Journal of Crystal Growth,* **287**, pp. 335–338.

113. Pejchal, J., Nikl, M. et al. (2009), *Journal of Physics D-Applied Physics,* **42**.

114. Autrata, R., Schauer, P. et al. (1978), *Journal of Physics E-Scientific Instruments,* **11**, pp. 707–708.

115. Robertson, J.M. and Vantol, M.W. (1980), *Applied Physics Letters,* **37**, pp. 471–472.

116. Robertson, J.M., Vantol, M.W. et al. (1981), *Philips Journal of Research,* **36**, pp. 15–30.

117. Babin, V., Gorbenko, V. et al. (2010), *Radiation Measurements,* **45**, pp. 415–418.

118. Zorenko, Y., Gorbenko, V. et al. (2008), *Physica Status Solidi (b),* **245**, pp. 1618–1622.

119. Babin, V., Gorbenko, V. et al. (2007), *Journal of Luminescence,* **127**, pp. 384–390.

120. Onderisinova, Z., Kucera, M. et al. (2015), *Journal of Luminescence,* **167**, pp. 106–113.

121. Kucera, M., Onderisinova, Z. et al. (2016), *Journal of Luminescence,* **169**, pp. 674–677.

122. Bachmann, V., Ronda, C. et al. (2009), *Chemistry of Materials,* **21**, pp. 2077–2084.

123. Kamada, K., Endo, T. et al. (2011), *Crystal Growth & Design,* **11**, pp. 4484–4490.

124. Kamada, K., Yanagida, T. et al. (2011), *Journal of Physics D: Applied Physics,* **44**, pp. 505104.

125. Zorenko, Y., Zorenko, T. et al. (2014), *Journal of Luminescence,* **156**, pp. 102–107.

126. Zorenko, Y., Gorbenko, V. et al. (2015), *Materials Research Bulletin,* **64**, pp. 355–363.

127. Przybylinska, H., Ma, C.-G. et al. (2013), *Applied Physics Letters,* **102**, pp. 241112.

128. Scott, G.B. and Page, J.L. (1977), *Journal of Applied Physics,* **48**, pp. 1342–1349.

129. Tachiya, M. and Mozumder, A. (1974), *Chemical Physics Letters,* **28**, pp. 87–89.

130. Huntley, D.J. (2006), *Journal of Physics: Condensed Matter,* **18**, pp. 1359–1365.

131. Cherepy, N.J., Kuntz, J.D. et al. (2010), *Proceedings of SPIE - The International Society for Optical Engineering*.

132. Dorenbos, P. (2013), *Journal of Luminescence,* **134**, pp. 310–318.

133. Fasoli, M., Vedda, A. et al. (2011), *Physical Review B,* **84**, pp. 081102(R).

134. Kamada, K., Yanagida, T. et al. (2012), *Journal of Crystal Growth,* **352**, pp. 88–90.

135. Prusa, P., Kamada, K. et al. (2013), *Radiation Measurements*, pp. 62–66.

136. Kamada, K., Kurosawa, S. et al. (2014), *Optical Materials,* **36**, pp. 1942–1945.

137. Prusa, P., Kucera, M. et al. (2013), *Optical Materials,* **35**, pp. 2444–2448.

138. Zorenko, Y., Gorbenko, V. et al. (2014), *Journal of Crystal Growth,* **401**, pp. 532–536.

139. Kucera, M., Nikl, M. et al. (2013), *Physica Status Solidi-Rapid Research Letters,* **7**, pp. 571–574.

140. van Eijk, C.W.E., Dorenbos, P. et al. (1994), *IEEE Transactions on Nuclear Science,* **41**, pp. 738–741.

141. Dujardin, C., Pedrini, C. et al. (1997), *Journal of Physics-Condensed Matter,* **9**, pp. 5229–5243.

142. Ogino, H., Yoshikawa, A. et al. (2006), *Journal of Crystal Growth,* **292**, pp. 239–242.

143. Swiderski, L., Moszynski, M. et al. (2009), *Ieee Transactions on Nuclear Science,* **56**, pp. 2499–2505.

144. Yanagida, T., Sato, M. et al. (2011), *Optical Materials,* **33**, pp. 413–418.

145. Kamada, K., Yanagida, T. et al. (2012), *Journal of Crystal Growth,* **352**, pp. 91–94.

146. Gorbenko, V., Zorenko, Y. et al. (2010), *Radiation Measurements,* **45**, pp. 461–464.

147. Zorenko, Y., Gorbenko, V. et al. (2010), *IEEE Transactions on Nuclear Science,* **57**, pp. 1335–1342.

148. Nikl, M., Pejchal, J. et al. (2006), *Applied Physics Letters,* **88**.

149. Dujardin, C., Pedrini, C. et al. (1998), *Journal of Physics-Condensed Matter,* **10**, pp. 3061–3073.
150. Zhuravleva, M., Novoselov, A. et al. (2008), *Ieee Transactions on Nuclear Science,* **55**, pp. 1476–1479.
151. Yanagida, T., Kamada, K. et al. (2010), *Nuclear Instruments and Methods in Physics Research A,* **623**, pp. 1020–1023.
152. Autrata, R., Schauer, P. et al. (1983), *Scanning,* **5**, pp. 91–96.
153. Mares, J.A., Beitlerova, A. et al. (2004), *Radiation Measurements,* **38**, pp. 353–357.
154. Mares, J.A., Nikl, M. et al. (2005), *Nuclear Instruments & Methods in Physics Research A,* **537**, pp. 271–275.
155. Chewpraditkul, W., Phunpueok, A. et al. (2013), *Physica Status Solidi (A) Applications and Materials Science,* **210**, pp. 1903–1908.
156. Wojtowicz, A.J., Glodo, J. et al. (1998), *Journal of Luminescence,* **79**, pp. 275–291.
157. Laguta, V.V., Nikl, M. et al. (2009), *Physical Review B,* **80**, pp. 045114.
158. Trummer, J., Auffray, E. et al. (2005), *Nuclear Instruments & Methods in Physics Research A,* **551**, pp. 339–351.
159. Zorenko, Y., Gorbenko, V. et al. (2007), *Radiation Measurements,* **42**, pp. 528–532.
160. Weber, M.J. (1973), *Journal of Applied Physics,* **44**, pp. 3205–3208.
161. Babin, V., Gorbenko, V. et al. (2011), *Journal of Physics D: Applied Physics,* **44**, pp. 315402.
162. Douissard, P.A., Martin, T. et al. (2014), *IEEE Transactions on Nuclear Science,* **61**, pp. 433–438
163. Felsche, J. (1973), The crystal chemistry of the rare earth silicates, in *Structure and Bonding,* p. 99–197, Springer.
164. Chiriu, D., Faedda, N. et al. (2007), *Physical Review B,* **76**.
165. Aitasalo, T., Hölsä, J. et al. (2005), *Optical Materials,* **27**, pp. 1511–1515.
166. Aitasalo, T., Hölsä, J. et al. (2004), *Optical Materials,* **26**, pp. 107–112.
167. Thibault, F., Pelenc, D. et al. (2008), *Optical Materials,* **30**, pp. 1289–1296.
168. Dorenbos, P., van Eijk, C.W.E. et al. (1994), *Journal of Physics: Condensed Matter,* **6**, pp. 4167.
169. Nikl, M., Laguta, V.V. et al. (2008), *Physica Status Solidi B-Basic Solid State Physics,* **245**, pp. 1701–1722.

170. Blahuta, S., Bessiere, A. et al. (2011), *Materials,* **4**, pp. 1224–1237.

171. Blahuta, S., Bessiere, A. et al. (2013), *IEEE Transactions on Nuclear Science,* **60**, pp. 3134–3141.

172. Zorenko, Y., Nikl, M. et al. (2011), *Optical Materials,* **33**, pp. 846–852.

173. Zorenko, Y., Gorbenko, V. et al. (2012), *Optical Materials,* **34**, pp. 1969–1974.

174. Kärner, T., Laguta, V.V. et al. (2014), *Journal of Physics D: Applied Physics,* **47**, pp. 065303.

175. Gorbenko, V., Krasnikov, A. et al. (2012), *Journal of Physics D-Applied Physics,* **45**, pp. 355304.

176. Rotman, S.R., Tuller, H.L. et al. (1992), *Journal of Applied Physics,* **71**, pp. 1209–1214.

177. Jary, V., Nikl, M. et al. (2012), *IEEE Transactions on Nuclear Science,* **59**, pp. 2079–2084.

178. Jary, V., Nikl, M. et al. (2011), *Optical Materials,* **34**, pp. 428–432.

179. Chewpraditkul, W., Wanarak, C. et al. (2013), *Optical Materials,* **35**, pp. 1679–1684.

180. Feng, H., Jary, V. et al. (2010), *Journal of Applied Physics,* **108**, pp. 033519.

181. Jarý, V., Mihóková, E. et al. (2014), *Journal of Physics D: Applied Physics,* **47**, pp. 365304.

182. Douissard, P.A., Cecilia, A. et al. (2012), *Journal of Instrumentation,* **7**, pp. P09016–P09016.

183. Cecilia, A., Jary, V. et al. (2014), *Radiation Measurements,* **62**, pp. 28–34.

184. Gerhardt, R., Kleine-Borger, J. et al. (1999), *Applied Physics Letters,* **75**, pp. 1210–1212.

185. Klimczak, M., Malinowski, M. et al. (2009), *Journal of Luminescence,* **129**, pp. 1869–1873.

186. Malinowski, M., Kaczkan, M. et al. (2001), *Journal of Luminescence,* **94**, pp. 29–33.

187. Piramidowicz, R., Lawniczuk, K. et al. (2008), *Journal of Luminescence,* **128**, pp. 708–711.

188. Nakielska, M., Kosko, J. et al. (2008), *Optical Materials,* **30**, pp. 759–762.

189. Malinowski, M., Nakielska, M. et al. (2007), *Spectroscopy Letters,* **40**, pp. 271–292.

190. Hlásek, T., Rubešová, K. et al. (2015), *Journal of Luminescence,* **164**, pp. 90–93.

Chapter 6

Luminescence of Pb- and Bi-Related Centers in Aluminum Garnet, Perovskite, and Orthosilicate Single-Crystalline Films

Svetlana Zazubovich,[a] Aleksei Krasnikov,[a] Yuriy Zorenko,[b,c] Vitali Gorbenko,[b,c] Vladimir Babin,[d] Eva Mihokova,[d] and Martin Nikl[d]

[a]*Institute of Physics, University of Tartu, 50411 Tartu, Estonia*
[b]*Institute of Physics, Kazimierz Wielki University in Bydgoszcz, 85090 Bydgoszcz, Poland*
[c]*Ivan Franko National University of Lviv, 79017 Lviv, Ukraine*
[d]*Institute of Physics, Czech Academy of Sciences, Cukrovarnicka 10, 16253 Prague, Czech Republic*
svea@ut.ee

6.1 Introduction

Single crystals (SCs) of lutetium–aluminum garnet $Lu_3Al_5O_{12}$ (LuAG), yttrium-aluminum garnet $Y_3Al_5O_{12}$ (YAG), lutetium–aluminum perovskite $LuAlO_3$ (LuAP), and yttrium-aluminum perovskite $YAlO_3$

Nanocomposite, Ceramic, and Thin Film Scintillators
Edited by Martin Nikl

ISBN 978-981-4745-22-2 (Hardcover), 978-981-4745-23-9 (eBook)
www.panstanford.com

(YAP), doped with Ce^{3+} and Pr^{3+} ions, have a high density and excellent mechanical and chemical stability, possess an intense and fast emission due to 5d–4f radiative transitions of Ce^{3+} and Pr^{3+} ions, and have been, thus, considered for fast scintillator applications (see, e.g., [1–11]). Some of these materials can be prepared also in the form of single-crystalline films (SCFs) of typical thickness of 1–50 μm and employed for imaging screens with high spatial resolution [12–14] with certain advantages with respect to classical powder phosphor-based screens [15] or the systems using thin cuts of their bulk SC analogues [16, 17].

Under excitation in the host absorption region, an intense slow intrinsic luminescence appears in the melt-grown SC of garnets, which arises from the excitons localized around the antisite $Lu_{Al}{}^{3+}$- or $Y_{Al}{}^{3+}$-related defects [16–22]. The presence of such defects in garnets and perovskites was predicted also by theoretical calculations [23–26]. The ability of such defects to create shallow electron traps was theoretically examined in YAP [27]. As the intrinsic emission bands are overlapped with the impurity absorption bands, the scintillation response of the Ce^{3+}- and Pr^{3+}-doped materials can be negatively affected due to the appearance of a considerable amount of slow decay components [16, 17, 24, 28]. In [29], the $Lu_{Al}{}^{3+}$ antisite defects in LuAG SC were identified with shallow electron traps, which are responsible for thermoluminescence glow curve peaks within 120–200 K and can thus be another reason for considerable amount of very slow decay components in the scintillation response of LuAG:Ce SC. The concentration of antisite defects is strongly suppressed in SCF grown by the liquid phase epitaxy (LPE) method, owing to that the Ce^{3+}- and Pr^{3+}-doped SCF show considerably better timing characteristics under excitation in the host absorption region with respect to their bulk analogues [10, 16, 17, 19, 28, 30–35].

Ce^{3+}-doped SC of rare earth oxyorthosilicates Lu_2SiO_5 (LSO), Y_2SiO_5 (YSO), Gd_2SiO_5 (GSO) were intensively studied as promising scintillation materials since the beginning of 1990 when the excellent scintillation characteristics of LSO:Ce were reported [36–38]. Indeed, LSO:Ce SC possesses high density (7.4 g/cm^3), short luminescence decay time (29–46 ns; ~40 ns under γ-ray excitation), high light yield (up to 32 000 photons/MeV), high stopping power (1.14 cm, R_{Moliere} = 2.07 cm), high effective atomic number (Z_{eff} = 66) necessary for efficient γ-ray detection, suitable emission wavelength

(~410 nm) compatible with photomultiplier detectors, relatively high radiation hardness, as well as good mechanical properties and chemical stability [5, 39–51]). Owing to that, this material was proposed for fast detection of γ- or X-rays in many applications such as medical imaging (positron emission tomography, X-ray mammography), nuclear physics, high-energy physics, and environmental monitoring (see, e.g., [37, 39, 44, 52, 53]).

LSO:Ce SCF prepared by the LPE method was found to have some advantages as compared with LSO:Ce SC, e.g., a larger (up to 1.45 times) light yield, smaller content of oxygen vacancies responsible for the afterglow, etc. [54–56]. Lower concentration of oxygen vacancies was reported also in LSO and YSO prepared by other low-temperature methods [49, 57]. However, at the preparation of SCF by the LPE method from the melt-solution based on the PbO-B_2O_3 flux, lead ions are introduced into the crystal lattice. Therefore, the study of luminescence characteristics of Pb-related centers and their influence on the scintillation characteristics of SCF is of great importance. Some aspects of absorption and luminescence of these centers in SCF of aluminum perovskites and garnets were reported in [13, 14, 58–61]. Luminescence characteristics of Pb-related centers in LSO and YSO were studied at 295 K in [54–56].

Luminescence of Bi^{3+}-related centers was studied in various materials (alkali halides, alkaline-earth oxides, sulphates, and phosphates, rare earth oxide-based materials, etc.) starting from 1960s (see, e.g., [62–82] and references therein). The Bi^{3+}-doped rare earth aluminate and gallate garnets and rare earth silicates, where a Bi^{3+} ion substitutes a trivalent rare earth ion, were considered as perspective materials for scintillators due to an intense and fast Bi^{3+}-related luminescence [78, 80, 81]. Besides, it was found that in SCF of $Y_3Al_5O_{12}$:Bi and $Lu_3Al_5O_{12}$:Bi, grown by the LPE method from the melt-solution based on a Bi_2O_3 oxide flux, a large and variable concentration of Bi^{3+} ions can be achieved [80, 81]. The interest in Bi^{3+}-doped compounds has recently re-appeared in connection with the preparation of phosphors for new generation of solid state light sources, so-called white LED [83–85]. Bi^{3+} ions are used to efficiently absorb light in near UV and transfer the energy to other emission centers in the green-red spectral range. Therefore, it is of interest to study the luminescence characteristics of various Bi^{3+}-doped materials in order to obtain information on the nature of luminescence centers, origin and structure of their relaxed excited states (RES), and on the processes taking place in the excited states.

This chapter reviews the detailed and systematic study of luminescence characteristics of Pb- and Bi-related centers in aluminum perovskites and garnets [86–92] and oxyorthosilicates [93–96] carried out by the time-resolved spectroscopy methods in the 4.2–400 K temperature range. The characteristics of the SCF with different Pb or Bi contents were compared with those of the corresponding SC and the substrates used at the SCF preparation. As a result of these studies, the spectral bands, arising from Pb- and Bi-related centers of different types, were identified and their RES structure and parameters defined. The processes of energy transfer from the host lattice to the impurity (Ce^{3+}, Pr^{3+}) ions and between various Pb-related centers and the impurity ions were studied, and the influence of various Pb-related centers on the luminescence characteristics of the impurity ions was clarified.

6.2 Sample Preparation and Characterization Methods

In the melt-grown bulk garnet, perovskite or silicate crystals mentioned above the strong evaporation of Bi_2O_3 and PbO oxides from the melt during the growth process at high (1750–2150°C) temperatures disables their doping when using classical Czochralski or Bridgman methods. However, these ions can be doped in SCFs of these compounds, grown by LPE from the Bi- and Pb- based or containing fluxes. The LPE method does enable to change the concentration of Pb^{2+}- and Bi^{3+}-doped ions in mentioned oxide compounds in wide concentration range of about 10^{-4} to 10^{-1} at.%, quite sufficient for both the basic research as well as application purposes [54–56, 80, 81].

The LuAG:Bi (series **a** and **b**) and YAG:Bi (series **a** and **b**) SCFs were grown onto (110) oriented YAG substrates from super-cooled melt-solution based on Bi_2O_3 oxide flux [81]. The concentration of crystal forming components with respect to the total content of melt-solution was 5.7 mol% (series **a**) and 6.8 mol% (series **b**). These values determine the different temperatures of saturation of respective melt-solutions T_S = 1023–1025°C and 1081–1086°C as well as the ranges of SCF growth temperature T_g = 967–993°C and 1018–1053°C, respectively (see Table 6.1).

The YSO:Bi (**a** and **b** series) and LSO:Bi (**c** series) SCFs were grown from super-cooled melt-solution based on Bi_2O_3 flux onto the undoped YSO and LSO substrates, respectively. For the growth of these SCF we used the equal-molar ($Y_2O_3(Lu_2O_3)/SiO_2$) ratio of the crystal-forming component in melt-solution. The concentration of crystal-forming components with respect to the total content of melt-solution was about 6.0 mol% (see Table 6.2).

Table 6.1 Growth conditions, luminescence, and scintillation properties of YAG:Bi (Y) and LuAG:Bi (L) SCFs prepared by LPE method from the Bi_2O_3 based flux grown onto YAG substrates

SCF Sa.	Bi content (at. %)	T_g (°C)	h (μm)	CL UV/ VIS (%)	K = ratio of UV/VIS CL intensity	N_{phels} (%)	FWHM (%)
Y a1	0.15	967	12	5.9/0.9	6.5	18.0	18.5
Y a2	0.13	993	19	9.6/1.2	8.0	20.6	13.7
Y b1	0.53	1018	29.5	2.2/1.1	2.0	12.0	13.9
Y b2	0.31	1034	22	6.5/1.7	3.8	15.9	18.1
Y b3	0.27	1051	30.5	7.5/1.8	4.2	20.5	14.9
L a1	0.18	987.5	12	6.6/2.1	3.0	15.1	20.0
L a2	0.05	991.5	21	11.3/1.3	8.7	8.4	27.4
L b1	0.12	1038	14.5	7.9 /1.5	5.2	11.2	19.4
L b2	0.10	1048	18.3	14.5/1.5	9.6	10.9	31.5
L b3	0.07	1053	8.2	25.5/1.2	21.1	7.0	31.3

Note: N_{phels} and FWHM were measured under 5.4857 MeV of ^{241}Am α-particles excitation and 3 μs shaping time.

T_g – temperature of growth; h – SCF thickness; CL intensity of ultraviolet (UV) and visible (VIS) bands with respect to CL intensity of YAP:Ce reference crystal; N_{phels} – relative photoelectron yield compared to that of YAP:Ce crystal in %; FWHM – energy resolution in %

The content of bismuth and lead ions in SCF of garnet, perovskite and orthosilicate compounds was determined by using a JEOL JXA-733 and DS-130 Akashi microanalyzers.

The absorption spectra of the SCF samples of Bi- and Pb-doped garnets, perovskites, and orthosilicates were measured using UV/

VIS/NIR absorption spectrophotometer Shimadzu UV-3101PC in the wavelength range 190–1200 nm at 300 K.

The steady-state emission and excitation spectra in the 80–300 K temperature range were measured using a setup, consisting of a deuterium DDS-400 lamp and two monochromators (SF-4 and SPM-1). The luminescence was detected by a photomultiplier (FEU-39 or FEU-79) with an amplifier and recorder. In the 4.2–300 K temperature range, the spectra were measured with the use of the computer-controlled setup consisting of the LOT-ORIEL xenon lamp (150 W), two monochromators (MDR-3 and ORIEL Corner Stone 1/8 m), and Hamamatsu 6240 photon counting system. The spectra were corrected for the spectral distribution of the excitation light, the transmission and dispersion of the monochromators, and the spectral sensitivity of the detectors.

Luminescence decay kinetics in the 10 μs–ms time range was measured at the latter setup but under excitation with a xenon flash lamp FX-1152 (EG&G) with the pulse duration of about 1 μs and maximum repetition frequency of 300 Hz. The detection of the signal and the registration of decay curves were performed with a photon counting head and a multichannel analyzer. Faster luminescence decays within the ns–μs range were measured by 5000M model spectrofluorometer Horiba Jobin Yvon equipped with the single grating excitation and emission monochromators. Excitation was performed by the nanosecond coaxial hydrogen-filled flashlamp 5000F and TBX-04 photomultiplier module was used for luminescence detection (IBH Scotland). The method of time-correlated single photon counting was employed. Deconvolution procedure (SpectraSolve software package) was applied to extract true decay times from the multiexponential approximation. These measurements were carried out in the 1.7–300 K temperature range.

The cathodoluminescence (CL) spectra were measured at 300 K with a set-up based on the SF-4A monochromator and photomultiplier (PMT) FEU-100 under the steady-state e-beam excitation (10 kV, 5 μA). Emission spectra were corrected for the spectral dependence of the detection part. The LY of CL was detected by the FDUK-2 photodiode in the spatial angle of 1 sr in comparison with standard YAG:Ce SCF sample with a LY of 250 μWt/sr.

Luminescence of Bi^{3+} and Pb^{2+} ions in some SCF was also investigated at 10–300 K range at the Superlumi station (HASYLAB at

DESY, Hamburg) under excitation by SR with an energy of 3.7–25 eV and pulse duration of 130 ns. The emission and excitation spectra were measured both in the integral regime and in the 1.2–6 ns and 150–200 ns time intervals (fast and slow components, respectively) in the limits of SR pulse regime with the repetition time of 200 ns. The decay kinetics of luminescence was measured in the time range of 0–200 ns at temperatures 10–300 K. These emission spectra were not corrected for the detection part spectral dependence.

Table 6.2 Growth conditions, luminescence, and scintillation properties of YSO:Bi (YS) and LSO:Bi (LS) SCFs grown onto YSO and LSO substrates, respectively

SCF samples	Bi content, (at. %)	h, μm	f_p, μm/min	T_g, (°C)	Sub-strate	K = Ratio of UV/VIS bands in CL intensity	N_{phels} (%)	FWHM, (%)
YS a1		1.0	0.06	985	YSO			
YS a2		23	1.15	964	YSO			
YS b6	0.11	10	0.15	975	YSO	11.5	12.2	22
YS b13	1.46	10	0.33	945	YSO	3.45	8.2	27
YS b14	1.87	24	0.8	948	YSO		1.8	49
LS c2	1.46	64	0.5	980	LSO	4.35		
LS c3	2.0	16	0.27	972	LSO			
LS c4	2.23	19	0.3	970	LSO	3.80		
LS c5	2.14	20	0.36	965	LSO			
YAG:Ce SC							40	

Note: N_{phels} and FWHM were measured under 5.4857 MeV of ^{241}Am a-particles excitation and 3 μs shaping time.
h – SCF thickness; f – growth rate; T_g – temperature of growth; K – ratio of cathodoluminescent (CL) intensity of the ultraviolet (UV) and visible (VIS) bands; N_{phels} – relative photoelectron light yield (LY) compared to that of YAP:Ce crystal in %; FWHM – energy resolution in %

N_{phels} photoelectron yield and energy resolution FWHM of Bi- and Pb-doped YAG, LuAG and YSO SCF were measured and compared with those of reference YAP:Ce bulk crystal scintillator (manufactured in CRYTUR Ltd., Turnov, Czech Republic) measured in the same conditions. Pulse height spectra were measured using different shaping times within 0.5–10 μs. We used ^{241}Am (energy 5.4857 MeV) radioisotope for α-particle excitation. Scintillation was detected by a HPMT (DEP PPO 475B type) which has maximum sensitivity in the range 200–400 nm (see [18] for details). N_{phels} and FWHM values of SCF samples as well as their other optical and luminescence characteristics were obtained at the same side of the substrate where the electron microprobe analysis was made.

6.3 Luminescence of Single Pb^{2+}-Based Centers in Aluminum Garnets and Perovskites

Emission spectrum of undoped perovskites and garnets consists of many relatively narrow bands located in UV and VIS spectral ranges. Detailed study of luminescence characteristics of many SCF with different lead contents carried out in [86–88] revealed the presence of considerable number of single Pb^{2+}-based centers only in LuAG SCF. Indeed, the comparison of the characteristics of LuAG SCF, LuAG SC, and LuAG:Pb ceramics allowed to conclude that the emission band peaking at 3.61 eV (FWHM = 0.47 eV) (Fig. 6.1a, curve 1) arises from single-Pb^{2+}-ion-based centers. Its dominant 4.75 eV excitation band (Fig. 6.2a, curve 1; see also Table 6.2) coincides with the absorption spectrum peak (curve 2) ascribed in [13, 14] to the $^1S_0 \rightarrow {}^3P_1$ transitions of a Pb^{2+} ion. Indeed, this emission is absent in the substrate. The intensity of only the 3.61 eV emission under 4.75 eV excitation increases proportionally with the increasing thickness of the SCF layer, i.e., with the total lead content in the SCF. Besides the 4.75 eV band, the narrow exciton-related band peaking at 7.12 eV and a weak 6.25 eV band were observed in the excitation spectrum of the 3.61 eV emission (Fig. 6.3a, curve 1). The positions of both the 3.61 eV emission band and the 4.75 eV excitation band and the value of the Stokes shift S=1.16 eV are close to those observed for the Pb^{2+} centres, e.g., in Al_2O_3 [58], $Y_3Ga_5O_{12}$ [97], and f.c.c. alkali halides [98, 99]. In this case, the absorption band, arising from the allowed $^1S_0 \rightarrow {}^1P_1$ transitions, should be located at ~1.5 eV higher energy. Therefore, the excitation band at 6.25 eV can be ascribed to these transitions. A relatively small intensity of this band can be due to the location of the $^1P_1$ energy level close to or inside the conduction band [97]. In LuAG:Ce SCF, the 3.61 eV emission band of Pb^{2+} centers is overlapped with the 3.6 eV absorption band of Ce^{3+} centres [7], therefore it looks like a "doublet" with the maxima at 3.75 eV and 3.40 eV (Fig. 6.1a, curve 2). The intensity of the 3.61 eV emission is constant up to ~ 100 K and then decreases (twice at ≈130 K) (the inset of Fig. 6.2, curve 1).

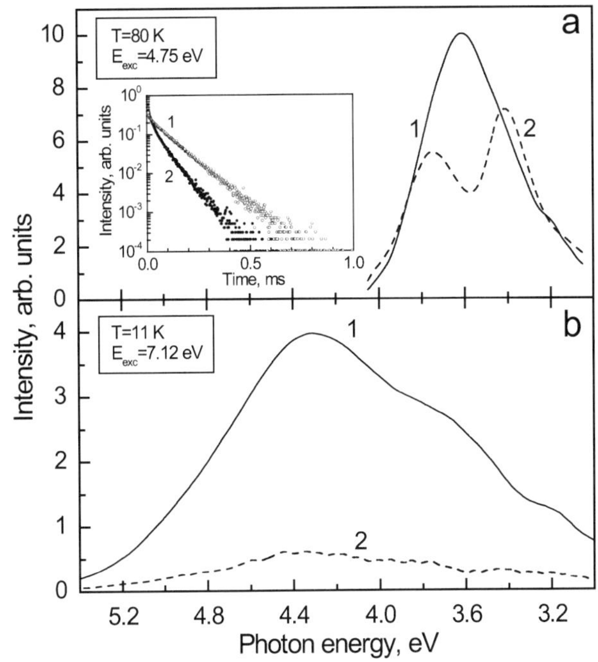

Figure 6.1 Emission spectra of LuAG SCF (curve 1) and LuAG:Ce SCF (curve 2) measured (a) at 80 K under E_{exc} = 4.75 eV and (b) at 11 K under E_{exc} = 7.12 eV. In the inset, normalized decay curves of the 3.61 eV emission measured at 4.2 K for LuAG SCF (curve 1) and LuAG:Ce SCF (curve 2); E_{exc} = 4.75 eV.

Decay kinetics of the 3.61 eV emission in LuAG SCF consists at 4.2 K of the fast submicrosecond component and the dominating slow exponential component with the decay time of 97 μs (the inset of Fig. 6.1, curve 1). This indicates to the existence of closely spaced radiative $^3P_1$ and metastable $^3P_0$ levels in the triplet RES of Pb^{2+}. Temperature dependence of the slow component decay time (Fig. 6.4a, curve 1) is similar to that characteristic for the triplet emission of Pb^{2+} centres in alkali halides (see, e.g., [98]) and allows to conclude that at $T < 20$ K, the radiative decay of the metastable minima of the triplet RES takes place. The shortening of the decay time at $T > 20$ K is caused by the thermally stimulated population of the emitting minimum and at $T > 100$ K, by the thermal quenching of the triplet emission. However, the slow component decay time at low temperatures is by about an order of magnitude shorter as compared with Pb^{2+}-doped alkali halides. This effect indicates a strong vibronic interaction in LuAG, which mixes the metastable $^3P_0$ and emitting $^3P_1$ levels of the triplet RES of Pb^{2+} centres and results in a larger probability of the radiative transitions from the metastable minima.

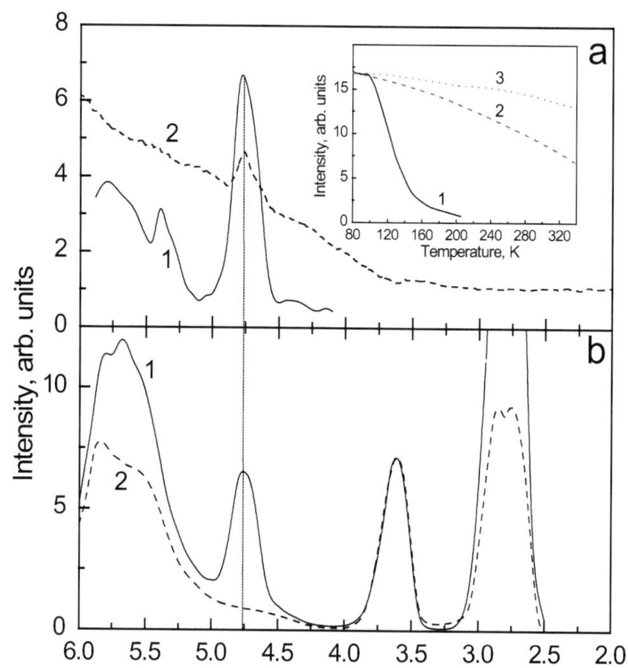

Figure 6.2 (a) Excitation spectrum of the 3.61 eV emission (curve 1) and absorption spectrum (curve 2) of LuAG SCF. (b) Normalized excitation spectra of the 2.46 eV emission of LuAG:Ce SCF (curve 1) and LuAG:Ce SC (curve 2). $T = 80$ K. In the inset, temperature dependence of the emission intensity measured under $E_{exc} = 4.75$ eV for the 3.61 eV emission of LuAG SCF (curve 1) and the 2.46 eV of LuAG:Ce SCF (curve 2), and under $E_{exc} = 2.8$ eV for the 2.46 eV emission of LuAG:Ce SCF (curve 3).

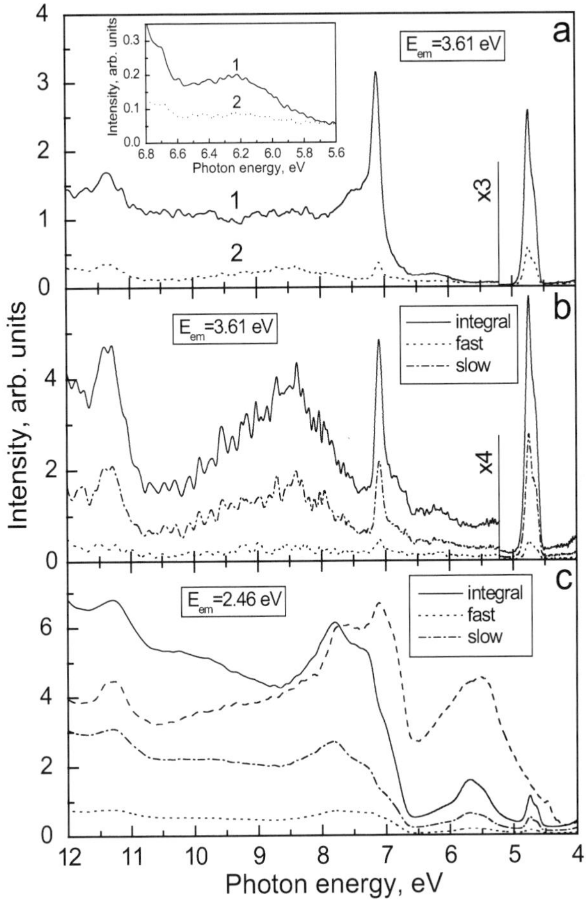

Figure 6.3 (a) Excitation spectra of the 3.61 eV emission measured at the same conditions for LuAG SCF (curve 1) and LuAG:Ce SCF (curve 2). In the inset, these spectra are shown in more detail. Time-resolved excitation spectra of LuAG:Ce SCF measured for (b) the 3.61 eV emission and (c) the 2.46 eV emission. For comparison, the integrated excitation spectrum of LuAG:Ce SC is shown (dashed line). T = 11 K.

According to [61], the UV emission band of Pb^{2+} centers in YAG SCF is located at ~3.7 eV and excited at 4.55 eV. However, in the 3.4–4.0 eV range, the emission spectrum of all the YAG SCF samples studied in [86–88] was found to consist of strongly overlapped weak emission bands [87]. No dependence of the emission intensity on the thickness of the SCF layer and its preparation temperature (consequently, on the lead content) was found. Similar emissions exist also in the substrates used, and their intensity is comparable with that in SCF or it is even larger than in SCF. For this emission range, no clear excitation band was observed in the 4.5–4.8 eV range (see, e.g., Fig. 6.5a). The existence of single Pb^{2+}-based centers appeared only in the presence of the 4.75 eV band in the excitation spectrum of the Ce^{3+} emission in YAG:Ce SCF (Fig. 6.5b, curve 1), where this band appears due to the $Pb^{2+} \rightarrow Ce^{3+}$ energy transfer. Indeed, in YAG:Ce SC, this band is absent (curve 2). The corresponding emission band is not evident due to its overlap with the Ce^{3+}-related 3.69 eV absorption band. Due to the $Pb^{2+} \rightarrow Ce^{3+}$ energy transfer, enabled by such an overlap, a slow decay component appears in the decay kinetics of the Ce^{3+} emission under excitation around 4.75 eV (compare curves 1 and 2 in the inset in Fig. 6.5a).

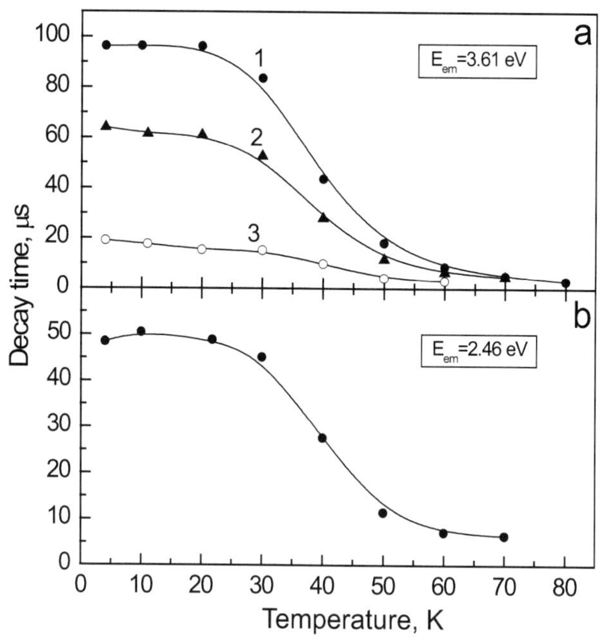

Figure 6.4 Temperature dependences of the slow component decay time of (a) the 3.61 eV emission of LuAG SCF (curve 1) and LuAG:Ce SCF (curves 2, 3) and (b) the 2.46 eV emission of LuAG:Ce SCF. E_{exc} = 4.75 eV.

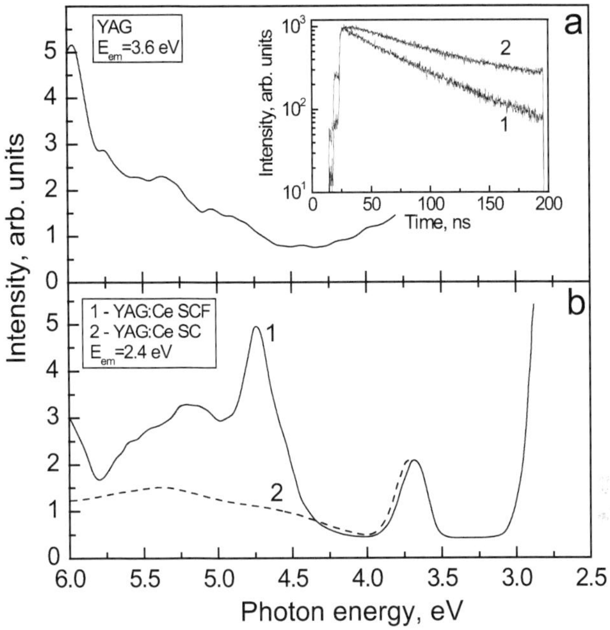

Figure 6.5 Excitation spectra of (a) the 3.6 eV emission of YAG SCF and (b) the 2.4 eV emission of YAG:Ce SCF (curve 1) and YAG:Ce SC (curve 2) with comparable Ce^{3+} contents (normalized at 3.69 eV); T = 80 K. In the inset, the decay curves of the 2.4 eV emission of YAG:Ce SCF under E_{exc} = 3.69 eV (curve 1) and E_{exc} = 4.75 eV (curve 2); T = 10 K.

According to [60], a weak 4.98 eV absorption band and the broad (FWHM ≈ 1 eV) emission bands, located at 3.65 eV and 2.14 eV, arise from Pb^{2+} ions in YAP SCF. In LuAP SCF, the 5.04 eV absorption band was ascribed to Pb^{2+} ions, and the emission bands of Pb^{2+} centers were concluded to be located at 3.45 eV, 3.18 eV, 2.39 eV, and 2.16 eV. However, in all the samples studied, no clear emission band was observed which could be ascribed to single Pb^{2+} centers [87]. No clear dependence of the emission intensity on the thickness of the SCF layer and its preparation conditions was found. No Pb-related band was found in this energy range in the excitation spectrum of the Ce^{3+} emission of YAP:Ce and LuAP:Ce SCF, where this band could appear due to the $Pb^{2+} \rightarrow Ce^{3+}$ energy transfer. Thus, the concentration of single Pb^{2+}-based centers in the YAP and LuAP SCF studied in [87] is negligible. The analysis of the data obtained for LuAP and YAP SCF in [60] allows one to suggest that only the 3.45–3.65 eV emission excited around 5.0 eV can arise from single Pb^{2+} centers (Table 6.3).

Table 6.3 Maxima positions (in eV) of the absorption (E_{abs}) and emission (E_{em}) bands of single and dimer Pb^{2+}-based centers [32, 33, 86, 87], Ce^{3+} centers [7, 32, 146], and Pr^{3+} centers [102, 147, 148]. T = 80 K.

	Single Pb^{2+}		Dimer Pb^{2+}		Ce^{3+}		Pr^{3+}	
SCF	E_{abs}	E_{em}	E_{abs}	E_{em}	E_{abs}	E_{em}	E_{abs}	E_{em}
LuAG	4.75	3.61	3.97	3.20	3.60	2.46	5.17	4.05
					2.8	2.22	4.35	3.81
YAG	4.75	3.70	3.95	3.22	3.69	2.25	5.21	3.94
					2.71	2.11	4.28	3.73
LuAP	5.04	3.45	3.85	3.15	4.52	3.38	~5.4	4.93
					4.25	3.30		4.35
					4.10			
YAP	4.98	3.65	3.87	3.15	4.52	3.51	~5.4	4.96
					4.25	3.27		4.43
					4.10			

6.4 Luminescence of Complex Pb-Related Centers in Aluminum Garnets and Perovskites

Under excitation around 3.85 eV, the 3.15 eV emission was observed in YAP and LuAP SCF (see the inset in Fig. 6.6a and [87]). In the decay kinetics of this emission, the component with the decay time of ~470 ns dominates at 80 K (Fig. 6.6). The luminescence with the mentioned characteristics was not observed in the corresponding single crystals; consequently, it can arise from lead-induced centers.

Similar (~3.2 eV) emission was observed under ~3.95 eV excitation also in aluminum garnets [87] (Fig. 6.7, curve 1). In LuAG SCF, this emission is relatively weak and overlaps the intense narrow 3.15 eV emission band excited at 3.4 eV (Fig. 6.7b) and ascribed in [100] to F^+ centers. This points to the existence of oxygen vacancies in the SCF under study. This conclusion was also confirmed by the TSL studies carried out in [87]. In the decay kinetics of the 3.2 eV emission, besides the dominating slow component with the decay time ~430–520 ns, a weak fast component (73–77 ns) was also detected (see, e.g., Fig. 6.8).

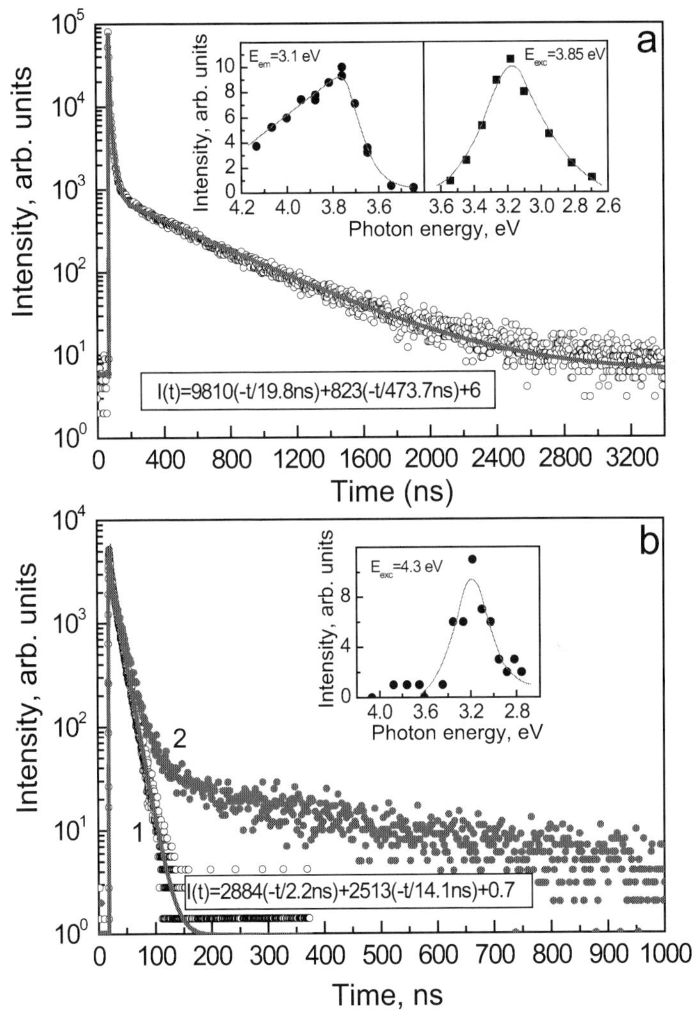

Figure 6.6 Decay curves measured at 77 K for (a) the 3.15 eV emission of LuAP SCF under E_{exc} = 3.85 eV and (b) the 3.4 eV (curve 1) and 3.15 eV (curve 2) emissions of LuAP SCF under E_{exc} = 4.3 eV. In the insets, the time-resolved excitation spectrum for E_{em} = 3.1 eV and emission spectrum under E_{exc} = 3.85 eV, both measured at t = 900 ns after the excitation pulse.

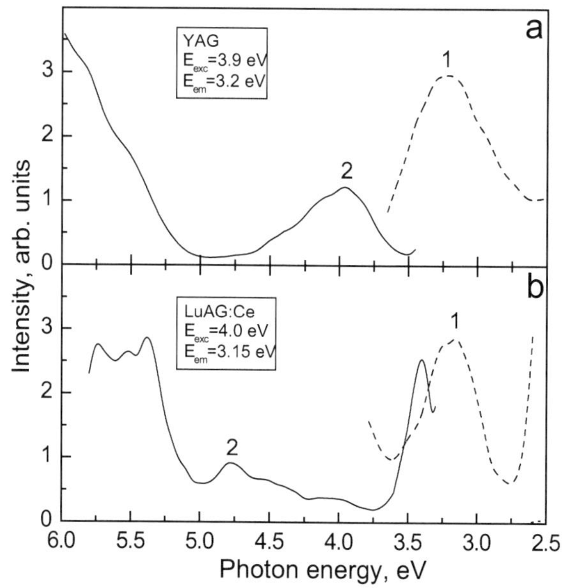

Figure 6.7 Emission spectra under 3.9–4.0 eV excitation (curve 1) and excitation spectra of the 3.15–3.20 eV emission (curve 2) measured for (a) YAG SCF and (b) LuAG:Ce SCF at 80 K.

Thus, the 3.15–3.20 eV emission with similar characteristics is observed in all the SCF studied (Table 6.3) as well as in LuAG:Pb ceramics [87] (Figs. 6.8a, 6.9). This emission was suggested to arise from dimer lead centers which was confirmed by the study of the emission intensity dependence on the lead content and quenching procedures carried out in [87] at the LuAG:Pb ceramics.

An intense visible emission observed in all the SCF studied was found to consist of strongly overlapped narrow (FWHM = 0.25–0.35 eV) components with the small Stokes shifts (S = 0.3–0.5 eV). Some of them can arise from aggregates of lead ions. Indeed, the increase of lead content in the LuAG:Pb ceramics was shown to result in the enhancement of the visible emission accompanied by the reduction of the 3.61 eV and 3.2 eV emissions (Fig. 6.9) [87]. Similar result was obtained at the study of the cathodoluminescence spectra of YAG SCF with strongly different lead contents [61]. No broad emission bands with the large Stokes shifts, which could be ascribed to the Pb-related localized excitons, were detected in the SCF studied.

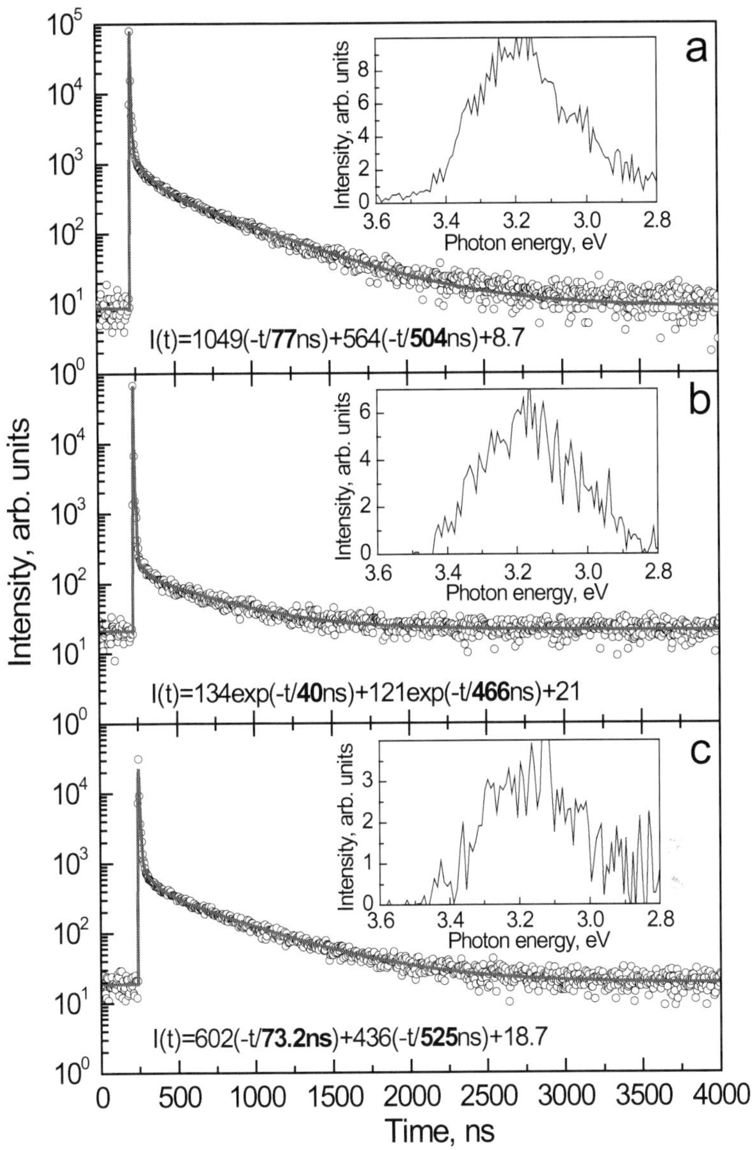

Figure 6.8 Decay curves measured at 80 K for the 3.15–3.20 eV emission of (a) the LuAG:1000 ppm Pb ceramics (b) LuAG:Ce SCF, and (c) YAG SCF. In the insets, the time-resolved emission spectra of the slow decay components are shown. E_{exc} = 3.9 eV.

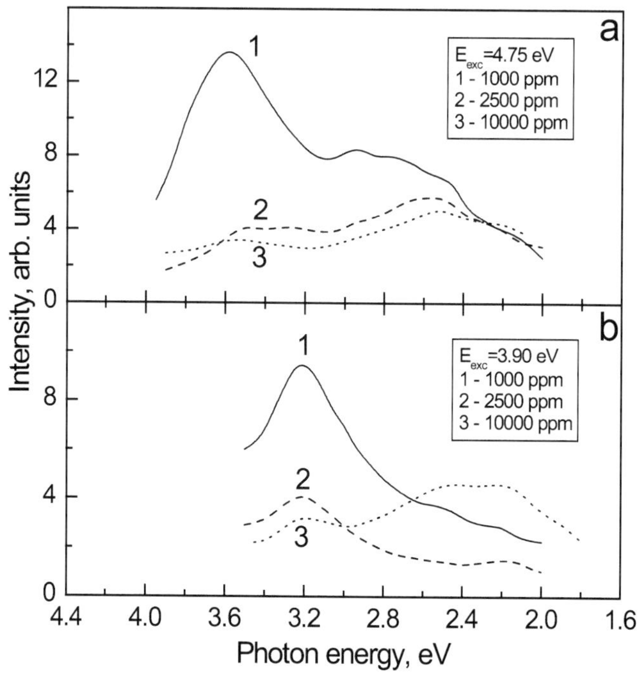

Figure 6.9 Emission sprectra of LuAG:Pb ceramics with different lead contents measured at the same conditions at 80 K under (a) E_{exc} = 4.75 eV and (b) E_{exc} = 3.9 eV.

6.5 Influence of Pb-Related Centers on Luminescence of Ce^{3+} and Pr^{3+} in Garnets and Perovskites

From the comparison of the positions of the absorption and emission bands of single and dimer lead centers with those of impurity Ce^{3+} and Pr^{3+} centers (Table 6.3), the following features were expected which can influence the luminescence characteristics of the Ce^{3+}- and Pr^{3+}-doped SCF: (i) the energy transfer from the excited Pb-related centers to impurity centers; (ii) the energy transfer from the excited impurity ions to the Pb-related centers; (iii) the reabsorption of the localized exciton emission in the absorption bands of impurity centers; (iv) the overlap of the emission bands of the Pb-related and impurity centers.

6.5.1 Energy Transfer from the Excited Pb-Related Centers to Impurity Ions

In LuAG:Ce and YAG:Ce SCF, the emission band of single Pb^{2+} centers is strongly overlapped with the 3.6–3.7 eV absorption band of Ce^{3+} centers (Table 6.3). Owing to that, an effective energy transfer from single Pb^{2+}-based centers to Ce^{3+} centers takes place [86, 88]. The appearance of the intense 4.75 eV band in the excitation spectrum of the Ce^{3+} emission in SCF (unlike SC) (Figs. 6.2b, 6.5b) supports the presence of such an energy transfer. Under 4.75 eV excitation, the decrease of the Ce^{3+} emission intensity with temperature in LuAG:Ce SCF (the inset of Fig. 6.2, curve 2) is more pronounced with respect to that observed under excitation in the Ce^{3+} bands (curve 3), and the slow component (~50 μs at 4.2 K) appears in the decay kinetics of the Ce^{3+} emission (Fig. 6.4b) besides the fast (52 ns) component characteristic for Ce^{3+} in LuAG SC [4]. In LuAG:Ce SCF, the slow component decay of the 3.61 eV emission accelerates and becomes distinctly non-exponential (see the inset in Fig. 6.1) in which the two-exponential fit was applied to get a quantitative description (Fig. 6.4a, curves 2 and 3). From these data, the multipole interaction mechanism [101] can be proposed for the $Pb^{2+} \rightarrow Ce^{3+}$ energy transfer.

No energy transfer from dimer lead centers to Ce^{3+} ions was detected in garnets [87, 89]. No energy transfer from both the single and dimer Pb^{2+} centers to Pr^{3+} ions was found in the SCF of Pr^{3+}-doped aluminum perovskites and garnets [89, 102].

6.5.2 Energy Transfer from Excited Ce^{3+} or Pr^{3+} Ions to Lead-Induced Centers

As the emission spectrum of Ce^{3+} centers in perovskites overlaps with the absorption band of dimer lead centers (Table 6.3), an energy transfer from Ce^{3+} to the dimers takes place. It results in a faster (~14 ns) decay of the Ce^{3+} emission in SCF (Fig. 6.6b, curve 1) as compared with SC (16–18 ns [103–105]), and in the appearance of the slow dimer-related emission (the inset) under excitation in the Ce^{3+}-related absorption region (around 4.3 eV where the emission of dimer centers is not directly excited) (curve 2). In the garnets, the influence of dimer lead centers on the characteristics of Ce^{3+} centers is smaller as the spectral bands of these centers are much less overlapped. However, in both the perovskite and garnet hosts, a dimer center will lower the scintillation efficiency because of its low intrinsic quantum efficiency (less than 10% at RT [87]) and enhanced ability to trap the migrating charge carriers created under band-to-band excitation and to become the recombination center (evidenced by CL spectra).

The energy transfer from Pr^{3+} centers to dimer lead centers, caused by the overlap of the Pr^{3+} emission band with the absorption band of dimers, was observed in LuAG:Pr and YAG:Pr SCF [102]. It resulted in the shortening of the Pr^{3+} emission decay time (from 18–19 ns in SC to 13–14 ns in SCF) and reduction of quantum efficiency of the Pr^{3+} emission. In perovskites, the emission bands of Pr^{3+} overlap the absorption band of single Pb^{2+} centers located, according to [32], around 5.0 eV (Table 6.3). Therefore, the energy transfer from Pr^{3+} to single Pb^{2+} centers is possible. However, this effect was not investigated, as the number of single Pb^{2+} centers in the perovskite SCF studied was negligible.

6.5.3 Reabsorption of the Localized Exciton Emission in the Absorption Bands of Ce^{3+} and Pr^{3+} Centers

Under excitation in the exciton absorption range, an intense broad (FWHM ≈ 0.9–1.0 eV) 4.2–4.3 eV emission band is observed in the SCF of perovskites and garnets [20, 33, 86] (see, e.g., Fig. 6.1b, curve 1). In the corresponding SC, this emission is absent. Taking into account the data of [20, 106], one can conclude that the 4.2–4.3 eV emission cannot arise neither from Pb^{2+} nor from Pt^{4+} ions. Most probably, it can be ascribed to an exciton localized around a Pt^{3+} ion incorporated into the SCF from the platinum crucible. The scintillation characteristics of the Ce^{3+}-doped perovskites and Pr^{3+}-doped garnets can be considerably influenced by a strong overlap of the 4f–5d absorption bands of Ce^{3+} and Pr^{3+} ions with the 4.2–4.3 eV emission band. For example, a strong suppression of this emission in LuAG:Ce SCF can be due to the reabsorption of the 4.2–4.3 eV emission by Ce^{3+} ions (curve 2). As the decay of this emission is slow [20, 86], this leads to the appearance of undesirable slow components in the scintillation decay of these materials. Because of that, the intensity ratios of the slow and fast emission components of the Ce^{3+}-related emission in LuAG:Ce SCF and LuAG:Ce SC (measured under excitation in the 7.1–7.8 eV energy range for the same time gates at 11 K) are close, despite the absence in SCF of antisite defects responsible for the slow decay of the Ce^{3+} emission in SC.

Strongly different excitation spectra observed in the host absorption region for the Pb^{2+}- and Ce^{3+}-related emissions in LuAG:Ce SCF (Figs. 6.3b and 6.3c) indicate to the competition between Pb^{2+} and Ce^{3+} ions in the energy transfer from the LuAG SCF lattice. Under the band-to-band excitation, the efficiency of the energy transfer from the host lattice to Pb^{2+} ions is relatively much smaller than to Ce^{3+} ions. However, under excitation in the exciton region, the energy transfer to Pb^{2+} ions is effective. A strong difference in the excitation spectra of the Ce^{3+} emission in the exciton region (around 7.12 eV) in LuAG:Ce SCF (containing Pb^{2+} centres) and LuAG:Ce SC (where Pb^{2+} centres are absent) (Fig. 6.3c) also indicates to the competition between Pb^{2+} and Ce^{3+} ions in the process of energy transfer from the exciton-related states. In LuAG:Ce SCF (Fig. 6.3b), the energy transfer from the exciton state to Pb^{2+} is strongly suppressed as compared with that in the undoped LuAG SCF (Fig. 6.3a). Under the band-to-band excitation, the competition between Pb^{2+} and Ce^{3+} ions is smaller.

6.5.4 Overlap of the Emission Bands of Lead-Induced and Impurity Centers

In Ce^{3+}-doped perovskites and Pr^{3+}-doped garnets, the emission bands of Pb-related and impurity centers are strongly overlapped (Table 6.3). As a result, in the spectral range characteristic for the Ce^{3+} and Pr^{3+} centers, besides the fast Ce^{3+}- and Pr^{3+}-related decay components, additional slow lead-related components appear in the luminescence decay.

6.6 Luminescence of Lead-Related Centers in Lutetium and Yttrium Oxyorthosilicates

Luminescence spectra of YSO and LSO SCF are similar and consist of a weak ultraviolet (UV) band and more intense complex blue (B) band [93, 94] (Fig. 6.10). Absorption spectra (Fig. 6.11a) indicate a huge concentration of Pb-related centers in the SCF studied. A close location of the absorption spectrum and the excitation spectrum of the B emission (Fig. 6.11b, dashed line) means that the centers responsible for the B emission strongly prevail in the SCF studied. Excitation spectra of the UV and B emissions are different.

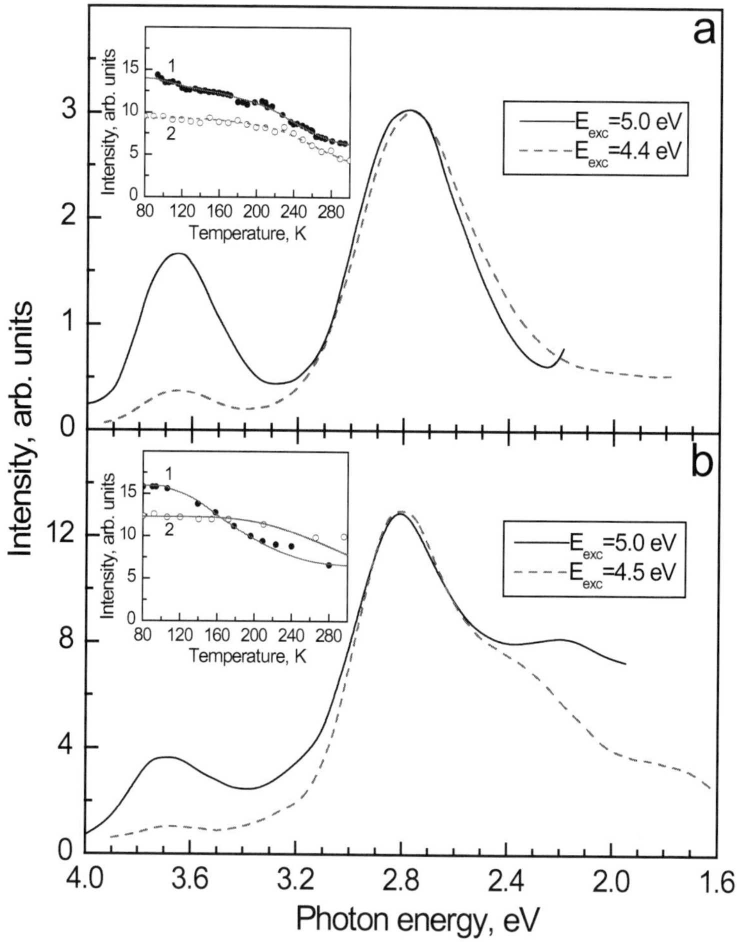

Figure 6.10 Emission spectra of Pb-related centers in (a) LSO SCF and (b) YSO SCF measured at 4.2 K. In the insets, temperature dependences of the intensities of the UV (curve 1) and B (curve 2) emissions.

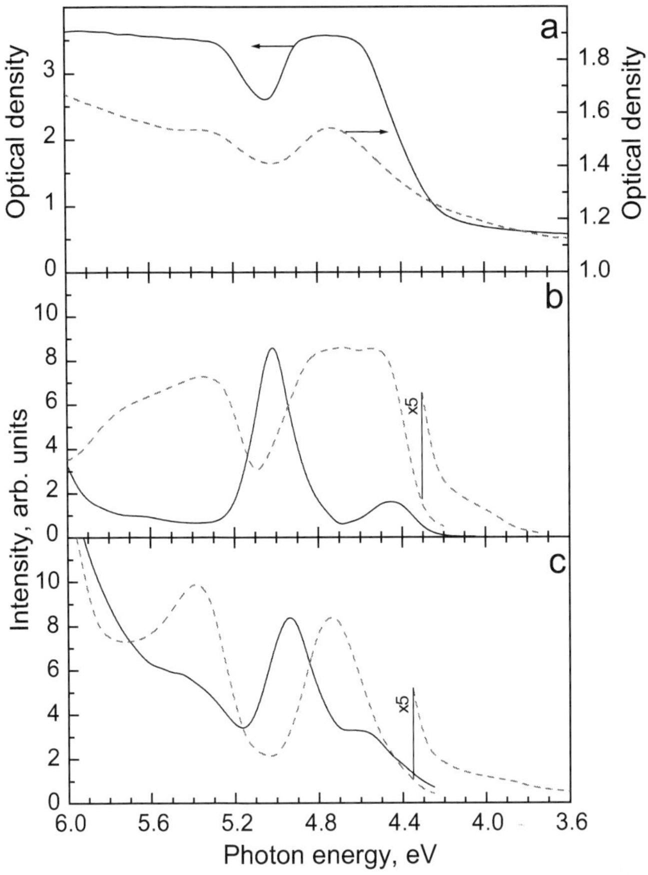

Figure 6.11 (a) Absorption spectra of LSO SCF with the thickness h = 10 μm (solid line) and h = 5.5 μm (dashed line) measured at 295 K. (b–c) Excitation spectra (normalized) of the UV (solid line) and B (dashed line) emissions of LSO SCF with h = 10 μm (b) and h = 5.5 μm (c) measured at 80 K.

6.6.1 Characteristics of the Ultraviolet Luminescence

The excitation spectrum of the UV emission (Fig. 6.11b, solid line) consists of the bands, located at E_{exc} > 6 eV, ~5.7 eV, 4.95–5.00 eV, and 4.45 eV. This spectrum is distorted due to a strong absorption (around 4.7 eV) of the centers responsible for the B emission. This is evident from the dependence of the positions of two components of the lowest-energy excitation band of the UV emission on the optical density (OD). Indeed, these components are located at 5.02 eV and 4.45 eV in the LSO SCF with larger OD (Fig. 6.11b) and at 4.92 eV and 4.55 eV, in the LSO SCF with smaller OD (Fig. 6.11c). Most probably, the above-mentioned components are two parts of the same excitation band located around 4.70–4.75 eV. In this case, the Stokes shift is ~1.1 eV (Table 6.4).

In the decay kinetics of the UV emission at 4.2 K, one slow exponential component is observed with the decay time of 109 μs in YSO and 126 μs in LSO. The decay time is constant up to 25 K and then decreases down to ~200 ns at RT (Figs. 6.12a,c). No fast (ns) component is observed.

The emission and excitation spectra of the UV emission (Table 6.4) are similar to those observed for the triplet emission of single Pb^{2+} centers in alkali halides (see, e.g., reviews [98, 99] and references therein) and LuAG SCF [86]. The lowest-energy (A) excitation band of this emission located around 4.70–4.75 eV arises from the transitions from the ground $^1S_0$ state to the triplet $^3P_1$ state of a Pb^{2+} ion. From the comparison of the electron transitions energies in a free Pb^{2+} ion and the suggested position of the A band, we expect that the C band, arising from the $^1S_0 \rightarrow {}^1P_1$ transitions, should be located at about 6.2–6.3 eV. The excitation band at about 5.7 eV should arise from the $^1S_0 \rightarrow {}^3P_2$ transitions (the B band).

The decay kinetics of the UV emission and temperature dependence of its decay time are also similar to those observed for Pb^{2+} centers in other hosts (see, e.g., [86, 98, 99]), suggesting similar structures of their triplet RES. At $T < 20$ K, where the slow component decay time is independent of temperature, the emission is caused by the radiative decay from the metastable level of the triplet RES of a Pb^{2+} center. As the UV emission intensity is practically constant up to 220 K (see the insets of Fig. 6.10), the decrease of its decay time at $T > 20$ K can be caused only by thermally stimulated transitions between the metastable and emitting levels of the triplet RES. The excited state dynamics of the luminescence centers responsible for the UV emission can be described within the phenomenological model sketched in Fig. 6.13a.

Table 6.4 Maxima positions (E_{em}), full widths at half maxima (FWHM), excitation bands maxima (E_{exc}), Stokes shifts (S), and slow component decay times (τ_{SC}) of Pb-related emissions in YSO and LSO SCF at 4.2 K, as well as the corresponding triplet RES parameters obtained in [93, 94] (for notation, see [93])

	E_{em} (eV)	FWHM (eV)	E_{exc} (eV)	S (eV)	τ_{SC} (μs)	D (meV)	$k_1 \times 10^4$ s^{-1}	$k_2 \times 10^6$ s^{-1}	$K \times 10^7$ s^{-1}	$K_x \times 10^7$ s^{-1}	E_x meV
YSO	3.67	0.34	>6 5.7 ~4.7	~1.08	109	21	0.9	19	6		
	2.86	0.40	5.25 4.75 4.2 3.9	1.04	77	9	1.3	1.9	0.1	10	120
	2.70	0.48	5.25 4.75 4.2 3.9	1.20	17	12	6	5	0.6	50	130
LSO	3.65	0.38	>6 5.7 ~4.7	~1.10	126	20	0.8	15	10		
	2.83	0.39	5.25 4.7 4.2 3.9	1.07	9.6	10	10.2	3	1	50	100
	2.72	0.41	5.25 4.7 4.2 3.9	1.18	21	9	4.3	1.3	1	1.5	65

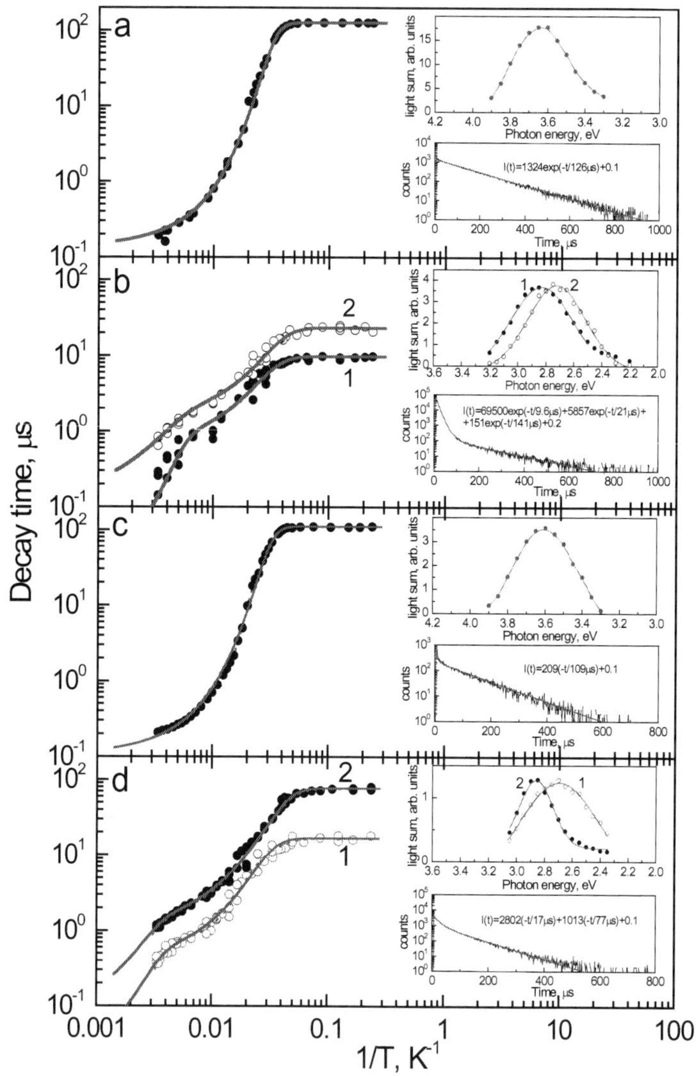

Figure 6.12 Temperature dependence of the decay times of different decay components measured for (a, c) the UV emission and (b, d) the B emission of Pb-related centers in LSO (a, c) and YSO (b, d). Symbols are experimental data, solid lines are the best fits to the data using the two excited-state level model described in the text. In the insets, decay curves and uncorrected time-resolved emission spectra (normalized) of the corresponding decay components measured at 4.2 K; (a, c) E_{exc} = 5.0 eV (b, d) E_{exc} = 4.55 eV.

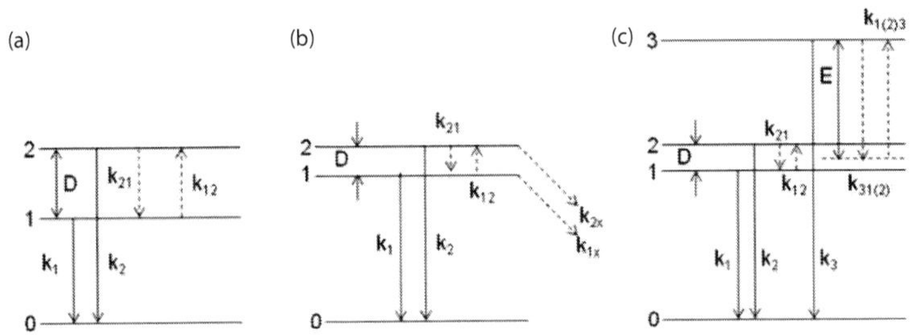

Figure 6.13 Energy level diagrams used for the description of the excited states dynamics of various Pb- and Bi-related luminescence centers in the SCF studied. For details, see the text.

The time evolution of the populations N_1, N_2 of the excited levels 1 and 2, respectively, can be described by the following rate equations:

$$dN_1/dt = -k_1N_1 - k_{12}N_1 + k_{21}N_2 - k_{1x}N_1 \quad (6.1)$$

$$dN_2/dt = -k_2N_2 - k_{21}N_2 + k_{12}N_1 - k_{2x}N_2,$$

where k_1, k_2, k_{12}, k_{21}, and $k_{1(2)x}$ are radiative transition rates from levels 1,2, non-radiative rates of phonon assisted transitions between the radiative level 2 and metastable level 1 and the quenching channel from the level 1(2), respectively. Non-radiative transitions between levels 1,2 can be written as

$$k_{21} = K(n+1),\ k_{12} = Kn,\ n = 1/[\exp(D/k_BT)-1] \quad (6.2)$$

Here K, n, D are the zero-temperature transition rate between the levels 1 and 2, the Bose–Einstein factor and energy spacing between the levels, respectively. Nonradiative quenching channel is considered in the usual barrier form:

$$k_{1(2)x} = K_{1(2)x}\exp(-E_{1(2)x}/k_BT) \quad (6.3)$$

with $K_{1(2)x}$ being a frequency factor and $E_{1(2)x}$ the height of the barrier.

The total emission $I(t)$ can be represented by the superposition of two exponential decays,

$$I(t) = \sum_{j=1}^{2} A_j \exp(-t/\tau_j)$$

and the two inverse decay times τ_j^{-1} are given by the eigenvalues τ^{-1} of the matrix of coefficients in Eq. 6.1. In the case of the UV emission no quenching channels are considered. The best fits of the model to the data are displayed as solid lines in Figs. 6.12a,c.

The parameters of the triplet RES corresponding to the UV emission are shown in Table 6.4. Note that the probability (k_1) of the radiative transition from the metastable level of Pb^{2+} centers in LSO and YSO is much larger as compared with the Pb^{2+}-doped alkali halides (~10^3–10^2 s^{-1} [98, 99]). This fact indicates a strong vibronic interaction in LSO and YSO (due to a strong covalent bond between Pb^{2+} and oxygen ions) mixing the metastable and emitting levels of the triplet RES of Pb^{2+} centers (for more details, see [99] and references therein). Similar effect was observed for Pb^{2+} centers in LuAG [86].

6.6.2 Characteristics of the Blue Luminescence

In the excitation spectrum of the B emission (Fig. 11b, dashed line), several bands are observed whose maxima positions are presented in Table 6.3. At $T < 15$ K, two components are observed in the decay kinetics of the B emission of YSO with the decay times of 77 μs and 17 μs (Fig. 6.12d). The time-resolved emission spectra of these components (see the inset) have the maxima at 2.86 eV and at 2.70 eV, respectively and their excitation spectra are different. In LSO, the time-resolved emission and excitation spectra indicate that the B emission band consists mainly of two components (2.83 eV and 2.72 eV) with the radiative lifetime values of 9.6 μs and 21 μs, respectively (Fig. 6.12b). The decay times of the both components decrease with the increasing temperature.

The phenomenological model described in item 5.1 was used to fit the temperature dependences of the B emission decay times. Taking into account the temperature dependence of the emission intensity (see the insets in Fig. 6.10), the quenching channel from the level 1 was employed in this case (Fig. 6.13b). The best fits are displayed as solid lines in Fig. 6.12b,d. The parameters of the fits are shown in Table 6.4.

The results obtained in [93, 94] allowed us to conclude that both the UV and the B emission of LSO and YSO SCF arise from Pb-related centers. Different characteristics of these emissions indicate that they arise from the Pb-related centers of different origin.

6.7 Luminescence of Bi^{3+}-Related Centers in Aluminum Garnets

The luminescence of Bi^{3+}-related centers in garnets was studied in [67, 75, 78–81]. In the emission spectrum of $Y_3Ga_5O_{12}$:Bi [78, 79], $Gd_3Ga_5O_{12}$:Bi [79], $Y_3Al_5O_{12}$:Bi [79–81], and $Lu_3Al_5O_{12}$:Bi [79, 81], two main bands were observed, located in the UV and VIS spectral ranges. The UV emission was ascribed to the radiative decay of the triplet RES of a single Bi^{3+} center. For a long time, the origin of the VIS emission was under discussion. It was ascribed to the Bi^{3+}-related bound exciton [75] or to pairs and/or clusters of Bi^{3+} ions [79–81]. The studies carried out in [90, 92] showed that both the UV and VIS emission bands of LuAG:Bi and YAG:Bi SCF surely arise from the presence of Bi^{3+} in the SCF, are of complex structure and have much different properties. The characteristics of their separate components are independent of the Bi^{3+} content (0.07, 0.183, and 1.34 at. % of Bi^{3+} in LuAG:Bi; 0.133 and 0.92 at. % of Bi^{3+} in YAG:Bi) and similar in LuAG:Bi and YAG:Bi (Fig. 6.14, Table 6.5). A large optical density (Fig. 6.15a) results in the distortion of the excitation spectra of both the UV and VIS emission bands (Fig. 15b,c).

6.7.1 Characteristics of the Ultraviolet Luminescence

At $T < 100$ K, the main UV emission band is located at 4.08 eV (FWHM = 0.24 eV) in LuAG:Bi and at 3.99 eV (FWHM = 0.24 eV) in YAG:Bi (see the inset in Fig. 6.14, solid line). In the excitation spectrum of this emission, wide bands at 5.95 eV and 4.63 eV in LuAG:Bi (Fig. 6.15b) and at 5.90 eV, 4.57 eV in YAG:Bi (Fig. 6.15c) are observed. As the temperature increases up to 150 K, the maximum of the UV emission band is shifting up to 4.19 eV in LuAG:Bi and 4.11 eV in YAG:Bi (FWHM = 0.27 eV) (see the inset in Fig. 6.14, dashed line). These data indicate to the presence of two components in the UV emission spectrum. The thermally stimulated redistribution of their intensities occurs around 115 K (the inset (a) in Fig. 6.15a) with the activation energy $E_a \approx 0.1$ eV. At $T > 180$ K, the emission band maximum gradually shifts back (to 4.15 eV at 300 K) (see the inset in Fig. 6.14, dotted line) which explains the reverse redistribution of the emission intensities shown in the inset (a) in Fig. 6.15a. In the temperature range of 4.2–150 K, the intensity of the UV emission decreases about twice and then remains practically constant (the inset (b) in Fig. 6.15a).

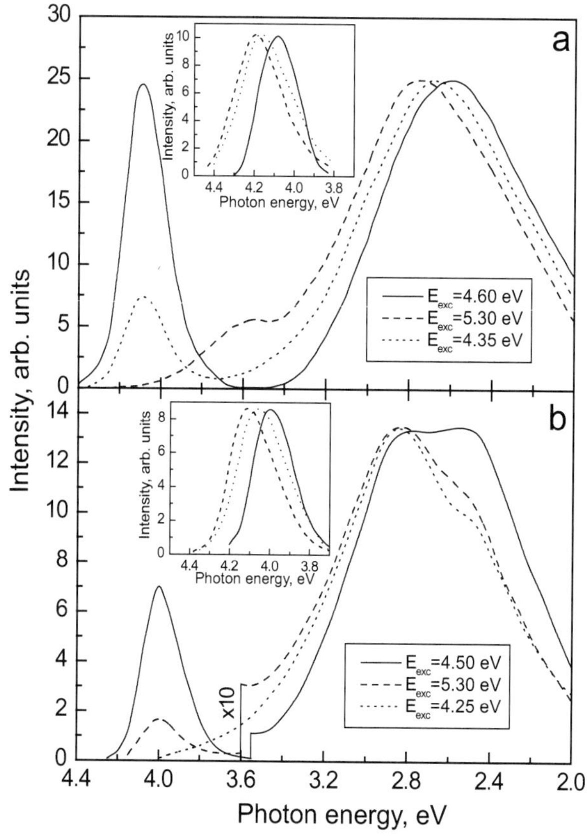

Figure 6.14 Emission spectra (normalized) of (a) LuAG:Bi and (b) YAG:Bi measured at 80 K under different excitations. In Fig. 6.1b, the spectra are distorted around 2.7 eV due to the reabsorption of the emission in the absorption band of Ce^{3+} centers. In the insets, the ultraviolet emission spectra (normalized) of (a) LuAG:Bi and (b) YAG:Bi measured at 80 K (solid line), 150 K (dashed line) and 300 K (dotted line).

In the decay kinetics of the UV emission at 4.2 K, mainly a slow component is observed. At $T < 80$ K, its decay time is constant ($\tau \approx$ 1 ms), but at $T > 100$ K, it exponentially decreases with increasing temperature (Figs. 6.16a and 6.17a). Thus, the decay kinetics of the

UV emission of LuAG:Bi and YAG:Bi is similar to that observed for the triplet emission of Bi^{3+} centers in some other hosts (see, e.g., [72, 77, 78, 82]). It can be described within the phenomenological model used in [78, 91, 93] and sketched in Fig. 6.13a (for more details see item 5.1).

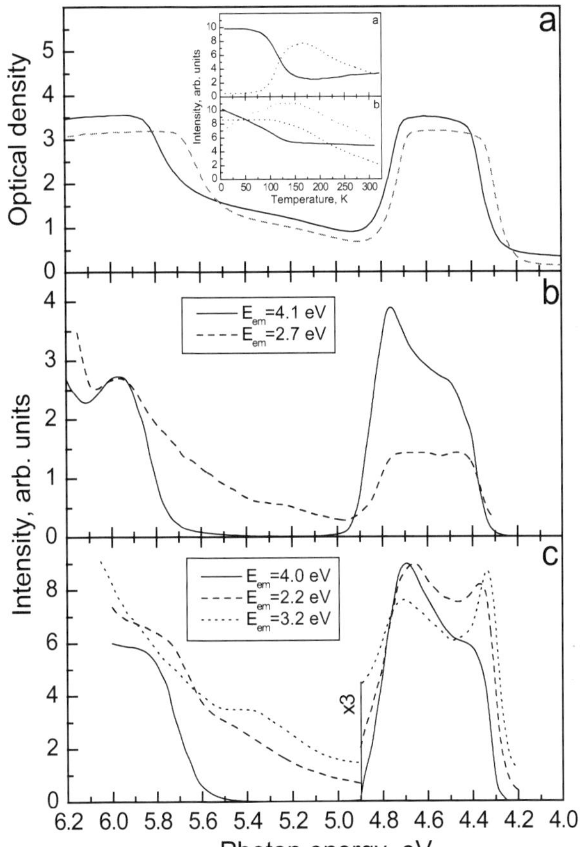

Figure 6.15 (a) Absorption spectra of LuAG:Bi (solid line) and YAG:Bi (dashed line) at 295 K. Excitation spectra (normalized) of (b) LuAG:Bi and (c) YAG:Bi measured at 80 K for different emission spectra regions. In the inset, temperature dependences of the LuAG:Bi emission intensities measured for (a) two opposite sides of the UV emission band at E_{em} = 3.95 eV (solid line) and E_{em} = 4.32 eV (dotted line) and (b) the maximum intensity of the UV (solid line), 2.6 eV (dotted line), and 2.75 eV (dashed line) emissions.

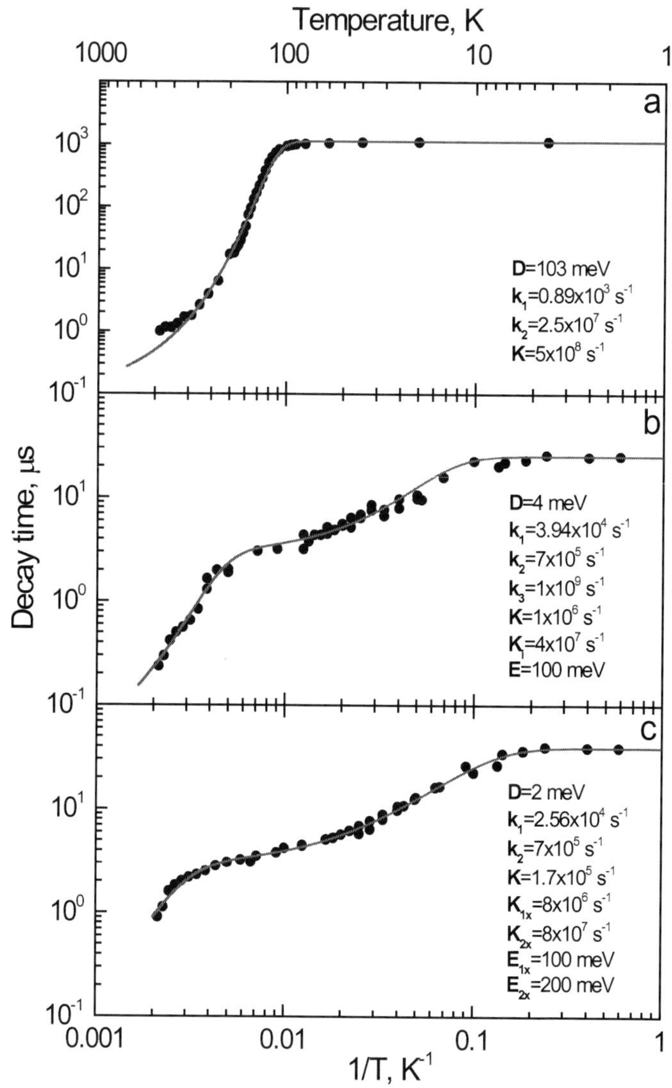

Figure 6.16 Temperature dependences of the decay times of (a) the UV emission (b) the 2.6 eV emission, and (c) the 2.75 eV emission of LuAG:Bi. The circles are experimental data. Solid lines are best fits of the two (a, c) or three (b) excited-state level models shown in Fig. 6.13 to the experimental data (for details see the text). The parameters of the fits are reported in the figures. (a) E_{exc} = 4.6 eV, E_{em} = 4.1 eV; (b) E_{exc} = 4.7 eV, E_{em} = 2.4 eV; (c) E_{exc} = 5.3 eV, E_{em} = 3.2 eV.

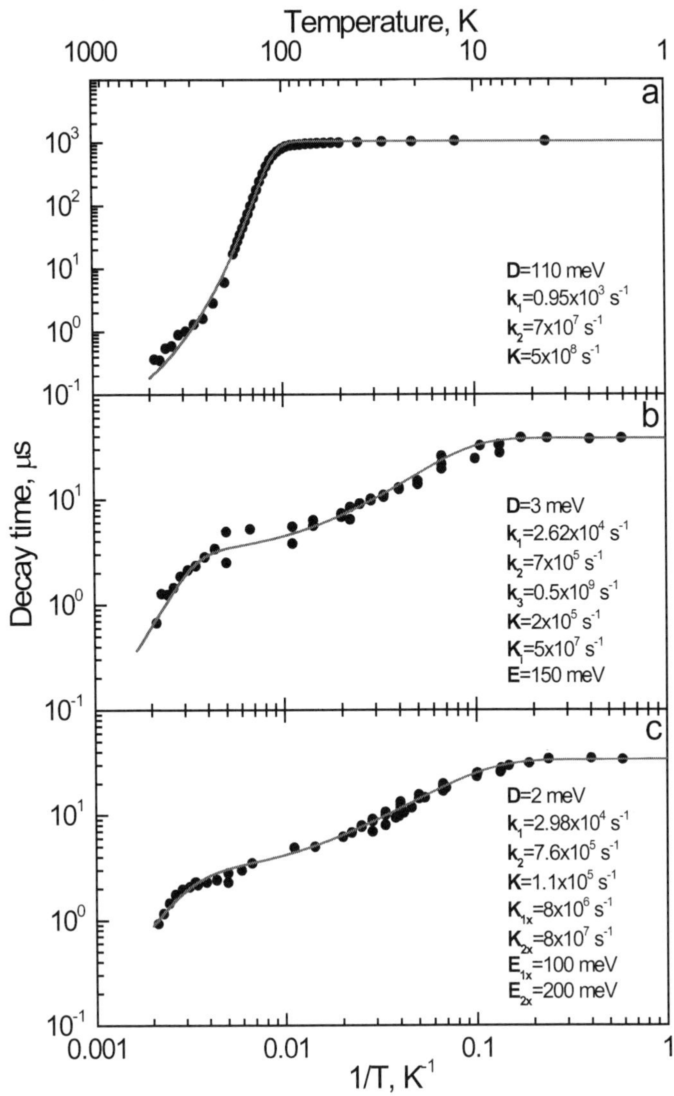

Figure 6.17 Temperature dependences of the decay times of (a) the UV emission (b) the 2.63 eV emission, and (c) the 2.75 eV emission of YAG:Bi (circles). Solid lines are best fits of the two (a, c) or three (b) excited-state level models shown in Fig. 6.13 to the experimental data (for details, see the text). The parameters of the fits are reported in the figures. (a) E_{exc} = 4.55 eV, E_{em} = 4.0 eV; (b) E_{exc} = 4.5 eV, E_{em} = 2.4 eV; (c) E_{exc} = 5.3 eV, E_{em} = 3.1 eV.

The best fit of the experimental $\tau(T)$ dependence is shown as a solid line in Figs. 6.16a and 6.17a for LuAG:Bi and YAG:Bi, respectively. The parameters of the triplet RES corresponding to the best fit are also reported in the figures (see also Table 6.6). The obtained data indicate that at $T < 80$ K, the contribution to the observed radiative decay comes only from the metastable minima (level 1) of the triplet RES of Bi^{3+}. At $T > 100$ K, the shortening of the UV emission decay time and the high-energy shift of the UV emission band with increasing temperature are both caused by thermally stimulated nonradiative transitions between the metastable (level 1) and radiative (level 2) minima of the triplet RES.

6.7.2 Characteristics of the Visible Luminescence

The VIS emission band consists of two components (Fig. 6.14, Table 6.5) with different excitation spectra (Fig. 6.15b,c). A strong lower-energy (~2.6 eV) emission (Fig. 6.14, solid line) is excited around 4.6 eV and 5.95 eV. Under excitation around 5.3 eV (dashed line) and 4.3 eV (dotted line), a weak higher-energy (2.75 eV) emission is observed. The temperature dependences of the two VIS emissions are different (compare dashed and dotted lines in the inset (b) in Fig. 6.15a). Their FWHM and Stokes shifts are about 3 times larger as compared with the UV bands (Table 6.5). In the 4.4–4.9 eV energy range, the excitation band of the ~2.6 eV emission (Fig. 6.15b,c, dashed line) is close to that of the UV emission (solid line), which allows one to assume that both these emissions arise from the same Bi^{3+} center. Under 4.6 eV excitation, the reduction of the UV emission (the inset (b) in Fig. 6.15a, solid line) is accompanied with the enhancement of the ~2.6 eV emission (dotted line), which takes place with an approximately same activation energy (6–9 meV). These data indicate the presence of thermally stimulated transitions between the excited states responsible for the UV and the ~2.6 eV emission. As the decrease in the UV emission intensity does not result in a shortening of its decay time at $T < 100$ K (Figs. 6.16a and 6.17a), one can conclude that the excited state, responsible for the ~2.6 eV emission, is thermally populated from the nonrelaxed triplet excited state of the Bi^{3+} ion.

According to [81], the increase in the VIS/UV ratio in the cathodoluminescence spectrum of LuAG:Bi with the increasing bismuth content is accompanied with a higher-energy shift of the VIS emission band. This can mean that the higher-energy component of the VIS emission has a stronger intensity dependence on the Bi^{3+} content as compared with the ~2.6 eV band. This fact allows one to assume that the 2.75 eV band can be connected with dimer Bi^{3+} centers.

Table 6.5 Characteristics of the ultraviolet and visible emissions of Bi^{3+}-related centers in LuAG:Bi and YAG:Bi SCF obtained in [90, 92]

Sample	T (K)	E_{em} (eV)	FWHM (eV)	E_{exc} (eV)	S (eV)	τ_{SC} (4.2 K) (μs)
LuAG:Bi	80	4.08	0.24	5.95 4.63	0.55	1100
	150	4.19	0.27	5.97 4.63	0.44	
	80	2.60	0.87	5.95 4.60	2.00	26
	80	2.75	0.90	5.20 4.35	1.60	40
YAG:Bi	80	3.99	0.24	5.90 4.57	0.58	1050
	150	4.11	0.27	5.90 4.57	0.46	
	80	2.63	0.85	5.90 4.55	1.92	38
	80	2.75	0.85	5.40 4.32	1.57	33

Note: The notations are the same as in Table 6.4.

The low-temperature decay kinetics of the both VIS emissions of LuAG:Bi and YAG:Bi is rather complicated and consist of up to four μs-components. No ns-component is observed under any excitation. The detailed study of time-resolved emission and excitation spectra of separate components carried out in [90–92] allows the conclusion that at 1.7 K, the decay time of the most intense component of the ~2.6 eV emission is about 26 μs in LuAG:Bi and 38 μs in YAG:Bi.

For the 2.75 eV emission, the corresponding decay times are 40 μs and 33 μs, respectively (Table 6.5). At $T < 4.2$ K, all the decay times remain constant. Temperature dependences of the decay times of each VIS emission are similar in both SCFs (Figs. 6.16b,c and 6.17b,c) and point to a small spin-orbit splitting energy of the corresponding triplet RES.

The decay kinetics of the VIS emissions was described within the phenomenological model [91]. In the case of the 2.75 eV emission, the thermal quenching from both (1 and 2, Fig. 6.13b) excited levels was taken into account (see Section 6.6.1). In the case of the ~2.6 eV emission, additional transitions to and from the singlet level 3 lying above the triplet levels of the RES were taken into account. The model is sketched in Fig. 6.13c, and the rate equations now read

$$\begin{aligned} dN_1/dt &= -k_1N_1 - k_{12}N_1 - k_{13}N_1 + k_{21}N_2 + k_{31}N_3 \\ dN_2/dt &= -k_2N_2 - k_{21}N_2 - k_{23}N_2 + k_{12}N_1 + k_{32}N_3 \\ dN_3/dt &= -k_3N_3 - k_{31}N_3 - k_{32}N_3 + k_{13}N_1 + k_{23}N_2, \end{aligned} \tag{6.4}$$

where analogous parameters have the same meaning as in Eq. 6.1, k_3 is the radiative transition rate from the singlet level 3, and $k_{3(2)1}$, $k_{1(2)3}$ are nonradiative rates of transitions between the levels 3(2) and 1:

$$k_{31} = K'(n'+1),\ k_{13} = K'n',\ n' = 1/[\exp(E/k_BT)-1] \tag{6.5}$$

Since $D << E$, we consider

$$k_{31} = k_{32},\ k_{13} = k_{23} \tag{6.6}$$

The best fit of the experimental $\tau(T)$ dependence is displayed as a solid line in Figs. 6.16b,c and 6.17b,c. The parameters used to obtain the best fit are reported in the figures. One can see that they are strongly different from those obtained for the UV emission (Figs. 6.16a and 6.17a).

The shortening of the decay time at further increase of the temperature in the case of the 2.75 eV emission is caused by thermal quenching of this emission at $T > 100$ K with the activation energies of about 100 and 200 meV, but in the case of the ~2.6 eV emission, which is not thermally quenched up to 160–175 K, by thermally stimulated transitions between the triplet and the singlet level separated by the energy distance of 100–150 meV [90–92].

Table 6.6 The values of the parameters of the triplet RES of Bi^{3+} centers obtained in [91, 95, 96] from the best fit to the experimental $\tau_{SC}(T)$ data. For details, see text.

SCF host	D (meV)	k_1 10^3 s^{-1}	k_2 10^8 s^{-1}	K 10^8 s^{-1}
LuAG	103	0.89	0.25	5
YAG	110	0.95	0.7	5
LSO	65	3.51	1	1
YSO	60	3.05	0.2	1

6.8 Luminescence of Bi^{3+}-Related Centers in Lutetium and Yttrium Oxyorthosilicates

For the first time, the luminescence of LSO:Bi and YSO:Bi SCF was detected in [95, 96]. The characteristics of Bi^{3+}-doped LSO and YSO SCF were found to be similar (Tables 6.7 and 6.8). In the emission spectrum of LSO:2.24% Bi, a strong UV band (3.455 eV), much weaker 3.3 eV band, and several visible bands are observed at 4.2 K (Fig. 6.18). The Bi content in this crystal is very large (Fig. 6.19a). In YSO:0.04% Bi, only the 3.56 eV band can surely be ascribed to Bi^{3+} centers because of a too small Bi content.

6.8.1 Characteristics of the Ultraviolet Luminescence

The characteristics of the UV emission of LSO:Bi and YSO:Bi are similar to those observed for the UV emission of single Bi^{3+} centers in other Bi-containing oxide compounds (see, e.g., [72, 75, 77–82, 90–92] and references therein). In the excitation spectrum of this emission in LSO:Bi, the 4.2 eV, 5.03 eV, and 5.95 eV bands are observed (Fig. 6.19b, Table 6.7), while in YSO:Bi, only the lowest-energy ≈4.5 eV band can be surely ascribed to Bi^{3+} centers (Table 6.8). As the temperature increases, the maximum of the UV emission band of LSO:Bi is gradually shifting to higher energies from 3.455 eV at $T < 50$ K (the inset in Fig. 6.19a, solid line) up to 3.525 eV (FWHM = 0.39 eV) at 130 K (dotted line). Similar temperature dependences are observed also in YSO:Bi.

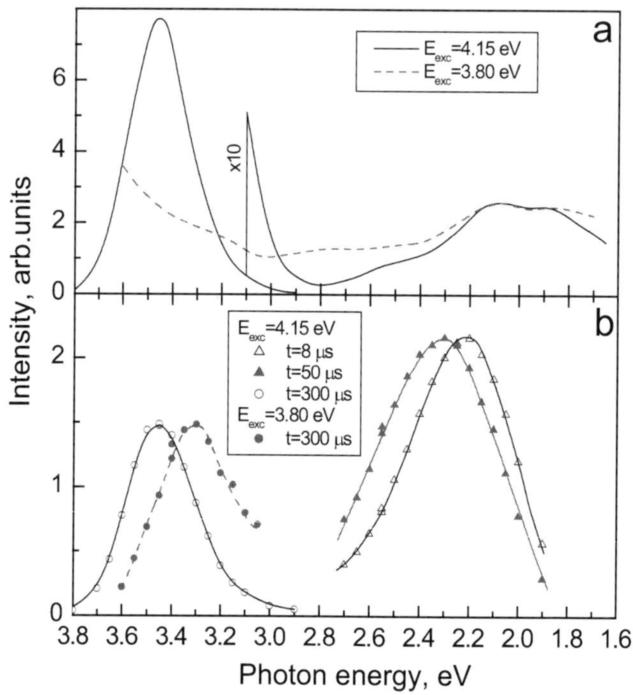

Figure 6.18 Normalized (a) steady-state and (b) time-resolved emission spectra of LSO:Bi measured at 4.2 K under excitation with E_{exc} = 4.15 eV and E_{exc} = 3.8 eV at different time instants (t) after the excitation pulse.

These data point to the presence of two components in the UV emission spectrum. Both the thermally stimulated redistribution of the intensities of the two emission components (Fig. 6.20a) and the largest shift of the UV emission band maximum (Fig. 6.20b, solid line) take place around 80 K. As the temperature increases, the FWHM of the emission band increases from 0.28 eV at 4.2 K up to 0.5 eV at 300 K (dashed line). The maximum intensity of the total UV emission remains constant up to 50 K and then decreases with increasing temperature (Fig. 6.20c). The quenching stage around 80 K well correlates with the temperature dependence of the maximum position of the UV band (Fig. 6.20b, solid line). It means that the non-radiative transitions occur from the upper level of the triplet RES.

In the decay kinetics of the UV emission, mainly the μs-component is observed at 4.2 K with the decay time of 292 μs in LSO:Bi (Fig.

6.21a) and 328 μs in YSO:Bi [96]. As the temperature increases, the decay time remains constant up to about 50 K in LSO:Bi and 60 K in YSO:Bi and then decreases exponentially. Similar temperature dependences were observed for the triplet emission of Bi^{3+} centers in some other hosts (see, e.g., [72, 77, 78, 82, 90–92]). Thus, the decay kinetics data clearly indicate that the UV emission of LSO:Bi and YSO:Bi arises from the triplet RES of Bi^{3+} ions.

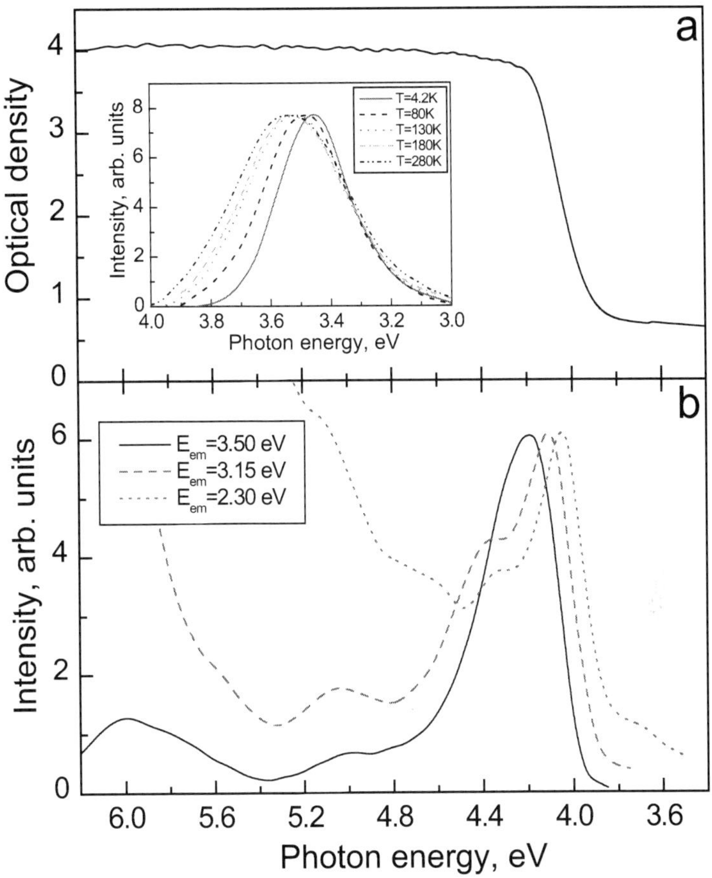

Figure 6.19 (a) Absorption spectrum of LSO:Bi at 295 K. In the inset, the normalized UV emission spectra of LSO:Bi measured at different temperatures indicated in the legend, E_{exc} = 4.15–4.07 eV. (b) Normalized excitation spectra measured at 4.2 K for different emission energies indicated in the legend.

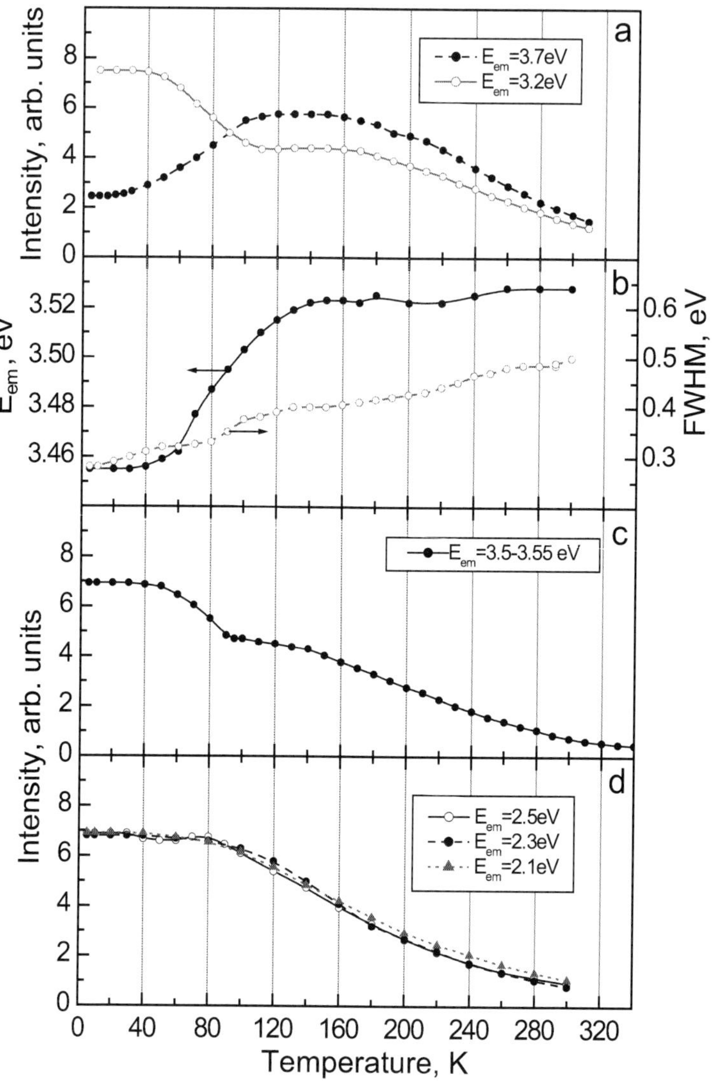

Figure 6.20 Temperature dependences of (a) the emission intensities measured for LSO:Bi under E_{exc} = 4.15 eV for two opposite sides of the UV emission band: at E_{em} = 3.2 eV (solid line) and E_{em} = 3.7 eV (dashed line); (b) the peak position (solid line) and FWHM (dashed line) of the UV emission band; (c, d) the maximum intensities of the UV emission (c) and the VIS emissions (d) under excitation in the A band maximum (4.15–4.07 eV) .

Table 6.7 The characteristics of some emission bands of Bi^{3+}-related centers in Lu_2SiO_5:Bi obtained at 4.2 K and 130 K in [95].

E_{em} (4.2 K) (eV)	FWHM (eV)	S (eV)	E_{em} (130 K) (eV)	FWHM (eV)	S (eV)	E_{exc} (4.2 K) (eV)	τ_{SC} (4.2 K) (μs)
3.455	0.28	0.745	3.525	0.39	0.675	5.95 5.03, 4.2	292
3.3	0.4	0.80				>5.9, 5.05 4.4, 4.1	298
2.2	1.0	1.85				4.35, 4.05	2.7
2.3	1.0	1.75				4.35, 4.05	16.8

Note: The notations are the same as in Table 6.4

Table 6.8 Comparison of the characteristics of the two (lower-temperature and higher-temperature) spectral components of the triplet luminescence of Bi^{3+} centers in $Lu_3Al_5O_{12}$ [90], $Y_3Al_5O_{12}$ [92], Lu_2SiO_5 [95], Y_2SiO_5 [96], KCl [77], and CaO [77].

Host	E_{em} (eV)	FWHM (eV)	S (eV)	E_{exc} (eV)	τ_{SC} (4.2 K) (μs)
$Lu_3Al_5O_{12}$	4.08 4.19	0.24 0.27	0.55 0.44	4.63	1100
$Y_3Al_5O_{12}$	3.99 4.11	0.24 0.27	0.58 0.46	4.47	1050
Lu_2SiO_5	3.455 3.525	0.28 0.39	0.745 0.675	4.20	292
Y_2SiO_5	3.56 3.63	0.33 0.43	0.94 0.87	4.50	328
KCl	2.68 2.46	0.38 0.41	1.13 1.35	3.81	2800
CaO	3.10 3.30	0.14 0.20	0.40 0.20	3.50	3250

Note: The notations are the same as in Table 6.4. E_{exc} is the position of the lowest-energy excitation band.

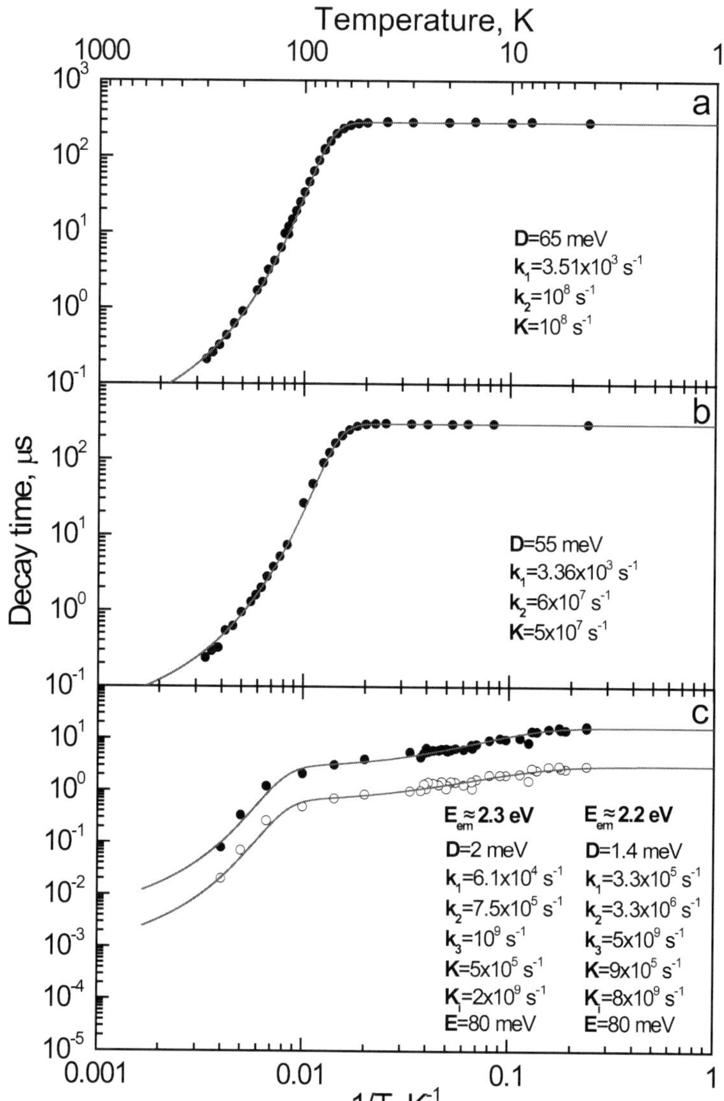

Figure 6.21 Temperature dependences of the μs-component decay times of the (a) 3.5 eV, (b) 3.3 eV, and (c) 2.2 eV (empty circles) and 2.3 eV (filled circles) emissions of LSO:Bi. E_{exc} = 4.15 eV (a, c) and E_{exc} = 3.9 eV (b). Solid lines are the best fits to the experimental data of the two (a, b) or three (c) excited-state level models shown in Fig. 6.13 (for details, see the text). The parameters of the fit are reported in the figures.

The decay kinetics of the UV emission of LSO:Bi can be described within the phenomenological model used for Bi^{3+} centers in $Y_3Ga_5O_{12}$ [78] and in $Y_3Al_5O_{12}$, $Lu_3Al_5O_{12}$ [91]. From the best fit of the experimental $\tau(T)$ dependence shown as a solid line in Fig. 6.21a, the characteristic parameters of the triplet RES are determined (see also Table 6.6). The obtained data indicate that at $T < 50$ K, the contribution to the observed radiative decay comes mainly from the metastable minima (level 1 in Fig. 6.13a) of the triplet RES of Bi^{3+}. At $T > 130$ K, the shortening of the decay time and the high-energy shift of the UV emission band with increasing temperature are both caused by thermally stimulated non-radiative transitions between the metastable (level 1) and radiative (level 2) minima of the triplet RES.

Besides the main ~3.5 eV emission of Bi^{3+} centers, much weaker ~3.3 eV band is observed in LSO:Bi under 3.8 eV excitation (Fig. 6.18a, dashed line), which most clearly appears in the time-resolved emission spectrum (Fig. 6.18b, filled circles). The excitation spectrum of this emission (Fig. 6.19b, dashed line) differs from that of the ~3.5 eV emission (solid line). Its shape indicates that in the excitation light absorption, this center competes with the ~3.5 eV emission centre and that the excitation spectrum of the ~3.3 eV emission is distorted due to an intense absorption arising from the main Bi^{3+} centers (Fig. 6.19a). At 4.2 K, the dominating decay component of the ~3.3 eV emission has the decay time of 298 μs. The temperature dependence of the decay time (Fig. 6.21b) is similar to that obtained for the ~3.5 eV emission (Fig. 6.21a). From the best fit of the experimental τ(T) dependence, the parameters of the corresponding triplet RES are determined. The obtained data indicate that the ~3.3 eV emission can also arise from the triplet RES of Bi^{3+}, but the centers responsible for the ~3.5 eV and ~3.3 eV emissions are different.

6.8.2 Characteristics of the Visible Luminescence

Unlike Bi^{3+}-doped aluminum garnets, the VIS emission of Bi-doped silicates is much weaker as compared with the UV emission (Fig. 6.18a). Due to that, in the steady-state emission spectra, no luminescence with the characteristics similar to those of the VIS emission in garnets was observed.

At 4.2 K, the decay curve of the VIS emission of LSO:Bi observed under 4.15 eV excitation consists of four components whose decay times are about 3 μs, 17 μs, 76 μs, and 384 μs [95]. The study of the time-resolved emission spectra of the separate decay components indicates that the emission spectra of the ~3 μs and ~17 μs components are broad (FWHM = 0.5–0.6 eV) bands with the maxima at about 2.2 eV and 2.3 eV, respectively (Fig. 6.18b, triangles), and the temperature dependences of their decay times (Fig. 6.21c) are similar to those observed for the VIS luminescence of Bi^{3+}-doped aluminum garnets (Figs. 6.16b,c and 6.17b,c). By analogy with [90–92], it is suggested that the ~3 μs and ~17 μs components could arise from the radiative decay of the triplet RES of excitons localized near Bi^{3+}-related centers. The light sums of these components are in, at least, two orders of magnitude smaller as compared with the light sum of the UV emission of the main Bi^{3+} centers. The decay kinetics of the both VIS emissions of LSO:Bi can be described within the phenomenological model used in [91] for the Bi^{3+}-related localized excitons in garnets and presented in Fig. 6.13c (see item 6.6.2). From the best fit of the experimental $\tau(T)$ dependences shown as solid lines in Fig. 6.21c, the characteristic parameters of the corresponding RES are determined.

6.9 Possible Models of Pb-Related Centers in Aluminum Garnets and Perovskites

In aluminum garnets, the coordination number for the impurity ion is CN = 8 in the Lu^{3+} (Y^{3+}) site and CN = 6, in the Al^{3+} site. The ionic radius of Pb^{2+} is 2.2 times larger than the ionic radius of Al^{3+}, so that Pb^{2+} ion can occupy only the Lu^{3+} or Y^{3+} site. The ionic radius of a Pb^{2+} ion is large (1.29 A) as compared also with the ionic radii of the substituting Lu^{3+} (0.98 A) or Y^{3+} (1.02 A) ions (Table 6.9). Besides, a Pb^{2+} ion has an effective negative charge with respect to the host crystal lattice. The charge and volume compensation can be achieved by small tetravalent impurity ions (e.g., Pt^{4+}, Pb^{4+}) or oxygen vacancies (V_O). In case Pb^{2+} ions and compensating defects are statistically distributed along the crystal lattice, single Pb^{2+} centers are mainly created. However, the close location of a Pb^{2+} ion and a compensating defect is more probable. Let us consider possible structures of Pb-related centers in the SCF studied.

Table 6.9 Ionic radii (R) of some trivalent rare earth ions as well as of Pb^{2+}, Bi^{3+}, and Pt^{4+} ions in the 6-fold/8-fold coordination [107] and the crystal structure of the corresponding oxyorthosilicates [108]

Ion	Lu^{3+}	Y^{3+}	Gd^{3+}	Ce^{3+}	Pb^{2+}	Pt^{4+}	Bi^{3+}
R (A)	0.86/ 0.98	0.90/ 1.02	0.94/ 1.05	1.01/ 1.14	1.19/ 1.29	0.63	1.03/1.17
Struct.	X2	X2	X1	X1	X1		

6.9.1 Single Pb^{2+}-Based Centers of the Type of $\{Pb^2 – Pt^{4+}\}$

The SCF studied are grown in a platinum crucible, and due to a chemical reaction between PbO and platinum metal at high temperatures, they contain Pt ions at a concentration, which can exceed the concentration of Pb^{2+} ions. In [97], it was concluded that in this case the charge and volume of a Pb^{2+} ion in the garnet lattice are both compensated by a much smaller Pt^{4+} ion located in the Al^{3+} site. However, ESR studies [106] have shown that trivalent Pt^{3+} ions are incorporated into the garnet lattice at a concentration of about several hundreds ppm which is comparable with the total Pt content in the SCF studied. Thus, the concentration of tetravalent Pt^{4+} ions in SCF should be small. In any case, a single Pb^{2+} ion associated with a Pt ion can be considered as a single Pb^{2+}-based center where a Pb^{2+} ion is only slightly perturbed by the compensating defect.

6.9.2 Single Pb^{2+}-Based Centers of the Type of $\{Pb^{2+}–Pb^{4+}\}$

In the SCF of garnets with a very large lead content, Pb^{4+} ions can also exist [97]. The Pb^{2+} and Pb^{4+} ions can be localized in dodecahedral positions of the garnet lattice, and in this case they have ionic radii 1.29 A and 0.94 A, respectively [107]. The sum of their ionic radii (2.33 A) exceeds the sum of the radii of two substituted Lu^{3+} (1.96 A) or Y^{3+} (2.03 A) ions, i.e., the total excess volume remains uncompensated. According to [97], a wide (FWHM $\approx$ 1 eV) absorption band of $Y_3Ga_5O_{12}$:Pb with a large lead content, arising from the charge-transfer transitions in the <111>-type {Pb^{2+}–Pb^{4+}} pairs, is located in the visible spectral range (at 2.23 eV). However, no such band was detected in the YAG and LuAG SCF studied. The corresponding visible emission was not detected either. Thus, the spectral bands located in the UV region cannot be connected with the above-mentioned <111>-type {Pb^{2+}–Pb^{4+}} pairs. In the {Pb^{2+}–Pb^{4+}} associates of the type of <100>, the influence of the Pb^{4+} ion on the Pb^{2+} ion states should be small, therefore, these centers can also be considered as single Pb^{2+}-based centers, where electronic transitions take place between the states of a Pb^{2+} ion only slightly perturbed by a Pb^{4+} ion. Thus, the small number of single Pb^{2+} centers in the SCF studied can be explained by a small concentration of the defects needed for the Pb^{2+} charge and volume compensation.

6.9.3 Dimer Lead Centers

The compensation of the Pb^{2+}-induced excess negative charge and excess volume can be achieved also by oxygen vacancies. The presence of F^+ centers in LuAG SCF and the TSL characteristics of SCF reported in [87] support the existence of oxygen vacancies in SCF, despite a low preparation temperature and air atmosphere during the SCF growth. The compensation with the aid of an oxygen vacancy V_O is the most preferable one in the case of formation of dimer lead centers close to V_O, where two Pb^{2+} ions are located in the nearest-neighboring dodecahedral Lu^{3+} or Y^{3+} sites. Due to a strong covalent bond between the two closely located Pb^{2+} ions, they prefer to form a quasimolecule-like $(Pb^{2+})_2$ center which should not be very sensitive to its surroundings. The characteristics of such dimer centers should differ noticeably from the characteristics of single Pb^{2+}-based centers.

Taking into account these considerations, the 3.15–3.20 eV emission of the SCF excited around 3.90 eV was assumed to arise arises from the dimer centers [87]. The relatively small Stokes shift (S = 0.7–0.8 eV) and the close location of the spectral bands of the dimer centers in perovskites and garnets (Table 6.3) can indicate to a relatively weak interaction of this center with its nearest environment. The data obtained indicate that the complex lead centers are more preferably produced in the SCF of LuAP, YAP, and YAG studied as compared with single Pb^{2+} centers.

Due to a high preparation temperature, the ceramics should contain a higher concentration of oxygen vacancies but they should be free from platinum. As the spectra of single and dimer lead centers in LuAG SCF and LuAG:1000 ppm Pb ceramics were found to be the same [87], this supports the suggestion made above that Pt ions do not influence the characteristics of the lead-induced centers studied.

Table 6.10 The rare earth (RE) coordination to the oxide (O) ions in oxyorthosilicates [108]

	X1 structure		X2 structure	
	Site 1	Site 2	Site 1	Site 2
Total coord. number	9	7	7	6
No. of only RE bound O	8	4	5	4
No. of Si and RE bound O	1	3	2	2
Av. Y-O distance (A)	2.46	2.39	2.43	2.28
Av. Lu-O distance (A)			2.32	2.22

6.10 The Origin of Pb-Related Centers in Lutetium and Yttrium Oxyorthosilicates

The oxyorthosilicates of the large (with ionic radius $R > 0.92$ A) trivalent rare earth (Re^{3+}) ions (e.g., GSO) always crystallize in a monoclinic X_1 type structure of the space group $P2_1/c$. The bulk single crystals of the oxyorthosilicates of the smaller Re^{3+} ions (e.g., LSO, YSO) have a monoclinic X_2 type structure of the space group C2/c [108, 109] (Table 6.9) (for more details, see also [49, 110–112]). However, the YSO and LSO powders prepared at the temperatures below 1190°C by the sol-gel method [113], hydrothermal method [114], or by solution combustion synthesis [49] have the X_1 type structure in case the temperature of the following annealing of the powder has not exceeded 1100°C [49, 113, 115]. Only their annealing at the temperatures above 1190°C results in the irreversible formation of the X_2 type structure [116]. The SCF studied were grown at $T < 1030$°C and, thus, should have the X_1 structure [49, 113–115, 117]. However, the X_2 structure of the SCF was reported in [54, 55] on the basis of the XRD measurements.

In both the X_1 and X_2 structures, there exist two crystallographically different Re^{3+} ion sites (site 1 and site 2) with the same (C_1) point symmetry but with different coordination numbers and Re-O distances [112] (Table 6.10). It is evident that the impurities with ionic radii larger than the substituted Re^{3+} ion should prefer to occupy the sites 1 where the coordination number and the Re-O distances are larger. As the ionic radius of Pb^{2+} is much larger as compared with that of Y^{3+} or Lu^{3+} (Table 6.9), one can conclude that Pb^{2+} ions should mainly occupy the crystal lattice sites 1, and the number of Pb^{2+} ions in the sites 2 should be negligible. Indeed, even in the case of smaller Ce^{3+} ions, ESR data [118] indicate that 95% of these ions are located in the sites 1. Therefore, the UV emission of Pb-related centers in YSO and LSO was ascribed in [93, 94] to the Pb^{2+} ions, substituting Y^{3+} or Lu^{3+} ions in the host lattice sites 1 of the X_2 structure (Pb1 centers). By analogy with Ce^{3+} [119], these centers can be considered as the PbO_7 clusters. The structure and parameters of the triplet RES responsible for the UV emission are similar to those observed for single Pb^{2+} centers in other hosts (see, e.g., [99]). A Pb^{2+} ion in YSO and LSO should be coupled to an additional defect, for example, an oxygen vacancy (whose presence appears in thermoluminescence [120] and EPR [57]) or a small Pt^{4+} ion (R = 0.63 Å [107]) to achieve both the charge and volume compensation.

The origin of the B emission is not yet clear. Some hypotheses were considered in [93]:

1. In [54, 55], the intense B emission was ascribed to the Pb^{2+} ions located in the lattice sites 2 of the X_2 structure (Pb2 centers). However, it is difficult to explain the preference of a large Pb^{2+} ion to be located in a smaller site 2, as even in the case of much smaller Ce^{3+} ions, the Ce1/Ce2 ratio is 95:5 [118]. Besides, much stronger electron–phonon interaction should be characteristic for a large Pb^{2+} ion in a smaller site 2. However, the Stokes shift and FWHM values of the UV and B bands are close (Table 6.4). The parameters of the RES, corresponding to the B emission (Table 6.4), are different from those associated with the UV emission, and they are not characteristic for the triplet RES of Pb^{2+} centers. Thus, the B

emission cannot be ascribed to the Pb^{2+} ions located in the sites 2 in the X_2 structure.

2. Like in aluminum perovskites and garnets, dimer lead centers could exist in YSO and LSO as well. However, due to the peculiarities of the crystal structure, close location of two Pb1 ions (formation of a dimer $(Pb^{2+})_2$ center) in oxyorthosilicates is not possible. Therefore, a dimer center can be considered as a Pb1 ion only slightly perturbed by another Pb1 ion and the luminescence characteristics of dimer and single centers should be similar. Besides, the number of dimer centers should be limited by the number of their charge and volume compensating oxygen vacancies. However, in the oxyorthosilicates prepared by low-temperature methods, the content of V_O is found to be very small [49, 54, 55, 57]. Strong difference in the characteristics of the UV and B emissions as well as a huge number of the centers responsible for a large absorption and the intense B emission in the SCF studied cannot be explained in the assumption that the B emission arises from dimer lead centers.
3. One could assume that the B emission arises from the radiative decay of excitons localized around Pb^{2+} ions. However, the FWHM and Stokes shift values for the emission of the excitons, localized around Tl^{+}, Pb^{2+}, and Bi^{3+} ions, are usually much larger than those characteristic for the triplet emission of the corresponding Tl^{+}, Pb^{2+}, and Bi^{3+} centers, and their RES parameters are strongly different as well (see, e.g., [90, 91, 95, 121] and references therein). But this is not the case of the B emission (see Table 6.4). As it was shown in [86, 87], the Pb^{2+}-related exciton emission is absent also in Pb^{2+}-doped aluminum garnets and perovskites. Unlike the Pb^{2+}-doped systems, intense exciton-like emission bands were observed in Bi^{3+}-doped garnets [90, 91]. However, in LSO:Bi SCF, the visible emission, ascribed to the exciton localized around Bi^{3+} ion, is very weak as compared with the intense UV emission of single Bi^{3+} centers [95]. It means that the formation of the localized exciton states in the silicates is much less probable than in garnets. Besides, the RES parameters obtained for the B emission in LSO:Pb (Table 6.4) are strongly different

from those obtained in [90, 91, 95] for the localized exciton emission in Bi^{3+}-doped garnets and silicates. These data allow us to conclude that the B emission of Pb-related centers in LSO cannot arise from the excitons localized near Pb^{2+} ions.

4. Besides the Pb^{2+} ions substituting for Lu^{3+} ions in the crystal lattice sites, luminescent compounds of different structure can in principle be accidently produced at the temperatures around 1000°C as external inclusions at the surface of the SCF from all or some of the oxides (Y_2O_3, SiO_2, PbO, B_2O_3) used at the preparation of the SCF (see, e.g., [122–124] and references therein). For example, in luminescent materials reported in [122], consisting of different oxides, including PbO, prepared at 700–1500 °C, the lead-related emission is observed at 2.58 eV and excited at 3.4–4.9 eV. The 2.43 eV emission is observed under 4.27 eV excitation of PbO in PbO/SiO_2 nanocomposite, while pure SiO_2 has blue emission (2.87 eV) [125]. In [126], the lead-related intrinsic luminescence of binary lead silicate glasses is found to be located at 2.75 eV (FWHM = 0.4–0.5 eV) and excited around 3.8–3.9 eV. The PbO nanoparticles have emission spectrum in a wide energy range with the maxima at about 2.95 eV, 2.75 eV, and 2.36 eV [127]. These selected examples indicate that some components of the complex B emission band could arise from the external morphologies of lead-rich compounds on the surface of films at the SCF preparation, especially as the PbO-based flux cannot be completely removed from the film surface at the end of the growth process, and the growth can continue under undefined conditions [128]. Detailed study of the surface-related luminescence properties of the SCF is necessary to exclude this possibility.
5. The lead oxyorthosilicate Pb_3SiO_5 is expected to crystallize in a monoclinic X_1 type structure (see Table 6.9). To explain the large intensity of the B emission, it was assumed in [93] that due to a large ionic radius of a Pb^{2+} ion, under certain conditions the local $X_2 \rightarrow X_1$ phase transition could take place. As a result, Pb-related centers would be produced in the YSO and LSO of the X_2 structure with the local X_1-type structure characteristic for Pb_3SiO_5. The location of Pb^{2+} ion in the X_1 structure appears preferable due to the larger coordination

number and Lu-O distance as compared with the X_2 structure (Table 6.10). Consequently, the number of these centers should increase as the Pb concentration increases. Indeed, the intensity ratio of the B and UV emission increases with the increasing lead content [54, 55]. Based on these data, it was suggested in [93] that the B emission of LSO and YSO SCF arises from the Pb-related centers, which can be considered as the molecular PbO_9 clusters with the local X_1-type structure of the space group $P2_1/c$, containing a Pb^{2+} ion and the surrounding oxygen ions. The cluster can be formed due to a strong covalent bond between Pb^{2+} and O^{2-} ions and a large ionic radius of Pb^{2+}, which stimulates the local polymorph $X_2 \rightarrow X_1$ transition. Similar processes were proposed for Pb-doped caesium halides (CsHal), where the perovskite $CsPbHal_3$ nanocrystals, embedded in the CsHal crystal lattice, were produced by annealing at appropriate temperatures (see, e.g., [129, 130] and references therein).

Due to an overlap of the Pb-related emission bands with the absorption bands of Ce^{3+} centers [40], an effective $Pb^{2+} \rightarrow Ce^{3+}$ energy transfer is possible, which results in the appearance of slow decay components and thermal quenching of the Ce^{3+} emission. Thus, the presence of lead ions in LSO:Ce and YSO:Ce SCF can negatively influence the scintillation characteristics of this material.

6.11 The Origin of Bi-Related Centers in Aluminum Garnets and Oxyorthosilicates

A trivalent Bi^{3+} ion replaces a trivalent Lu^{3+} or Y^{3+} ion in the garnet and oxyorthosilicate crystal lattice (see, e.g., [131]). As no charge and volume compensation is needed, a large concentration of single Bi^{3+} centers can be achieved in these systems.

As the ionic radius of Bi^{3+} in oxyorthosilicates is larger than the ionic radius of Y^{3+} or Lu^{3+} (Table 8), the location of a Bi^{3+} ion in the larger site 1 with the larger Lu–O or Y–O distance seems to be much preferable as compared with the site 2. Therefore, we conclude that the intense UV emission of Bi^{3+} centers arises mainly from the Bi1 centers. By analogy with the Ce2 emission in these silicates (see, e.g., [48]), the emission band of Bi2 centers should be located at lower energies and excited at the lower-energy side of the A absorption band of Bi1 centers. In LSO:Bi, this emission is expected to appear in the 3.0–3.3 eV energy range and being excited around 3.8–4.1 eV. In principle, the weak ~3.3 eV emission, which has the characteristics similar to those of the UV emission of Bi1 centers, could arise from Bi2 centers. The larger Stokes shift and FWHM of this emission band as compared with the main UV band (Table 6.7) indicate to a stronger electron–phonon interaction, which should be characteristic for a Bi^{3+} ion in a smaller Lu2 site.

According to [132], the energies of electronic transitions from the ground $^1S_0$ level to the excited $^3P_1$, $^3P_2$, and $^1P_1$ levels of a free Bi^{3+} ion (E_{free}) are 9.41 eV, 11.96 eV, and 14.21 eV, respectively. In the crystal, the absorption bands, labeled as A, B, and C, correspond to the electronic transitions to these levels. The lowest-energy A absorption band, corresponding to the $^1S_0 \rightarrow {}^3P_1$ transition, is located around 4.63 eV in LuAG:Bi and 4.57 eV in YAG:Bi, i.e., the transition energy in the crystal (E_{crys}) is about twice as small as in a free Bi^{3+} ion. Taking into account the fact that the E_{free}/E_{crys} ratio increases with the increasing E_{free} according to the approximate equation $E_{free}/E_{crys} = 1 + kE_{free}$, experimentally found in [133] and confirmed to be valid for all the ns^2-ion-doped alkali halide crystals, the positions of the B and C bands of the Bi^{3+} center can be estimated. One can assume that the ~5.95 eV band in LuAG:Bi and ~5.9 eV band in YAG:Bi arise from the $^1S_0 \rightarrow {}^1P_1$ transitions. In that case, the B absorption band should be located at about 5.2 eV. The same consideration for Bi-doped silicates allows the conclusion that the lowest-energy absorption (excitation) band located at 4.2 eV in LSO:Bi and 4.5 eV in YSO:Bi, arises from the $^1S_0 \rightarrow {}^1P_1$ transitions, the ~5.95 eV band in LSO:Bi, from the allowed $^1S_0 \rightarrow {}^1P_1$ transitions, and the weak 5.03 eV band, from the $^1S_0 \rightarrow {}^3P_2$ transitions. A small intensity of the 5.95 eV excitation band can be caused by the location of the corresponding excited level close or inside the conduction band.

The characteristics of the VIS emissions in the systems studied (e.g., much larger Stokes shifts and FWHM, smaller values of the slow component decay times at 4.2 K, the increase of the decay times with the decreasing temperature down to 4.2 K) are strongly different from the characteristics of the UV emission which clearly indicates to different origins of these emissions. Let us consider the origin and structure of the energy levels responsible for the UV and VIS emission bands of a Bi^{3+}-related center in the systems studied.

6.11.1 The Centers Responsible for the Ultraviolet Luminescence

The data obtained at the luminescence study clearly indicate that the UV emission of Bi-doped garnets and oxyorthosilicates arises from the triplet RES of the Bi^{3+} center. At low temperatures, the UV emission arises from the radiative decay of the metastable minima, related to the $^3P_0$ level of a free Bi^{3+} ion. At higher temperatures, the radiative decay of the thermally populated higher-energy radiative excited state, related to the $^3P_1$ level of a free Bi^{3+} ion, also takes place. The electronic transitions between the ground state ($^1S_0$) and the $^3P_1$-related excited state are partly allowed due to the mixing of the triplet $^3P_1$ state with the singlet $^1P_1$ state by the spin–orbit interaction. The radiative transitions from the $^3P_0$-related state can occur due to the mixing of the $^3P_1$- and $^3P_0$-related states by the vibronic interaction with the non-totally symmetric vibrations or by the hyperfine interaction (see, e.g., [99] and references therein). As the only stable Bi isotope ^{209}Bi has a nuclear spin of $I = 9/2$, in Bi^{3+}-doped or Bi^{3+}-containing compounds with a weak vibronic interaction, mainly the hyperfine interaction can be expected to be responsible for the radiative decay of the $^3P_0$-related level. The influence of the hyperfine interaction on the probability of the radiative $^3P_0 \rightarrow {}^1S_0$ transitions in Bi^{3+}-doped alkali-earth oxides was considered in [70].

Hitherto, two models have been proposed for the description of the RES of the ns^2-ion-doped ionic crystals with strongly different electron–phonon and spin–orbit interactions. The systems with a *strong spin–orbit interaction and a very weak electron–phonon interaction* should be considered in the RES model, proposed by Seitz [134], which takes into account the spin–orbit interaction as the main one in the RES. In this model, the excited states of the luminescence center originate from the $^3P_0$, $^3P_1$, $^3P_2$, and $^1P_1$ levels of a free ns^2 ion, which are split in the crystal field of the corresponding symmetry. For the generated energy levels, the Jahn–Teller effect is taken into account as a perturbation. The totally symmetric $^3P_0$ state is not generated, therefore it cannot be active in the Jahn–Teller effect. The configuration coordinates (q) of the $^3P_0$ and $^1S_0$ minima in this model should coincide.

For the systems with a *strong electron–phonon interaction*, the RES theory was developed in [135]. In this theory, the interaction of impurity optical electrons with non-totally symmetric vibrations is considered as the main one in the RES, while the spin–orbit, hyperfine, and other interactions are taken as small perturbations. As a result, the Jahn–Teller minima of different symmetries can be formed on the adiabatic potential energy surface of the singlet (^1P) and triplet (^3P) excited states. Due to the spin–orbit interaction, each Jahn–Teller minimum of the triplet RES is split into the upper emitting level and the lower metastable level. The applicability of this model was confirmed by the systematic experimental study of luminescence characteristics of ns^2-ion-doped alkali halide crystals (see, e.g., review [99]). It was also found that in the Tl^+, Pb^{2+}, Bi^{3+} centers with a strong spin–orbit interaction, each metastable minimum of the triplet RES can lie not exactly under the corresponding emitting minimum, like in the Ga^+, In^+, Ge^{2+}, Sn^{2+} centers with a weak spin–orbit interaction, but it can be shifted towards smaller q values with respect to the emitting minimum. Due to that, the energy barriers for the thermally stimulated transitions between the metastable minima of various orientations can be much lower than between various emitting minima (see, e.g., [77, 99, 136] and the references therein).

A free Bi^{3+} ion is characterized by the largest spin–orbit interaction energy (ξ) among all ns^2 ions (ξ = 1.074 eV [76]). The luminescence characteristics of two Bi^{3+}-doped crystals with strongly different electron–phonon interaction (KCl:Bi and CaO:Bi) were compared in [77]. It was concluded that the theoretical model [135] is still valid in the case of KCl:Bi but it is not valid in the case of CaO:Bi due to a very weak electron–phonon interaction in the latter system. The comparison of the values of the Stokes shifts and FWHM of emission bands presented in Table 6.8 indicate that the Bi^{3+} center in garnets and oxyorthosilicates can be considered as an intermediate case between Bi^{3+}-doped CaO and KCl. The data presented in Tables 6.6 and 6.8 allow also the conclusion that the electron–phonon interaction in the silicates is noticeably stronger than in the garnets. For example, the larger probability (k_1) of the radiative decay of the metastable minima of the triplet RES points to the stronger electron–phonon interaction in the silicates. The smaller value of the spin–orbit splitting energy (D) of the triplet RES of Bi^{3+} center in the silicates as compared with the garnets can be just due to the stronger suppression of the spin–orbit interaction by the electron–phonon interaction.

6.11.2 The Centers Responsible for the Visible Luminescence

The origin of the VIS emission in garnets and oxyorthosilicates is surely different from that of the UV emission. As it was mentioned above, two explanations have been proposed for the VIS emission in garnets. In [79–81], this emission was ascribed to Bi^{3+} pairs or clusters. In [75], it was ascribed to impurity trapped excitons. The detailed experimental studies and decay kinetics modeling carried out in [90–92, 95, 96] allowed the conclusion that the *~2.6 eV emission* arises from the radiative decay of an exciton localized near a Bi^{3+} ion (ex^0Bi^{3+}). A similar interpretation was proposed for the well-known visible luminescence of a CsI:Tl scintillator [121, 137] and also of some other Tl^+- and Pb^{2+}-doped caesium halides (see, e.g., [121, 138–140] and references therein), as well as for Pb^{2+}-doped $CdCl_2$ crystals [141]. Indeed, large values of the Stokes shifts and FWHM, as well as the low-temperature luminescence decay kinetics, namely, relatively short decay times at 4.2 K and especially their temperature dependences, pointing to a very small spin–orbit splitting energy of the triplet RES as compared with that in a ns^2 ion, are characteristic for the radiative decay of the triplet exciton state. The RES structure and parameters obtained for the ex^0Bi^{3+} emission are characteristic for various localized exciton states in aluminum garnets (see, e.g., [20]). At $T \leq 4.2$ K, this emission is due to the radiative decay of the metastable minima of the triplet state of the localized exciton. The shortening of the decay time with increasing temperature is caused by thermally stimulated nonradiative transitions between the metastable and radiative minima of the triplet RES. The energy distance between them is $\sim10^{-3}$ eV. Around 30–50 K, the thermal equilibrium between these minima is reached. The shortening of the decay time at further increase of the temperature is caused by thermal quenching of the emission or by thermally stimulated transitions between the triplet and singlet localized exciton states separated by ~10^{-2} eV.

The processes of the localized exciton states creation under excitation in the impurity-induced absorption bands have been considered for CsI:Tl [138] and CsI:Pb [140]. It was suggested that under excitation in the impurity-induced absorption bands, an electron transfer occurs from a halogen ion to an impurity ion, resulting in the creation of an electron impurity center and a self-trapped hole. A fast tunneling recombinations in the close pairs of the optically created electron and hole centers result in the appearance of the localized exciton emission (see also [142]). In LuAG:Bi, the localized exciton states can also be produced as a result of photostimulated electron transfer processes (see, e.g., [67]). As a result, electron (Bi^{2+}) and hole (Bi^{4+} or O^-) centers can be created. The formation of Bi^{2+} and Bi^{4+} centers is possible due to a variable valence of Bi [132]. The possibility of formation of Bi-related hole centers was mentioned in [75]. The formation of O^--type hole centers in LuAG was detected by the ESR method in Ref. [11]. The subsequent immediate electron–hole recombination close to a Bi^{3+} ion results in the formation of an exciton localized near the Bi^{3+} ion. In Tl^+ and Pb^{2+}-doped caesium halides, the optically created stable electron and hole centers were detected in the TSL and EPR studies of the UV-irradiated crystals [138, 143, 144]. However, after the UV irradiation of LuAG:Bi and YAG:Bi SCFs in the 4.5–6.0 eV energy range at 80–300 K, no TSL is observed. It means that the fast radiative recombination of the optically created electron and hole centers considerably dominates in the SCFs studied.

The large Stokes shift and FWHM of the *2.75 eV emission* in LuAG:Bi and YAG:Bi (Table 6.5) and the parameters of the corresponding RES (Figs. 6.16c, 6.17c) indicate that this emission is also of an exciton origin. The increase of the VIS/UV emission intensity ratio accompanied with the higher-energy shift of the VIS emission band observed in [81] allows the suggestion that the 2.75 eV emission arises from an exciton localized near a dimer $\{Bi^{3+}–Bi^{3+}\}$ center. The studies of ns^2-ions-doped alkali halides (see, e.g., [136, 145]) and aluminum perovskites and garnets [87] showed that the spectral bands of dimer impurity centers are usually noticeably shifted to lower-energies with respect to the bands of the corresponding single centers. Furthermore, their luminescence decay time is shorter. However, the detailed study of luminescence characteristics of LuAG:Bi and YAG:Bi SCF with different bismuth contents carried out in [91] allows the conclusion that the dimer centers do not appear in the luminescence of the SCFs studied. Probably, the formation of the exciton, localized near these centers, is much preferable process comparing to the radiative decay of the dimer centers themselves. Similar effect was observed, e.g., in CsI:Tl, where the strong yellow emission of an exciton localized near a Tl^+ ion considerably prevails over the much weaker UV emission of Tl^+ centers [121, 137].

On the basis of the experimental data obtained, a schematic configuration coordinate diagram of the triplet excited state of the Bi^{3+}-related center is presented in Fig. 6.22. A large spin–orbit splitting energy results in the appearance in the UV emission spectrum of two separate emission components, arising from the radiative decay of the metastable and the emitting triplet RES minima. The excitation in the A absorption band results in the electronic transitions into the $^3P_1$ state. A very small lightsum of the fast component in the decay kinetics of the UV emission means that the initial population of the $^3P_1$-related RES is comparatively small. It points to the population of the metastable $^3P_0$-related minimum mainly from the non-relaxed $^3P_1$ state. Similar features were observed in many other Bi^{3+}-doped oxide-based materials (see, e.g., [68, 69, 72, 77, 78]). The same effect was detected also in the In^+- and Sn^{2+}-doped alkali halides and explained by the peculiarities of vibrational relaxation processes in the triplet excited state of the impurity ions [99]. In the systems with a weak electron–phonon interaction, this effect can be caused by a suitable mutual location of the corresponding adiabatic potential energy surfaces (e.g., by the intersection of the $^3P_1$- and $^3P_0$-related potential surfaces below the energy level where the $^1S_0 \rightarrow {}^3P_1$ transitions take place). As the temperature increases, the $^3P_1$-related level of Bi^{3+} centers becomes thermally populated from the $^3P_0$-related level with an activation energy of $\sim 10^{-1}$ eV which corresponds to the spin–orbit splitting energy of the triplet RES of a Bi^{3+} ion.

A small (0.03 eV, see Table 6.5) shift of the excitation band of the ex^0Bi^{3+} emission with respect to that of the Bi^{3+} emission can be explained by the mutual location of the corresponding nonrelaxed levels and different relaxation ways into the corresponding excited states minima (Fig. 6.22). Note that in CsI:Tl, the excitation band of the Tl^+-related localized exciton emission is also slightly shifted with respect to the A band in the excitation spectra of the A_T and A_X emission bands of Tl^+ centers [121]. Like in CsI:Tl, thermally stimulated transitions take place between the states, responsible for the Bi^{3+} and ex^0Bi^{3+} emissions, which indicate that both these bands arise from the same Bi^{3+} center.

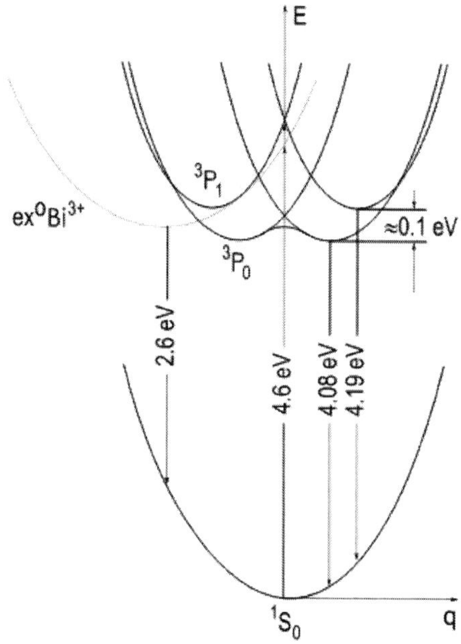

Figure 6.22 Schematic configuration coordinate diagram of the Bi^{3+}-related triplet excited state in LuAG:Bi. Electronic transitions between the ground state, corresponding to the $^1S_0$ level, and the excited states, corresponding to the $^3P_1$ and $^3P_0$ levels of a free Bi^{3+} ion and the state of the exciton localized near Bi^{3+} ion (ex^0Bi^{3+}) are indicated by arrows.

6.12 Scintillation Characteristics of Single-Crystalline Films

Nowadays the fast development of microimaging techniques for applications under the X-ray or synchrotron radiation sources in micro-tomography and industry strongly requires the SCF scintillating screens with spatial resolution in the µm or even in the sub-µm range [12, 15, 149]. For such a purpose the screens emitting the UV light [150–152] are of advantage due to diffraction limit. Contradictory requirements of small thickness (to achieve high spatial resolution) and high X-ray absorption (bigger thickness needed) especially materials of high density and effective atomic number are intensively searched as, e.g., the Lu-based LuAG, LSO and LuAP compounds [150, 151, 152] with the density values ρ = 6.67, 7.4 and 8.34 g/cm^3, respectively, and effective atomic number Z_{eff} = 59, 65.2, and 62, respectively.

Given the fact that Pb^{2+} emission in all these matrices is significantly quenched at room temperature and Coulombic compensation is necessary to balance non-equal Lu^{3+} (Y^{3+}) cation charge state upon divalent lead ion embedding, the Pb^{2+} dopant cannot be used for such an application, and, as described in the details above, its presence in LPE films leads always to the decrease of luminescence and scintillation efficiency.

On the contrary, the Bi^{3+} ions can be used as activators in LuAG, LSO and LuAP hosts due to favorable charge state and their dominant intrinsic $^1S_0$ -$^3P_1$ radiative transition in UV range. An example is provided in Fig. 6.23 where cathodoluminescence spectra are shown.

The N_{phels} photoelectron yields of LuAG:Bi and YAG:Bi SCFs under excitation by α–radiation of ^{241}Am (5.4857 MeV) are summarized in Table 6.1 [81]. Highest N_{phels} of LuAG:Bi SCFs is realized at Bi concentration of 0.18 at. % and optimum concentration can be estimated of about 0.2–0.3 at.%. Fluctuations in N_{phels} value in sets of YAG:Bi and LuAG:Bi SCF samples can be caused by different content of Bi^{3+} ions and some irregularities of SCFs as well. However, the stable Bi^{3+} UV emission position and halfwidth point to its regular embedding into crystal structure. All the 6s^2 ions (Pb^{2+}, Bi^{3+}) are very sensitive to the crystal environment due to unshielded character of the optical transition.

In LuAG:Bi we observe also the systematic increase of N_{phels} values with increasing Bi^{3+} concentration, while the total CL intensity is decreasing (see Table 6.1). Such feature indicates reduction of very slow decay components in the scintillation response. It is frequently observed in scintillators due to shortening of mean free path of migrating free charge carriers before the emission center is reached.

Figure 6.24 shows the N_{phels} photoelectron yield (a) and FWHM resolution (b) time dependence of Bi^{3+}-doped LuAG and YAG SCFs measured in the shaping time range 0.5–10 μs under excitation by

α-radiation of ^{241}Am (5.4857 MeV) [81]. 90% of N_{phels} value at the long time limit is reached within 3.5 and 3.8 μs for LuAG:Bi and YAG:Bi SCFs, respectively. Finally, for the best performing Bi-doped LuAG and YAG SCFs the maximum N_{phels} yield is about 15% and 21%, respectively, with respect to that of YAP:Ce standard crystal scintillator, using the shaping time of 3 μs [81] (see Table 6.1). Photoelectron yield in the Bi-doped YSO and LSO SCF's shows several times lower values [153] compared to the above described garnet hosts which excludes practical applications.

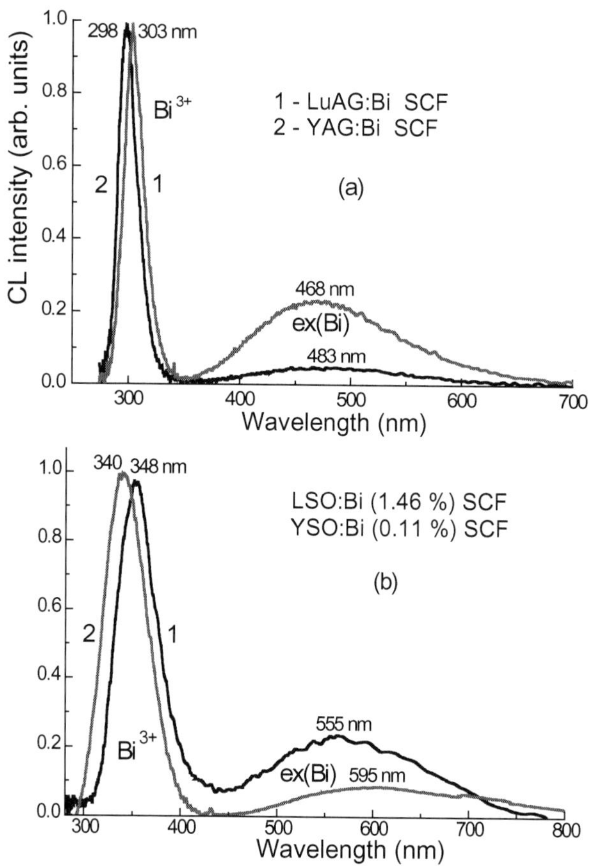

Figure 6.23 (a) CL spectra of LuAG:Bi (1) and YAG:Bi (2) SCFa . Reprinted from Ref. 81 with permission, © IOP Publishing. All rights reserved. (b) CL spectra of LSO:Bi (1) and YSO:Bi (b2) SCFs. T = 300 K.

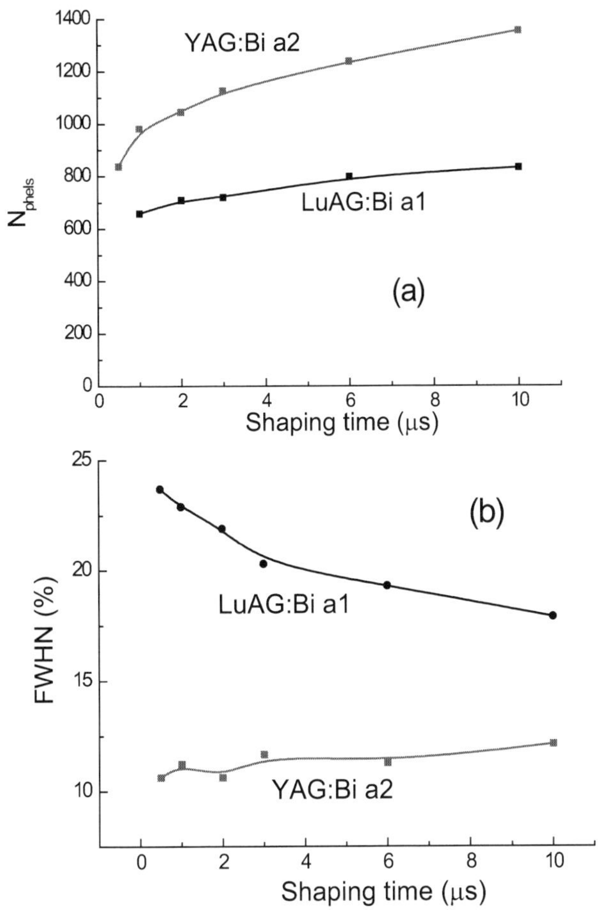

Figure 6.24 N_{phels} photoelectron yield (a) and energy resolution FWHM (%) (b) of LuAG:Bi and YAG:Bi SCFs, measured in the shaping time range 0.5–10 μs (values 0.5, 1, 2, 3, 6 and 10 μs) under excitation by α-radiation of ^{241}Am (5.4857 MeV).

6.13 Conclusions

The luminescence spectrum of Pb-related centers in the SCF studied consists of the bands of different origin. The characteristics of the highest-energy UV emission, arising from the radiative decay of the triplet RES of a single Pb^{2+}-ion-based center, are similar in all the SCF studied. In perovskites, the intensity of this emission is negligible and in YAG, it is considerably smaller than in LuAG and silicates. Relative concentration of the single Pb^{2+}-ion-based centers in the SCF studied is relatively small.

The characteristics of the 3.15–3.20 eV emission, ascribed to the dimer lead centers of the type of {Pb^{2+} - oxygen vacancy - Pb^{2+}}, are also similar in all the perovskite and garnet SCF studied. The dominating blue emission of the LSO and YSO SCF is suggested to arise from the molecular PbO_9 clusters with the local X_1-type structure, containing a Pb^{2+} ion and the surrounding oxygen ions, embedded into the crystal lattice of the X_2 structure. No Pb^{2+}-related localized exciton emission is detected in the SCF studied.

A considerable negative influence of various Pb-related centers on the scintillation characteristics of the Ce^{3+}- and Pr^{3+}-doped SCF, depending on the type of the center and on the host material, was revealed. As the substitution of trivalent Y^{3+} or Lu^{3+} ions by relatively large divalent Pb^{2+} ions is possible only in case both the charge and the volume of a Pb^{2+} ion are effectively compensated, in order to decrease the number of undesirable Pb-related centers and to improve their scintillation characteristics, the number of the corresponding compensating defects in the SCF should be considerably reduced.

In Bi^{3+}-doped garnets and silicates, the coexistence of the Bi^{3+}-ion-related energy levels and the levels of an exciton, localized near the Bi^{3+} ion, is found in the triplet RES of the luminescence center. The radiative decay of the emitting and metastable minima, arising from the $^3P_1$- and $^3P_0$-related levels of a free Bi^{3+} ion, results in the appearance of the UV emission. The excitation bands of the UV emission, located around 4.2–4.6 eV, 5.0–5.3 eV and 5.9–6.0 eV, arise from the $^1S_0 \rightarrow {}^3P_1$, $^1S_0 \rightarrow {}^3P_2$, and $^1S_0 \rightarrow {}^1P_1$ transitions of a free Bi^{3+} ion, respectively. The dominant UV emission of Bi^{3+} centers in silicates arises from Bi1 centers. The weak ≈3.3 eV emission of LSO:Bi is assumed to arise from Bi2 centers. The radiative decay of excitons, localized around single Bi^{3+} and dimer Bi^{3+}–Bi^{3+} centers, results in the appearance of the VIS emissions. Unlike the Bi^{3+}-doped aluminum garnets, the emission of the Bi^{3+}-related localized excitons in the silicates is very weak.

Thermally stimulated transitions between the metastable and emitting minima of the triplet state of both the Bi^{3+} ion and the exciton localized near the Bi^{3+} ion, between the states of the Bi^{3+} ion and the localized exciton, and between the triplet and singlet states of the localized exciton reveal themselves in the temperature dependences of the luminescence spectra and decay kinetics.

The phenomenological models are proposed to describe the excited-state dynamics of all the luminescence centers studied. Application of the two- or three-excited level models on the temperature evolution of the ultraviolet and visible luminescence decay times allowed determination of characteristic parameters of the corresponding relaxed excited states (the energy separations between the excited states and the rates of the radiative and non-radiative transitions from these states).

Owing to the presence at room temperature of the intense and relatively fast luminescence of Bi^{3+}-related centers, especially in aluminium garnets, these materials can be perspective for scintillator-oriented applications.

Acknowledgments

The work was supported by the projects of the Estonian (No. 8678), Czech (GA CR P204/12/0805), Polish (NCN No 2012/07/B/ST5/02376 and NCBR No. NANOLUX2014/ID 286) and Ukrainian (No. SF-20F and No. Fk 64/34) Science Foundations, the Institutional Research Funding IUT02-26 of the Estonian Ministry of Education and Research, and the NATO project CBP.NUKR.CLG984305.

References

1. Weber, M. J. (1973). Nonradiative decay from 5d states of rare earths in crystals, *Solid State Commun.*, **11**, pp. 741–744.
2. Autrata, R., Schauer, P., Kvapil, J., and Kvapil, J. (1978). A single crystal of YAG-new fast scintillator in SEM, *J. Phys. E: Sci. Instrum.*, **11**, pp. 707–708.
3. Moszynski, M., Ludziewski, T., Wolski, D., Klarma, W., and Norlin, L. O. (1994). Properties of the YAG-Ce scintillator. *Nucl. Instrum. Methods Phys. Res., Sect. A*, **345**, pp. 461–467.
4. Lempicki, A., Randles, M. H., Wisniewski, D., Balcerzyk, M., Brecher, C., and Wojtowicz, A. J. (1995). $LuAlO_3$:Ce and other aluminate scintillators, *IEEE Trans. Nucl. Sci.*, **42,** pp. 280–284.

5. Rodnyi, P. A. (1997) *Physical Processes in Inorganic Scintillators* (CRC Press LLC, Boca Raton, USA).
6. Nikl, M., Mihokova, E., Mares, J. A., Vedda, A., Martini, M., Nejezchleb, K., and Blazek, K. (2000). Traps and timing characteristics of LuAG:Ce^{3+} scintillator, *Phys. Status Solidi B,* **181**, pp. R10–R12.
7. Blazek, K., Krasnikov, A., Nejezchleb, K., Nikl, M., Savikhina, T., and Zazubovich, S. (2004). Luminescence and defects creation in Ce^{3+}-doped $Lu_3Al_5O_{12}$ crystals, *Phys. Status Solidi B,* **241,** pp. 1134–1140.
8. Nikl, M., Ogino, H., Krasnikov, A., Beitlerova, A., Yoshikawa, A., and Fukuda, T. (2005). Photo- and radioluminescence of Pr-doped $Lu_3Al_5O_{12}$ single crystal, *Phys. Status Solidi A,* **202**, pp. R4–R6.
9. Ogino, H., Yoshikawa, A., Nikl, M., Krasnikov, A., Kamada, K., and Fukuda, T. (2006). Growth and scintillation properties of Pr-doped $Lu_3Al_5O_{12}$ crystals, *J. Cryst. Growth,* **287**, pp. 335–338.
10. Nikl, M. (2006). Scintillation detectors for x-rays, *Meas. Sci. Technol.,* **17**, pp. R37–R54.
11. Nikl, M., Laguta, V. V., and Vedda, A. (2008). Complex oxide scintillators: Material defects and scintillation performance, *Phys. Stat. Sol. B,* **245**, pp. 1701–1722.
12. Koch, C., Raven, P., Spanne, P., and Snigirev, A. (1998). X-ray imaging with submicrometer resolution employing transparent luminescent screens, *J. Opt. Soc. Am. A,* **15**, pp. 1940–1951.
13. Zorenko, Yu. Gorbenko, V., Konstankevych, I., Grinev, B., and Globus, M. (2002). Scintillation properties of $Lu_3Al_5O_{12}$:Ce single-crystalline films, *Nucl. Instrum. Methods Phys. Res., Sect. A,* **486**, pp. 309–314.
14. Zorenko, Yu., Konstankevych, I., Gorbenko, V., and Yurchyshyn, P. (2002). Preparation, luminescence properties and application of scintillators based on $Lu_3Al_5O_{12}$:Ce single-crystal films, *J. Appl. Spectrosc.,* **69**, pp. 769–776.
15. Koch, A. and Riekel, C. (1996). X-ray video camera with 10 μm spatial resolution, *Rev. Sci. Instrum.* , **67**, 1737–1740.
16. Zorenko, Yu. Gorbenko, V., Voloshinovskii, A., Stryganyuk, G., Mikhailin, V., Kolobanov, V., Spassky, D., Nikl, M., and Blazek, K. (2005). Exciton-related luminescence in LuAG:Ce single crystals and single crystalline films, *Phys. Status Solidi A*, **202,** pp. 1113–1119.
17. Zorenko, Yu., Gorbenko, V., Konstankevych, I., Nikl, M., Mares, J. A., Blazek, K., Nejezchleb, K., Grinyov, B., Mikhailin, V., Kolobanov, V., and Spassky, D. (2005). Scintillators based on YAG:Ce and LuAG:Ce single

crystalline films: non-trivial advantages over bulk crystal analogues, *Proc. of the 8 Int. Conf. on Inorganic Scintillators and their Use in Science and Industrial Applications*, pp. 205–211.

18. Babin, V., Blazek, K., Krasnikov, A., Nejezchleb, K., Nikl, M., Savikhina, T., and Zazubovich, S. (2005). Luminescence of undoped LuAG and YAG crystals, *Phys. Status Solidi C*, **1**, pp. 97–100.

19. Zorenko, Yu., Gorbenko, V., Konstankevych, I., Voloshinovskii, A., Stryganyuk, G., Mikhailin, V., Kolobanov, V., and Spassky, D. (2005). Single-crystalline films of Ce-doped YAG and LuAG phosphors: advantages over bulk crystals analogues, *J. Lumin.*, **114**, pp. 85–94.

20. Babin, V., Bichevin, V., Gorbenko, V., Kink, M., Makhov, A. , Maksimov, Yu., Nikl, M., Stryganyuk, G., Zazubovich, S., and Zorenko, Yu. (2011). Time-resolved spectroscopy of exciton-related states in single crystals and single crystalline films of $Lu_3Al_5O_{12}$ and $Lu_3Al_5O_{12}$:Ce, *Phys. Status Solidi B*, **248**, pp. 1505–1512.

21. Zorenko, Yu., Voloshinovskii, A., Gorbenko, V., Zorenko, T., Nikl, M., and Nejezchleb, K. (2007). Intrinsic luminescence of $YAlO_3$ perovskites, *Phys. Status Solidi C*, **4**, pp. 963–967.

22. Babin, V., Gorbenko, V., Kondakova, I., Kärner, T., Laguta, V. V., Nikl, M., Zazubovich, S., and Zorenko, Yu. (2011). Time-resolved spectroscopy of exciton states in single crystals and single crystalline films of $YAlO_3$ and $YAlO_3$:Ce, *J. Phys. D: Appl. Phys.*, **44**, 315402 (8 pp).

23. Kuklja M. M. (2000). Defects in yttrium aluminium perovskite and garnet crystals: atomistic study, *J. Phys.: Condens. Matter*, **12**, pp. 2953–2967.

24. Stanek, C. R., McClellan, K. J., Levy, M. R., and Grimes, R. W. (2006). Extrinsic defect structure of $RE_3Al_5O_{12}$ garnets, *Phys. Status Solidi B*, **243**, pp. R75–R77.

25. Stanek, C. R., McClellan, K. J., Levy, M. R., Milanese, C., and Grimes, R. W. (2007). The effect of intrinsic defects on garnet scintillator performance, *Nucl. Instrum. Methods Phys. Res., Sect. A*, **579**, pp. 27–30.

26. Stanek, C. R., McClellan, K. J., Levy, M. R., and Grimes, R. W. (2006). Defect behavior in rare earth $REAlO_3$ scintillators, *J. Appl. Phys.*, **99**, 113518.

27. Singh, D. J. (2007). Antisite defects and traps in perovskite $YAlO_3$ and $LuAlO_3$: Density functional calculations, *Phys. Rev. B*, **76**, 214115.

28. Nikl, M. (2005). Energy transfer phenomena in the luminescence of wide band-gap scintillators, *Phys. Status Solidi A*, **202**, pp. 201–206.

29. Nikl, M., Mihokova, E., Pejchal, J., Vedda, A., Zorenko, Yu., and Nejezchleb, K. (2005). The antisite LuAl defect-related trap in $Lu_3Al_5O_{12}$:Ce single crystal, *Phys. Status Solidi B*, **242,** pp. R119–R121.

30. Zorenko, Yu., Voloshinovskii, A., Konstankevych, I., Kolobanov, V., Mikhailin, V., and Spassky, D. (2004). Luminescence of excitons and antisite defects in the phosphors based on garnet compounds, *Radiat. Meas.*, **38**, pp. 677–680.

31. Zorenko, Yu., Voloshinovskii, A., Savchyn, V., Vozniak, T., Nikl, M., Nejezchleb, K., Mikhailin, V., Kolobanov, V., and Spassky, D. (2007). Exciton and antisite defect-related luminescence in $Lu_3Al_5O_{12}$ and $Y_3Al_5O_{12}$ garnets, *Phys. Status Solidi B*, **244**, pp. 2180–2189.

32. Zorenko, Yu., Gorbenko, V., Konstankevych, I., Voznjak, T., Savchyn, V., Nikl, M., Mares, J.A., Nejezchleb, K., Mikhailin, V., Kolobanov, V., and Spassky, D. (2007). Peculiarities of luminescence and scintillation properties of YAP:Ce and LuAP:Ce single crystals and single crystalline films, *Radiat. Meas.*, **42,** pp. 528–532.

33. Zorenko, Yu., Gorbenko, V., Voznjak, T., Mikhailin, V., Kolobanov, V., Spassky, D., and Nikl, M. (2008). Intrinsic and Ce^{3+}-related luminescence in single crystalline films and single crystals of LuAP and LuAP:Ce perovskites, *IEEE Trans. Nucl. Sci.*, **55**, pp. 1192–1196.

34. Prusa, P., Cechak, T., Mares, J. A., Nikl, M., Beitlerova, A., Solovieva, N., Zorenko, Yu. V., Gorbenko, V. I., Tous, J., and Blazek, K. (2008). The α-particle excited scintillation response of the liquid phase epitaxy grown LuAG:Ce thin films, *Appl. Phys. Lett.*, **92**, 041903.

35. Prusa, P., Nikl, M., Mares, J. A., Kuchera, M., Nitsch, K., and Beitlerova, A. (2009). The α-particle excited scintillation response of YAG:Ce thin films grown by liquid phase epitaxy, *Phys. Status Solidi A*, **206**, pp. 1494–1500.

36. Melcher, C. L. and Schweitzer, J. S. (1992). Cerium-doped lutetium oxyorthosilicate: a fast, efficient new scintillator, *IEEE Trans. Nucl. Sci.*, **39**, pp. 502–505 (see also US Patents No. 4958080, 1990; Nos. 5025151, 5660627, 1992).

37. Melcher, C. L. and Schweitzer, J. S. (1992). A promising new scintillator: cerium-doped lutetium oxyorthosilicate, *Nucl. Instrum. Methods Phys. Res., Sect. A*, **314**, pp. 212–214.

38. Melcher, C. L., Manente, R. A., Peterson, C.A., and Schweitzer, J. S. (1993). Czochralski growth of rare earth oxyorthosilicate single crystals, *J. Cryst. Growth*, **128**, pp. 1001–1005.

39. Melcher, C. L., Schmand, M., Eriksson, M., Eriksson, L., Casey, M., Nutt, R., Lefaucheur, J. L., and Chai, B. (2000). Scintillation properties of LSO:Ce boules, *IEEE Trans. Nucl. Sci.,* **47**, pp. 965–968.

40. Suzuki, H., Tombrello, T. A., Melcher, C. L., and Schweitzer, J. S. (1992). UV and gamma-ray excited luminescence of cerium-doped rare-earth oxyorthosilicates, *Nucl. Instrum. Methods Phys. Res., Sect. A,* **320**, pp. 263–272.

41. Ren, G., Qin, L., Lu, S., and Li, H. (2004). Scintillation characteristics of lutetium oxyorthosilicate (Lu_2SiO_5:Ce) crystals doped with cerium ions, *Nucl. Instrum. Methods Phys. Res., Sect. A,* **531**, pp. 560–565.

42. Kobayashi, M., Ishii, M., and Melcher, C. L. (1993). Radiation damage of a cerium-doped lutetium oxyorthosilicate single crystal, *Nucl. Instrum. Methods Phys. Res., Sect. A,* **335**, pp. 509–512.

43. Moses, W. W. (2000). Scintillator requirements for medical imaging, *Proc. of the Fifth Int. Conf. on Inorganic Scintillators and Their Applications,* SCINT99, pp. 11–21.

44. Moses, W. W. (2001). Trends in PET imaging, *Nucl. Instrum. Methods Phys. Res., Sect. A,* **471**, pp. 209–214.

45. Lempicki, A., Glodo, J. (1998). Ce-doped scintillators: LSO and LuAP, *Nucl. Instrum. Methods Phys. Res., Sect. A,* **416**, pp. 333–344.

46. Dorenbos, P., Bos, A. J. J., and van Eijk, C. W. E. (2002). Photostimulated trap filling in Lu_2SiO_5:Ce^{3+}, *J. Phys.: Condens. Matter,* **14**, L99–L101.

47. Cooke, D. W., Bennett, B. L., Muenchausen, R. E., Lee, J.-K., and Nastasi, M. A. (2004). Intrinsic ultraviolet luminescence from Lu_2O_3, Lu_2SiO_5 and Lu_2SiO_5:Ce^{3+}, *J. Lumin.,* **106**, pp. 125–132.

48. Suzuki, H., Tombrello, T. A., Melcher, C. L., and Schweitzer, J. S. (1993). Light emission mechanism of Lu_2 (SiO_4) O: Ce, *IEEE Trans. Nucl. Sci.,* **40**, pp. 380–383.

49. Yukihara, E. G., Jacobsohn, L. G., Blair, M. W., Bennett, B. L., Tornga, S. C., and Muenchausen, R. E. (2010). Luminescence properties of Ce-doped oxyorthosilicate nanophosphors and single crystals, *J. Lumin.,* **130**, pp. 2309–2316.

50. Melcher, C. L., Spurrier, M. A., Eriksson, L., Eriksson, M., Schmand, M., Givens, G., Terry, R., Homant, T., and Nutt, R. (2003). Advances in the scintillation performance of LSO: Ce single crystals, *IEEE Trans. Nucl. Sci.,* **50**, pp. 762–766.

51. Wojtowicz, A., Drozdowski, W., Wisniewski, D., Lefaucher, J-L., Galazka, Z., Gou, Z., Lukasiewicz, T., and Kisielewski, J. (2006). Scintillation

properties of selected oxide monocrystals activated with Ce and Pr, *Opt. Mater.*, **28**, pp. 85–93

52. Casey, M. E., Eriksson, L., Schmand, M., Andreaco, M. S., Dahlbom, M., Nutt, R., and Paulus, M. (1997). Investigation of LSO crystals for high resolution positron emission tomography, *IEEE Trans. Nucl. Sci.*, **44**, pp. 1109–1113.
53. van Eijk, C. W. E. (2003). Inorganic scintillators in medical imaging detectors, *Nucl. Instrum. Methods Phys. Res., Sect. A,* **509**, pp. 17–25.
54. Zorenko, Yu., Nikl, M., Gorbenko, V., Mares, J. A., Savchyn, V., Voznyak, T., Solsky, I., Grynyov, B., Sidletskiy, O., Kurtsev, D., Beitlerova, A., and Kucerkova, R. (2010). Growth and luminescent properties of Lu_2SiO_5 and Lu_2SiO_5:Ce single crystalline films, *IOP Conf. Ser.: Mater. Sci. Eng.*, **15**, 012010.
55. Zorenko, Yu., Nikl, M., Gorbenko, V., Savchyn, V., Voznyak, T., Kucerkova R., Sidletskiy, O., Grynyov, B., and Fedorov, A. (2011). Growth and luminescent properties of Lu_2SiO_5 and Lu_2SiO_5:Ce single crystalline films, *Opt. Mater.,* **33**, pp. 846–852.
56. Zorenko, Yu., Gorbenko, V., Savchyn, V., Voznyak, T., Gorbenko, V. V., Nikl, M., Mares, J. A., Sidletskiy, O., Grinyov, B., Fedorov, A., Fabisiak, K., and Paprocki, K. (2012). Scintillation and luminescent properties of undoped and Ce^{3+} doped Y_2SiO_5 and Lu_2SiO_5 single crystalline films grown by LPE method, *Opt. Mater.,* **34**, pp. 1969–1974.
57. Cooke, D. W., Blair, M. W., Smith, J. F., Bennett, B. L., Jacobsohn, L. G., McKigney, E. A., and Muenchausen, R. E. (2008). EPR and luminescence of F+ centers in bulk and nanophosphor oxyorthosilicates, *IEEE Trans. Nucl. Sci.,* **55**, pp. 1118–1122.
58. Zorenko, Yu., Gorbenko, V., Konstankevych, I., Pashkovsky, M., Globus, M., Grinyov, B., Tarasov, V., Dorenbos, P., van Eijk, C. W. E., and van Loef, E. (2000). Scintillators on the base of single crystalline films of Al_2O_3-Y_2O_3 system oxides, *Proc. of the Fifth Int. Conf. on Inorganic Scintillators and Their Applications*, pp. 476–481.
59. Zorenko, Yu. (2006). Luminescence of Sc^{3+} and La^{3+} isoelectronic impurities in $Lu_3Al_5O_{12}$ single-crystal films, *Opt. Spectrosc.,* **100**, pp. 572–580.
60. Zorenko, Yu., Gorbenko, V., Mihokova, E., Nikl, M., Nejezchleb, K., Vedda, A., Kolobanov, V., and Spassky, D. (2007). Single crystalline film scintillators based on Ce- and Pr-doped aluminium garnets, *Radiat. Meas.*, **42,** pp. 521–527.

61. Zorenko, Yu., Gorbenko, V., Voznyak, T., and Zorenko T. (2008). Luminescence of Pb^{2+} ions in YAG:Pb single-crystalline films, *Phys. Status Solidi B*, **245,** pp. 1618–1621.

62. Lushchik, Ch. B., Lushchik, N. E., and Muuga, I. A. (1963). Band spectra of crystals activated with mercury-like ions (I), *Trudy IFA AN EstSSR* **23,** pp. 22–37.

63. Zazubovich, S. G., Lushchik, N. E., and Lushchik, C. B. (1963). Optical structure of luminescence centres in ionic crystals activated by mercury-like ions, *Opt. Spektrosk.*, **15,** pp. 381–388.

64. Lushchik, Ch., Gindina, R., Zazubovich, S., and Lushchik, N. (1970). Nature of luminescence centers in ionic crystals, *Czech. J. Phys.*, **20B**, pp. 585–604.

65. Boulon, G., Jacquier, B., and Gaume-Mahn, F. (1973). Evolution du niveau $^3P_1$ des centres isoelectroniques du mercure en fonction de l'environnement, *J. Solid State Chem.*, **7**, pp. 349–352.

66. Blasse, G., Braam, W. M., and Heerschop, M. (1977). Influence of crystal structure on the luminescence of ions with s2 configuration, *J. Solid State Chem.*, **20**, pp. 63–65.

67. Lacklison, D. E., Scott, G. B., and Page, J. L. (1974). Absorption spectra of Bi^{3+} and Fe^{3+} in $Y_3Ga_5O_{12}$, *Solid State Commun.*, **14,** pp. 861–863.

68. Hughes, A. E. and Pells, G. P. 1975). The luminescence spectra of Bi^{3+} ions in MgO and CaO, *Phys. Status Solidi B*, **71,** pp. 707–718.

69. Ellervee, A. F. (1977). Luminescence of Ph^{2+} and Bi^{3+} centres in alkali-earth sulphides and oxides, *Phys. Status Solidi B*, **82,** pp. 91–98.

70. Zavt, G. S. and Ellervee, A. F. (1979). Pb^{2+} and Bi^{3+} impurity centres in alkali-earth oxides vibronic spectra, lattice dynamics, and electron-phonon interaction, *Phys. Status Solidi B*, **94,** pp. 757–768.

71. Wolfert, A. and Blasse, G. (1985). Luminescence of Bi^{3+}-doped crystals of Cs_2NaYBr_6 and $Cs_2NaLaCl_6$, *J. Solid St. Chem.*, **59,** pp. 133–142.

72. Wolfert, A. and Blasse, G. (1985). Luminescence of Bi^{3+}-activated LaOBr, a system with emission from different states, *J. Lumin.* **33**, pp. 213–226.

73. Boulon, G. (1987) Spectroscopy of post-transition metal ions, In *Spectroscopy of Solid-State Laser-Type Materials*, ed. Di Bartolo, B., Plenum Press: New York, pp. 223–266.

74. Blasse, G. (1988). Luminescence of inorganic solids: From isolated centres to concentrated systems, *Prog. Solid State Chem.*, **18,** pp. 79–171.

75. Ilmer, M., Grabmaier, B. C., and Blasse, G. (1994). Luminescence of Bi^{3+} in gallate garnets, *Chem. Mater.* **6,** pp. 204–206.

76. Kang, J. G., Yoon, H. M., Chun, G. M., Kim, Y. D., and Tsuboi, T. (1994). Spectroscopic studies of Bi^{3+} colour centres in KCl single crystals, *J. Phys.: Condens. Matter,* **6,** pp. 2101–2116.

77. Aceves, R., Barboza Flores M., Maaroos, A., Nagirnyi, V., Perez Salas, R., Tsuboi, T., Zazubovich, S., and Zepelin, V. (1996). RES structure of Bi^{3+} centres in KCI: Bi, S and CaO:Bi crystals, *Phys. Status Solidi B,* **194,** pp. 619–631.

78. Nikl, M., Novoselov, A., Mihokova, E., Polak, K., Dusek, M., McClune, B., Yoshikawa, A., and Fukuda, T. (2005). Photoluminescence of Bi^{3+} in $Y_3Ga_5O_{12}$ single-crystal host, *J. Phys.: Condens. Matter,* **17,** pp. 3367–3375.

79. Setlur, A. A. and Srivastava, A. M. (2006). The nature of Bi^{3+} luminescence in garnet hosts, *Opt. Mater.,* **29,** pp. 410–415.

80. Zorenko, Yu., Gorbenko, V., Voznyak, T., Vistovsky, V., Nedilko, S., and Nikl, M. (2007). Luminescence of image ions in image single crystalline films, *Radiat. Meas.,* **42,** pp. 882–886.

81. Zorenko, Yu., Mares, J. A., Kucerkova, R., Gorbenko, V., Savchyn, V., Voznyak, T., Nikl, M., Beitlerova, A., and Jurek, K. (2009). Optical, luminescence and scintillation characteristics of Bi-doped LuAG and YAG single crystalline films, *J. Phys. D.: Appl. Phys.,* **42,** 075501.

82. Donker, H., Yamashita, N., Smit, W. M. A., and Blasse, G. (1989). Luminescence decay times of the Sb^{3+}, Pb^{2+}, and Bi^{3+} ions in alkaline-earth sulfides, *Phys. Status Solidi B,* **156,** pp. 537–544.

83. Chen, L., Chen, K. J., Lin, Ch. Ch., Chu, C. I., Hu, S. F., Lee, M. H., and Liu, R. S. (2010). Combinatorial approach to the development of a single mass YVO4:Bi^{3+},Eu^{3+} phosphor with red and green dual colors for high color rendering white light-emitting diodes, *J. Comb. Chem.,* **12**, pp. 587–594.

84. Mu, Z., Hu, Y., Chen, L., and Wang, X. (2011). Enhanced luminescence of Dy^{3+} in $Y_3Al_5O_{12}$ by Bi^{3+} co-doping, *J. Lumin.,* **131**, pp. 1687–1691.

85. Xia, Z., Chen, D., Yang, M., and Ying, T. (2010). Synthesis and luminescence properties of YVO4:Eu^{3+},Bi^{3+} phosphor with enhanced photoluminescence by Bi^{3+} doping, *J. Phys. Chem. Solids,* **71**, pp. 175–180.

86. Babin, V., Gorbenko, V., Makhov, A., Mares, J. A., Nikl, M., Zazubovich, S., and Zorenko, Yu. (2007). Luminescence characteristics of Pb^{2+} centres

in undoped and Ce^{3+}-doped $Lu_3Al_5O_{12}$ single-crystalline films and $Pb^{2+} \rightarrow Ce^{3+}$ energy transfer processes, *J. Lumin.*, **127**, pp. 384–390.

87. Babin, V., Bichevin, V., Gorbenko, V., Makhov, A., Mihokova, E., Nikl, M., Vedda, A., Zazubovich, S., and Zorenko, Yu. (2009). Luminescence of dimer lead centers in aluminium perovskites and garnets, *Phys. Status Solidi B*, **246**, pp. 1318–1326.

88. Babin, V., Gorbenko, V., Makhov, A., Nikl, M., Zazubovich, S., and Zorenko, Yu. (2007). The role of Pb^{2+} ions in the luminescence of LuAG:Ce single crystalline films, *Phys. Stat. Sol. C*, **4**, 797–800.

89. Babin, V., Gorbenko, V., Krasnikov, A., Makhov, A., Nikl, M., Zazubovich, S., and Zorenko, Yu. (2010). Influence of lead-related centers on luminescence of Ce^{3+} and Pr^{3+} centers in single crystalline films of aluminium perovskites and garnets, *Radiat. Meas.*, **45**, pp. 415–418.

90. Babin, V., Gorbenko, V., Krasnikov, A., Makhov, A., Nikl, M., Polak, K., Zazubovich, S., and Zorenko, Yu. (2009). Peculiarities of excited state structure and photoluminescence in Bi^{3+}-doped $Lu_3Al_5O_{12}$ single-crystalline films, *J. Phys.: Condens. Matter*, **21**, 415502.

91. Babin, V., Gorbenko, V., Krasnikov, A., Makhov, A., Mihokova, E., Nikl, M., Zazubovich, S., and Zorenko, Yu. (2012). Origin of Bi^{3+}-related luminescence centres in $Lu_3Al_5O_{12}$:Bi and $Y_3Al_5O_{12}$:Bi single crystalline films and the structure of their relaxed excited states, *Phys. Status Solidi B*, **249**, pp. 1039–1045.

92. Babin, V., Gorbenko, V., Krasnikov, A., Makhov, A., Nikl, M., Zazubovich, S., and Zorenko, Yu. (2010). Photoluminescence of $Lu_3Al_5O_{12}$:Bi and $Y_3Al_5O_{12}$:Bi single crystalline films, *Radiat. Meas.*, **45**, pp. 331–335.

93. Gorbenko, V., Krasnikov, A., Mihokova, E., Nikl, M., Zazubovich, S., and Zorenko, Yu. (2012). Luminescence of lead-related centres in single crystalline films of Lu_2SiO_5, *J. Phys. D: Appl. Phys.*, **45**, 355304.

94. Babin, V., Gorbenko, V., Krasnikov, A., Mihokova, E., Nikl, M., Zazubovich, S., and Zorenko, Yu. (2013). Luminescence and origin of lead-related centers in single crystalline films of Y_2SiO_5 and Lu_2SiO_5, *Radiat. Meas.*, **56**, pp. 124–128.

95. Gorbenko, V., Krasnikov, A., Mihokova, E., Nikl, M., Zazubovich, S., and Zorenko, Yu. (2013). Photoluminescence and excited state structure of Bi^{3+}-related centers in Lu_2SiO_5:Bi single crystalline films, *J. Lumin.*, **134**, pp. 469–476.

96. Babin, V., Gorbenko, V., Krasnikov, A., Mihokova, E., Nikl, M., Zazubovich, S., and Zorenko, Yu. (2013). Photoluminescence and excited state

structure in Bi^{3+}-doped Y_2SiO_5 single crystalline films, *Radiat. Meas.*, **56**, pp. 90–93.

97. Scott, G. B. and Page, J. L. (1977). Pb valence in iron garnets, *J. Appl. Phys.*, **48**, pp. 1342–1349.

98. Jacobs, P. W. M. (1991). Alkali halide crystals containing impurity ions with the ns2 ground-state electronic configuration, *J. Phys. Chem. Solids*, **52**, pp. 35–67.

99. Zazubovich, S. (1994). Polarization spectroscopy of ns2 impurity ions in alkali halides, *Int. J. Mod. Phys. B*, **8**, pp. 985–1031.

100. Babin, V., Laguta, V. V., Maaroos, A., Makhov, A., Nikl, M., and Zazubovich, S. (2011). Luminescence of F^+-type centers in undoped $Lu_3Al_5O_{12}$ single crystals, *Phys. Status Solidi B*, **248**, pp. 239–242.

101. Inokuti, M. and Hirayama, F. (1965). Influence of energy transfer by the exchange mechanism on donor luminescence, *J. Chem. Phys.*, **43**, pp. 1978–1989.

102. Gorbenko, V., Krasnikov, A., Nikl, M., Zazubovich, S., and Zorenko, Yu. (2009). Luminescence characteristics of LuAG:Pr and YAG:Pr single crystalline films, *Opt. Mater.*, **31**, pp. 1805–1807.

103. Weber, M. J. (1973). Optical spectra of Ce^{3+} and Ce^{3+}-sensitized fluorescence in $YAlO_3$, *J. Appl. Phys.*, **44**, pp. 3205–3208.

104. Mareš, J. A., Nikl, M., Pedrini, C., Moine, B., and Blazek, K. (1992). A study of fluorescence emission of Ce^{3+} ions in $YAlO_3$ crystals by the influence of doping concentration and codoping with Nd^{3+} and Cr^{3+}, *Mater. Chem. Phys.*, **32**, pp. 342–348.

105. Mareš, J. A., Nikl, M., Chval, J., Dafinei, I., and Kvapil, J. (1995). Fluorescence and scintillation properties of $LuAlO_3$:Ce crystal, *Chem. Phys. Lett.*, **241**, pp. 311–316.

106. Laguta, V. V., Slipenyuk, A. M., Glinchuk, M. D., Bykov, I. P., Zorenko, Yu., Nikl, M., Rosa, J., and Nejezchleb, K. (2007). Paramagnetic impurity defects in LuAG:Ce thick film scintillators, *Radiat. Meas.*, **42**, pp. 835–838.

107. Weber, M. J. (2003). *Handbook of Optical Materials* (CRC Press, Boca Raton).

108. Henry, N. F. M. and Lonsdale, K. (1952). *International Tables for X-ray Crystallography* **1** (Kynoch Press, Birmingham UK).

109. Felsche, J. (1973). The crystal chemistry of the rare earth silicates, In *Structure and Bonding, Volume* 13, Springer-Verlag: Berlin, pp. 99–197.

110. Brandle, C. D., Valentino, A. J., and Berkstresser, G. W. (1986). Czochralski growth of rare-earth orthosilicates (Ln_2SiO_5), *J. Cryst. Growth,* **79**, pp. 308–315.

111. Chiriu, D., Faedda, N., Lehmann, A. G., Ricci, P. C., and Anedda, A. (2007). Structural characterization of $Lu_{1.8}Y_{0.2}SiO_5$ crystals, *Phys. Rev. B,* **76**, 054112.

112. Gustafsson, T., Klintenberg, M., Derenzo, S. E., Weber, M. J., and Tomas, J. O. (2001). Lu_2SiO_5 by single-crystal X-ray and neutron diffraction, *Acta Crystallogr., Sect. C: Cryst. Struct. Commun.*, **57**, pp. 668–669.

113. Aitasalo, T., Hölsä, J., Lastusaari, M., Niittykoski, J., and Pelle, F. (2005). Delayed luminescence of Ce^{3+} doped X1 form of Y_2SiO_5, *Opt. Mater.,* **27**, pp. 1511–1515.

114. Cooke, D. W., Lee, J.-K., Bennett, B. L., Groves, J. R., Jacobsohn, L. G., McKigney, E. A., Muenchausen, R. E., Nastasi, M., Sickafus, K. E., Tang, M., Valdez, J. A., Kim, J.-Y., and Hong, K. S. (2006). Luminescent properties and reduced dimensional behavior of hydrothermally prepared Y_2SiO_5:Ce nanophosphors, *Appl. Phys. Lett.*, **88**, 103108.

115. Ito, J. and Johnson, H. (1968). Synthesis and study of yttrialite, *Am. Mineral.,* **53**, pp. 1940–1952.

116. Aitasalo, T., Hölsä, J., Lastusaari, M., Legendziewicz, J., Niittykoski, J., and Pelle, F. (2004). Delayed luminescence of Ce^{3+} doped Y_2SiO_5, *Opt. Mater.,* **26**, pp. 107–112.

117. Coetsee, E., Terblans, J. J., Ntwaeaborwa, O. M., and Swart, H. C. (2009). Luminescent mechanism of Y 2SiO 5:Ce phosphor powder, *Phys. B,* **404**, pp. 4426–4430.

118. Pidol, L., Guillot-Noöl, O., Kahn-Harari, A., Viana, B., Pelenc, D., and Gourier, D. (2006). EPR study of Ce^{3+} ions in lutetium silicate scintillators $Lu_2Si_2O_7$ and Lu_2SiO_5, *J. Phys. Chem. Solids,* **67**, pp. 643–650.

119. Andriessen, J., Sobolev, A., Kuznetsov, A., Merenga, H., Dorenbos, P., and van Eijk, C. W. E. (1996). Theoretical investigation of the 4f and 5d levels of cerium in LSO, studied HF-LCAO and the embedded cluster scattered wave method, *Proc. Int. Conf. on Inorganic Scintillators and their Applications,* SCINT95, pp. 130–137.

120. Vedda, A., Nikl, M., Fasoli, M. M., Mihokova, E., Peichal, J., Dusek, M., Ren, G., Stanek, C. R., McClellan, K. J., and Byler, D. D. (2008). Thermally stimulated tunneling in rare-earth-doped oxyorthosilicates, *Phys. Rev. B,* **78**, 195123.

121. Nagirnyi, V., Stolovich, A., Zazubovich, S., Zepelin, V., Mihokova, E., Nikl, M., Pazzi, G. P., and Salvini, L. (1995). Peculiarities of the triplet relaxed excited-state structure and luminescence of a CsI: Tl crystal, *J. Phys.: Condens. Matter*, **7**, pp. 3637–3653.

122. Halbritter, G., Kiss, A., Kleinschmit, P., and Volker, W. (1985). Lead activated material and process for its production, U.S. Patent 4578212.

123. Pisarski, W. A., Pisarska, J., Dominiak-Dzik, G., and Ryba-Romanovski, W. (2004). Visible and infrared spectroscopy of Pr $^{3+}$ and Tm $^{3+}$ ions in lead borate glasses, *J. Phys.: Condens. Matter*, **16**, pp. 6171–6184.

124. Pisarska, J. (2009). Optical properties of lead borate glasses containing Dy^{3+} ions, *J. Phys.: Condens. Matter*, **21**, 285101.

125. Zhou, T., Wang, N., Li, Ch., Yan, H., and Xiao, D. (2010). Sulfide sensor based on room temperature phosphorescence of PbO SiO_2 nanocomposite, *Anal. Chem.*, **82**, pp. 1705–1711.

126. Zatsepin, A. F., Zhidkov, I. S., Kukharenko, A. I., Zatsepin, D. A., Andronov, M. P., and Cholakh, S. O. (2012). An intrinsic luminescence in binary lead silicate glasses, *Opt. Mater.*, **34**, pp. 807–811.

127. Shah, M. A. (2010). Lead oxide (PbO) nanoparticles prepared by a new technique for biomedical applications, *Int. J. Biomed. Nanosci. Nanotechnol.*, **1**, pp. 3–9.

128. Zorenko, Yu., Gorbenko, V., Savchyn, V., Voznyak, T., Grinyov, B., Sidletskiy, O., Kurtsev, D., Fedorov, A., Baumer, V., Nikl, M., Mares, J. A., Beitlerova, A., Prusa, P., and Kucera, M. (2011). Growth and luminescent properties of Lu_2SiO_5:Ce and $(Lu_{1-x}Gdx)_2SiO_5$:Ce single crystalline films, *J. Cryst. Growth*, **337**, pp. 72–80.

129. Nikl, M., Nitsch, K., Polak, K., Pazzi, G. P., Fabeni, P., Citrin, D. S., and Gurioli, M. (1995). Optical properties of the Pb^{2+}-based aggregated phase in a CsCl host crystal: Quantum-confinement effects, *Phys. Rev. B*, **51**, 5192.

130. Aceves, R., Babin, V., Barboza Flores, M., Fabeni, P., Maaroos, A., Nikl, M., Nitsch, K., Pazzi, G. P., Perez Salas, R., Sildos, I., Zazubovich, N., and Zazubovich, S. (2001). Spectroscopy of $CsPbBr_3$ quantum dots in CsBr: Pb crystals, *J. Lumin.*, **93**, pp. 27–41.

131. Fischer, P., Haelg, W., Stoll, E., and Segmueller, A. (1966). X-ray and neutron diffraction study of the substitutional disorder in the yttrium-iron-gallium garnets, *Acta Crystallogr.*, **21**, pp. 765–769.

132. Landolt-Börnstein (1950) *Zahlenwerte und Funktionen, I.Band, Atom- und Molekularphysik, I Teil, Atome und Ionen*, ed. Eucken, A., Springer-Verlag, p. 439.

133. Lushchik, N. E. and Lushchik, Ch. B. (1960). Spectroscopy of the luminescence centres in alkali-halide crystals activated by homologous series of ions, *Opt. Spektrosk.,* **8**, pp. 839–846.

134. Seitz, F. (1938). Interpretation of the properties of alkali halide-thallium phosphors, *J. Chem. Phys.*, **6,** pp. 150–162.

135. Hizhnyakov, V. V. and Kristoffel, N. N. (1984). Jahn-Teller mercury-like impurities in ionic crystals, In *The Dynamical Jahn-Teller Effect in Localized Systems,* eds. Perlin, Yu. and Wagner, M., Elsevier: Amsterdam, pp. 383–438.

136. Nagirnyi, V., Soovik, T., Vaino, P., Zazubovich, S., and Jaanson, N. (1991). Triplet relaxed excited state structure and luminescence of thallium-doped alkali halides, *Phys. Status Solidi B,* **164,** pp. 493–502.

137. Nagirnyi, V., Zazubovich, S., Zepelin, V., Nikl, M., and Pazzi, G. P. (1994). A new model for the visible emission of the CsI: Tl crystal, *Chem. Phys. Lett.,* **227,** pp. 533–538.

138. Babin, V., Kalder, K., Krasnikov, A., and Zazubovich, S. (2002). Luminescence and defects creation under photoexcitation of CsI:Tl crystals in Tl^{+}-related absorption bands, *J. Lumin.*, **96**, pp. 75–85.

139. Babin, V., Krasnikov, A., Nikl, M., Nitsch, K., Stolovich, A., and Zazubovich, S. (2003). Luminescence and relaxed excited state origin in CsI:Pb crystals, *J. Lumin.,* **101,** pp. 219–226.

140. Babin, V., Kalder, K., Krasnikov, A., Nikl, M., Nitsch, K., and Zazubovich, S. (2002). Defect creation under UV irradiation of CsI :Pb crystals in Pb^{2+}-induced absorption bands investigated by luminescence methods, *Phys. Status Solidi B,* **234,** pp. 689–700.

141. Moine, B., Pedrini, C., and Ghiordanescu, V. (1994). An emission and kinetic study of the impurity-trapped exciton in $CdCl_2$:Pb^{2+}, *J. Phys.: Condens. Matter,* **6,** pp. 4093–4098.

142. Fu, C.-R., Chen, L.-F., and Song, K. S. (1999). Self-trapped excitons in pure and Na- and Tl-doped caesium halides and the recombination luminescence, *J. Phys.: Condens. Matter,* **11,** pp. 5517–5532.

143. Zazubovich, S., Aceves, R., Barboza Flores, M., Fabeni, P., Kärner, T., Pazzi, G. P., Perez Salas, R., and Jaanson, N. (1997). Exciton 4.29 and 3.65 eV luminescence in CsI:Tl and CsI:Pb, *J. Phys.: Condens. Matter,* 9, pp. 7249–7256.

144. Babin, V., Krasnikov, A., and Zazubovich, S. (2003). Charge-transfer processes in doped alkali halides, *Radiat. Eff. Defects Solids,* **158**, pp. 227–230.

145. Tsuboi, T. and Jacobs, P. W. M. (1991). Optical studies of s2-ion dimer centers in alkali halide crystals, *J. Phys. Chem. Solids*, **52,** pp. 69–80.

146. Krasnikov, A., Savikhina, T., Zazubovich, S., Nikl, M., Mares, J. A., Blazek, K., and Nejezchleb, K. (2005). Luminescence and defects creation in Ce^{3+}-doped aluminium and lutetium perovskites and garnets, *Nucl. Instrum. Methods Phys. Res., Sect. A,* **537**, pp. 130–133.

147. Drozdowski, W., Lukasiewicz, T., Wojtowicz, A. J., Wisniewski, D., and Kisielewski, J. (2005). Thermoluminescence and scintillation of praseodymium-activated $Y_3Al_5O_{12}$ and $LuAlO_3$ crystals, *J. Cryst. Growth*, **275**, pp. e709–e714.

148. Pejchal, J., Nikl, M., Mihokova, E., Mares, J. A., Yoshikawa A., Ogino, H., Schillemat, K. M., Krasnikov, A., Vedda, A., Nejezchleb, K., and Mucka, V. (2009). Pr $^{3+}$-doped complex oxide single crystal scintillators, *J. Phys. D: Appl. Phys.*, **42**, 055117.

149. Martin, T. and Koch, A. (2006). Recent developments in X-ray imaging with micrometer spatial resolution, *J. Synchrotron Radiat.*, **13**, pp. 180–194.

150. Zorenko, Yu., Gorbenko, V., Nikl, M., Mares, J. A., Martin, T., and Douissard, P.-A. (2010). Development of novel UV emitting single crystalline film scintillators, *IEEE Trans. Nucl. Sci.*, **57**, pp. 1335–1342.

151. Zorenko, Yu., Gorbenko, V., Savchyn, V., Voznyak, T., Nikl, M., Mares, J. A., and Winnacker, A. (2010). The luminescence of ZnO ceramics, *Radiat. Meas.*, **45**, pp. 441–443.

152. Zorenko, Yu., Gorbenko, V., Savchyn, V., Voznyak, T., Sidletskiy, O., Grinyov, B., Nikl, M., Mares, J. A., Martin, T., and Douissard, P.-A. (2012). Single crystalline film scintillators based on the orthosilicate perovskite and garnet compounds, *IEEE Trans. Nucl. Sci.*, **59**, pp. 2260–2268.

153. Zorenko, Yu., Gorbenko, V., Zorenko, T., Malinowski, P. , Jary, V., Kucerkova, R., Beitlerova, A., Mares, J. A., Nikl, M., and Fedorov, A. (2014). Luminescent and scintillation properties of Bi^{3+} doped Y_2SiO_5 and Lu_2SiO_5 single crystalline film, *J. Lumin.*, **154**, pp. 525–530.

154. Zorenko, Yu., Voznyak, T., Gorbenko, V., Zorenko, T., Voloshinovski, A., Vistovsky, V., Nikl, M., Nejezchleb, K., Kolobanov, V., and Spasski, D. (2008). Luminescence spectroscopy of excitons and antisite defects in $Lu_3Al_5O_{12}$ single crystals and single-crystal films, *Opt. Spectrosc.*, **104**, pp. 75–87.

Chapter 7

ZnO-Based Phosphors and Scintillators: Preparation, Characterization, and Performance

Daniel Nižňanský,[a,b] Jakub Růžička,[a] Alena Beitlerová,[c] Jindřich Houžvička,[d] Petr Horodyský,[d] Václav Tyrpekl,[a,b] Ivo Jakubec,[b] Akira Yoshikawa,[e,f] and Martin Nikl[c]

[a]*Faculty of Science, Department of Inorganic Chemistry, Charles University, Hlavova 2030, 12840 Prague, Czech Republic*

[b]*Institute of Inorganic Chemistry of ASCR, Řež u Prahy, Czech Republic*

[c]*Institute of Physics, Czech Academy of Sciences, Cukrovarnicka 10, 16253 Prague, Czech Republic*

[d]*CRYTUR spol. S r.o., Palackeho 175, 51101 Turnov, Czech Republic*

[e]*Institute for Materials Research, Tohoku University, 2-1-1 Katahira, Aoba-ku, Sendai, 980-8577, Japan*

[f]*New Industry Creation Hatchery Center, 6-6-10 Aoba, Aramaki, Aoba-ku, Sendai, Miyagi 980-8579, Japan*

nikl @fzu.cz

In this chapter, a review of R&D activity in the field of ZnO-based scintillators is presented. Two preparation methods for undoped and Ga-doped ZnO powder phosphor which provide material with high intensity of fast excitonic luminescence in near-UV are

Nanocomposite, Ceramic, and Thin Film Scintillators
Edited by Martin Nikl

ISBN 978-981-4745-22-2 (Hardcover), 978-981-4745-23-9 (eBook)
www.panstanford.com

developed, while the slow visible emission is suppressed. XRD and SEM experiments are employed to characterize the structure and morphology of prepared materials and radioluminescence is the main tool to characterize their scintillation performance. An attempt to prepare ZnO:Ga ceramics is presented as well.

7.1 Introduction

The research on zinc oxide (ZnO) started already in the 1930s and its initial period has been reviewed, e.g., in [1]. ZnO belongs to IIb–VI semiconductor group and crystallizes preferentially in the hexagonal wurtzite structure. Its direct bandgap and rather wide forbidden gap of 3.4 eV determine its optical and luminescence properties. Luminescence spectrum is dominated by the exciton-based luminescence in near-UV region. At room temperature (RT) it is peaks at about 375–380 nm and has 400 ps decay time; see the recent review [2] for spectral characteristics and [3] for the decay time measurements. Temperature dependence of the UV exciton luminescence is shaped by the thermal quenching energy barriers due to thermal disintegration of the donor-bound exciton (10–13 meV) and of the free exciton itself (60 meV) [4, 5]. Its intensity decreases at RT approximately 10 times respect to T = 10 K. In addition, there is a complex visible luminescence in ZnO arising at various lattice and structural defects. Up to 10 different emission bands within 1.75–2.6 eV were resolved at T = 10 K [6]. In the formation of some of them the zinc and oxygen vacancies may play the role [7]. A focused review of defects in bulk ZnO samples has been published recently [8]. Temperature dependence of visible luminescence (bands at 2.10 and 2.43 eV) decreases with temperature more rapidly than the UV one mentioned above. At RT the decrease of about two orders of magnitude was found respect to T = 10 K and thermal quenching energy barriers of 15 and 200 meV were calculated [4].

In the search for superfast scintillators to be used in various time-of-flight or subnanosecond timing requiring applications the family of cross-luminescence halide materials was considered first; see the review in [9]. However, owing to inconveniently placed spectra in fluoride compounds in VUV-deep UV spectral region and low light yield in general their practical figure-of-merit appeared too low. In the last decade the Wannier exciton emission from direct bandgap semiconductors was reconsidered for above mentioned applications and CuI, HgI_2, PbI_2, ZnO:Ga, and CdS:In compounds in powder form were studied, especially for the scintillation decay characteristics under ps X-ray pulsed excitation [10]. Advantage of the whole family of direct gap semiconductors is a favourable combination of a smaller gap, providing higher number of electron–hole pairs (excitons) after initial conversion stage in scintillation mechanism, and an UV–vis emission center based on Wannier exciton which provides subnanosecond radiative lifetimes due to microscopic superradiance effect [3, 11]. On the other hand, Stokes shift of such emission centers is necessarily low (typically below 0.1 eV) and it prevents their usage in the bulk form due to enhanced reabsorption effect (see Fig. 7.1). Among the above-mentioned semiconducting materials, ZnO:Ga has shown the best combination of subnanosecond decay time and emission intensity values at RT. Ga-doping was applied in ZnO already in the 1960s [12, 13]. According to Derenzo et al. [10] and Lehmann [12, 13], the Ga states create a degenerated donor band around the edge of conduction band, and that results in a low-energy shift of the UV emission peaking around 390–395 nm at RT. Thus, the superfast UV emission in ZnO:Ga is intensified. Furthermore, its Stokes shift increases from about 60 meV in the undoped ZnO (given by the binding energy of free exciton) up to 200 meV along with increasing Ga concentration up to 1–2 × 10^{20} cm^{-3} [14]. The Stokes shift increase beneficially contributes to diminishing the above-mentioned undesirable reabsorption effects. It is worth noting the recent study [15] which, on the basis of picosecond time-resolved absorption measurement, presents relatively slow disappearance (in the order of hundreds of nanoseconds) of free electrons from conduction band of an undoped ZnO single crystal after a band-to-band excitation. The effect is ascribed to electron trapping at shallow electron traps, as a result of which only a very small part (of about 5%) of created electron–hole pairs can promptly recombine and give rise to excitonic emission with subnanosecond decay times. The beneficial effect of Ga doping can then be understood as due to excess free electrons provided to the conduction band. The excess free electrons can fill the mentioned electron traps and disable the trapping effect.

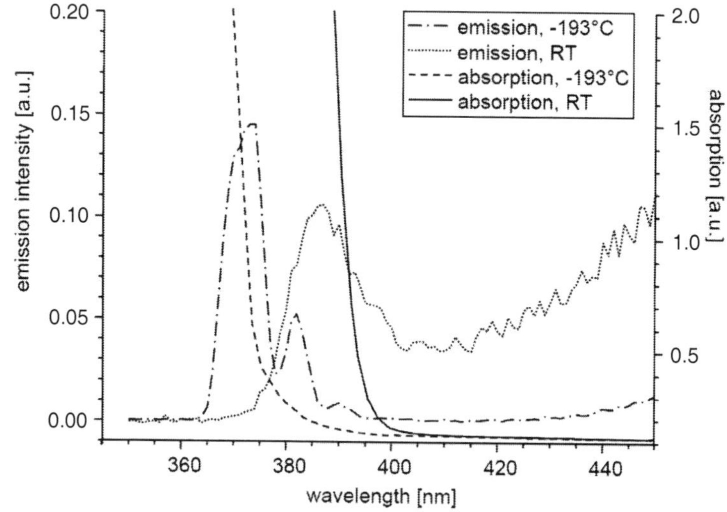

Figure 7.1 Radioluminescence and absorption spectra of the 1 mm thick plate of undoped ZnO at RT around the band edge. Reprinted with permission from [16].

Although the scintillation characteristics were measured also for the ZnO bulk crystals [17, 18] or undoped thin films [19, 20], the mainstream of R&D was focused on the Ga-doped powders and thin ZnO film systems [10, 14, 16, 21–24] due to their above-mentioned Ga-induced scintillation efficiency increase in the UV region and diminished reabsorption problem. An important improvement in the scintillation efficiency of the Ga-doped ZnO was reported recently [22, 25], namely, when the annealing in Ar-H_2 atmosphere at a temperature of about 800°C is applied as the last step of the post-preparation treatment of the Zn vacancy containing ZnO powder, the intensity of UV emission increases dramatically. The effect is interpreted as the UV radiative transition caused by the recombination of the Ga^{3+} donor band electrons and the holes trapped at shallow acceptors. The latter are ascribed to H^+ ions localized in Zn vacancies. Proposed interpretation is consistent with the characteristics of hydrogen centers in ZnO studied in [26, 27].

Let us briefly survey the applications which are the driving force for the ZnO-based scintillator development. As mentioned above, due to the reabsorption problem, only powder screens, thin films of any "optically thin" arrangement can be used for practical purposes.

ZnO-based scintillators suit particularly well as the alpha particle detectors, e.g., in a deuterium–tritium (D-T) neutron generator [28] or in other environmental, radiation safety, instrumental, or industrial applications [24, 29–31]. Large-area ZnO films are tested as fast cathodoluminescence screens [19]. Systematic research work is devoted to obtaining even faster response from quenched ZnO and it is accomplished by suitable doping [32–34]. Although scintillation efficiency is decreased, the response as fast as 15 ps has been reported recently [34]. Such a material can be used, e.g., for the free electron laser beam diagnostics or other superfast EUV or soft X-ray pulse sources.

Many different methods were applied to prepare ZnO in the bulk, thin film, nanopowder, ceramic [28], or (nano)composite forms. High-quality single crystals up to several inches in size were prepared by hydrothermal method [35], a high-pressure direct melting technique has been used as well [17, 36]. Solvothermal growth methods were applied to obtain atomically flat thin films a few micrometers in size (see review in [37]), while pulsed laser deposition, epitaxial growth, or sputtering were also used for scintillator thin film manufacturing [19, 20, 24, 38]. $Mg_xZn_{1-x}O$ nanopowders with an average grain size of 30 nm and with intense UV exciton luminescence were synthesized by the thermal decomposition method from water solution of acetate precursors [39]. ZnO nanoparticles were recently prepared by innovative radiation [40] and photochemical [41] methods as well. SiO_2/ZnO nanocomposites were prepared by sol-gel [42] or thermal annealing [43] methods. In the latter case the microscopic superradiance effect was observed that consists in substantial radiative lifetime shortening of the excitonic emission in ZnO quantum dot down to 65 ps at a temperature of 4.3 K in comparison with the value of 322 ps found in bulk ZnO.

Let us concentrate on the different methods that can be used for high-quality powder material preparation and that were reported in literature.

(a) The citrate (autocombustion) method [21, 42] is based on the preparation of mixed solution of all the used cations in the form of nitrates. This solution is then stabilised by citric acid (complexation reaction) and evaporated to the form of gel material. Further heating of this material leads to autocombustion and to the formation of very fine powder formed from the high-purity mixed oxide.

(b) Co-precipitation methods [44, 45] again employ the mixed solution of all the required cations in the form of nitrates, chlorides, or sulphates. After that, the precipitation follows using organic basic compounds, ammonium carbonate, or ammonia. The resulting intermediate product consists generally of hydrated carbonates, hydrated oxides, or hydroxides of the respective cations. Such intermediated products can be converted into oxide by heat treatment. Another possibility consists in the co-precipitation of the respective cations by organic acid-forming, poorly soluble compounds. Oxalic acid is very often used as a reactant. This acid does not form the insoluble Ga^{3+} compound, but a low amount of gallium cations can be easy incorporated into zinc oxalate by adsorption mechanisms. The resulting precipitate can be decomposed into oxides by hydrogen peroxide or simply by the heat treatment of the oxalates.

(c) The zinc oxide doping of Ga_2O_3 by solid state reaction is based on the diffusion of Ga^{3+} cations into ZnO structure using high annealing temperatures [21, 25, 45, 46, 47, 48]. This reaction is generally carried out by the joint heating of both oxides either in open air or in a closed fused quartz tube.

(d) Sublimation (vapor phase transport) [2, 6, 15, 26, 43] of Ga:ZnO in the vacuum at higher temperatures takes advantage of the ability of powder ZnO to resublime in a closed tube. When an evacuated tube or a tube filled with the inert gas is used, the sublimation can be observed already at the temperature of 750°C. The resulting product consists of the very fine powder which can be resublimed again. Its color depends on reducing environment in the closed tube. The more reduction atmosphere the darker the observed colour of the final product (changing from white to yellow, then to grey, and finally to black in a very strong reduction atmosphere).

The aim of this work was to find a preparation procedure that would lead to the material with the most intense exciton luminescence compared to the visible defect-based one peaking at the wavelength higher than 420 nm. Furthermore, we focused on the intensity of the emission output from the material (the influence of reabsorption, crystal size, and its agglomeration) and the influence of these properties on the preparation procedure, the calcinations temperatures, and the morphology of the final material. The economic aspects of industrial application were taken into account as well. The possibility of ceramic materials preparation was studied for the most efficient powders. Scintillation performance was checked by routine (X-ray excited) radioluminescence measurements..

7.2 Experimental

7.2.1 Sample Preparation

On the basis of bibliographic searching we chose several procedures of the powder material preparation. The ZnO system was the subject of many research works. For example, the group of Weber from Lawrence Berkeley National Laboratory published works on both the traditional ceramic method of preparation (subsequent multiple grinding and heat treatment of the commercial ZnO and Ga_2O_3) and the low temperature method based on the reaction of zinc nitrate and hexamethylenetetramine [22]. The classic method generally suffers from residual non-reacted reactants and inhomogeneities due to the solid state reaction of thermodynamically stable oxides. Low-temperature preparation in the water solution leads to the presence of OH group in the final material which can have spurious effects on luminescence efficiency. In this chapter we present alternative preparation methods in order to avoid the above-mentioned inconveniencies. Our aim was to prepare reactive oxides of both ZnO and doping elements which can then react at lower temperature giving more homogeneous material. With respect to the maximum possible purity of the final material and in order to lower the risk of the contamination by the other metal cations, the following preparation procedures and thermal treatments were chosen:

Citrate (autocombustion) method (CIT)

On the basis of our previous experience with similar oxide systems preparation, we chose the following amounts of reactants: 0.5 g of citric acid for 1 g of resulting ZnO. The starting solution containing the respective nitrates (the Ga amount ranging from 0 to 2 mass per cent) was heated up to the water evaporation, followed by autocombustion reaction. The preparation of co-doped samples with Ga^{3+} and Li^{+} (molar ratio of 1:1; total sum of Ga^{3+} and Li^{+} ranged from 0.1 to 2 mass per cent) were carried out using the same way in order to compensate for the charge of Ga^{3+}. The samples were denoted as **CIT B** with indication of the type and concentration of an eventual doping element.

Co-precipitation methods (CP)

The first group of the samples is based on increasing the pH of solution,

$$Zn^{2+} + 2OH^{-} \rightarrow Zn(OH)_2$$

$$Zn(OH)_2 \rightarrow ZnO + H_2O,$$

and creation of hydroxide mixtures of basic salts of Zn^{2+} and Ga^{3+} which include traces of other ions from solution (NH_4^{+}, OH^{-}, O^{2-}, NO_3^{-}, CO_3^{2-}, SO_4^{2-}, Cl^{-}). For the increasing pH of the solution we used ammonia, urea, and ammonium carbonate or bicarbonate. These samples were calcinated and analyzed, but the intensity of luminescence was very low, probably due to the large contamination by other anions.

The second group of the samples, where the co-precipitation method was employed, used the oxalic acid as reactant.

$$Zn^{2+} + C_2O_4^{2-} \rightarrow ZnC_2O_4$$

And the resulting zinc oxalates were decomposed thermally or even chemically.

$$2ZnC_2O_4 + O_2 \rightarrow 2ZnO + 4CO_2$$

$$ZnC_2O_4 + H_2O_2 \rightarrow ZnO + H_2O + 2CO_2$$

These samples were heat-treated and denoted as **CP**.

The incorporation of gallium nitrate into the ZnO or ZnC_2O_4 precipitate (IK)

The suspension of pure zinc oxide or oxalate was doped by gallium nitrate solution. The solvent was evaporated and the resulting solid product was grinded and progressively heated. The *decomposition of zinc oxalate* and gallium nitrate took place during the heating and

Ga^{3+} was incorporated into ZnO structure. This method combines the precipitation method with the doping using solid state reaction. The samples doped by heat treatment in corundum crucible in the air were proven to have higher luminescent effectiveness than samples treated in the vacuum. Moreover, this procedure can lead to the lower production costs. The doping was done in suspension with gallium nitrate solution; this mixture was subsequently dried. The reaction was carried out at lower temperature and no expensive mechanical homogenization was necessary. The samples containing 0 to 2 mass percent of Ga were prepared and denoted as **IK**.

Zinc oxide doping using Ga_2O_3 by solid state reaction

This method is based on the diffusion of Ga^{3+} cations into the ZnO structure using the high annealing temperatures. We studied the preparation procedure based on simultaneous heating of both oxides, both in the air and in the closed fused quartz tube.

Pure ZnO prepared by different makers and methods was mixed with the appropriate amount of Ga_2O_3 and then heat-treated in the evacuated quartz tube at the temperatures higher than 1000°C. However, this procedure suffered some inconveniences due to the undesirable partial sublimation of the SiO_2 from quartz tube into the resulting material and solid state reaction of ZnO with SiO_2 surface. These effects lead to powder contamination and quartz tube destruction.

Sublimation of Ga:ZnO in vacuum at higher temperatures

This method was used for the powder prepared by citrate method. The resulting product consisted of the very fine powder and it was denoted as **ZnO sublimate**. The color of the sample changed from yellow-white in the case of the sample treated in the presence of oxygen to black in the case of the sample prepared under *reduction atmosphere*. All the materials were retreated in air and reduced at the temperature of 800°C, just like powders prepared by other methods. But the resulting luminescence efficiency was very low.

Heat treatment of the samples

The prepared samples were heat-treated by different methods. The heat pre-treatment of the sample put into corundum crucible and annealed in chamber furnace at a temperature of 800–1100°C proved to be the most efficient procedure. These conditions lead to ZnO crystal growth and Ga^{3+} diffusion into zinc oxide structure. Then, the heat treatment in the reduction atmosphere at a temperature of 800°C followed [25]. This procedure was realized in a quartz tube placed into tubular furnace under controlled atmosphere using a mix of the 5% H_2 and argon. The use of N_2 instead of argon did not lead to good results.

7.2.2 Characterization Methods

Powder XRD spectra were measured at ambient temperature using Phillips X-pert diffractometer with Cu Kα-radiation.

Thermal analysis (TG and DTA) measurements were carried out using SETARAM device and NETZSCH STA (QMS) 409/429-403 coupled to the mass spectroscopy unit.

SEM was accomplished by Scanning Electron Microscope PHILIPS XL 30 CP and by Tescan Proxima SEM system equipped with SE, BSE, and CL detectors. Cathodoluminescence (CL) detector is sensitive in spectral range 350–650 nm.

Radio- and photoluminescence characterization was performed using model 5000M Spectrofluorometer, Horiba Jobin, Yvon, equipped with single-grating monochromators and photon-counting detectors. In the case of *radioluminescence* the X-ray tube (40 kV, 15 mA) Seifert GmbH was used as excitation source. In case of photoluminescence the D_2 continuous lamp and microsecond pulse xenon flashlamp were used for the spectra and slow *decay kinetics* measurements, respectively. *Scintillation decay* was measured using the ps X-ray pulse source excitation and streak camera detection (see the details in [49]). All measurements were performed at RT.

7.3 Experimental Results

The sample prepared by the incorporation of gallium nitrate into the precipitate (IK) was analyzed using thermal analysis TG-DTA coupled to mass spectroscopy analyzer. The results are represented in the Fig. 7.2. From the TG curve we can conclude that the decomposition of the zinc oxalate consists of two steps. The first step corresponds to the release of the structural water (endo-effect with the onset at a temperature of 173 °C on the DTA curve). The 19% weight loss corresponds to two water molecules and that is in good agreement with the formula of zinc oxalate ($ZnC_2O_4.2H_2O$). The second step corresponds to the release of carbon dioxide which occurs at a temperature of 410°C. No other mass change takes place above the temperature of 430°C up to 1100°C. Zinc oxide is already formed.

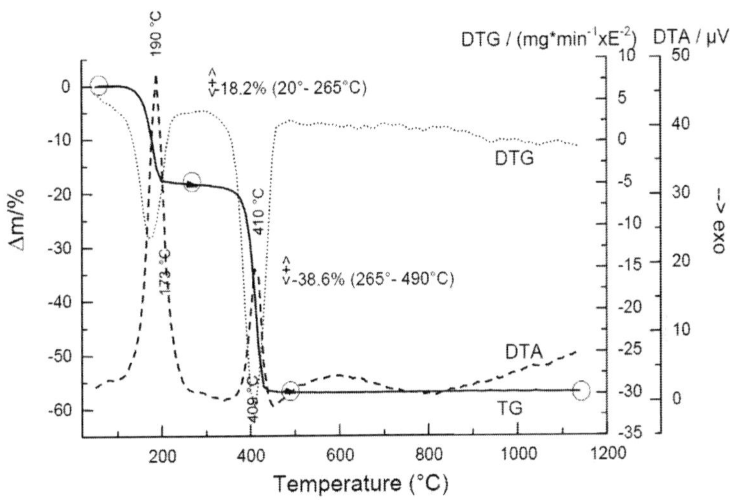

Figure 7.2 Thermal analysis of the zinc oxalate prepared by co-precipitation. TG curve corresponds to the thermogravimetric analysis. DTG curve represents its derivative, and DTA curve represents differential thermal analysis.

The *XRD analysis* was carried out in order to characterize the phase composition and also to estimate the crystallite size from the line broadening.

Two samples prepared by precipitation method (IK) were analyzed using XRD (sample heat-treated at 450°C and at 900°C). The results are shown in Fig. 7.3. It shows that in agreement with the thermal analysis (Fig. 7.2) the decomposition of zinc oxalate is already finished at the temperature of 450°C; only the ZnO phase is present and other phases are not observed. The sample heat-treated at 900°C was measured using XRD in order to confirm that no phase transition between 450°C and 900°C took place.

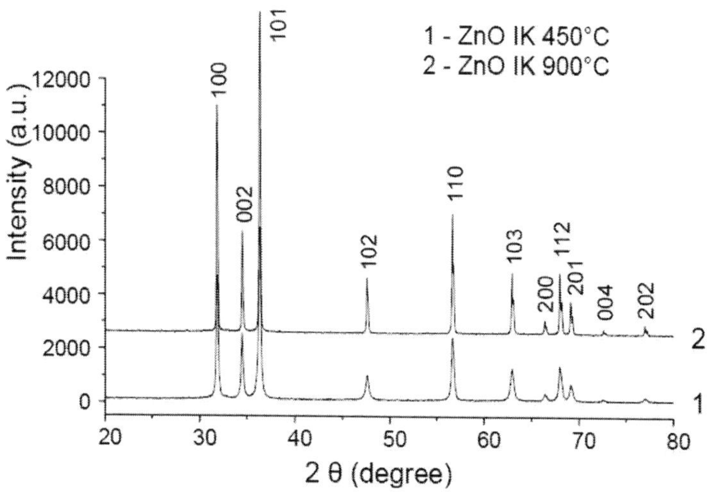

Figure 7.3 XRD of the sample prepared by precipitation method (IK) heat-treated at 450°C and at 900°C.

The same XRD analysis was carried out for the sample prepared by citrate method (CIT B). Figure 7.4 shows the results. The measurement proved that only the ZnO phase is present above the temperature of 450°C and no phase transition took place in temperature range between 450°C and 900°C. We can observe only peak narrowing due to the crystal growth (Fig. 7.3). This phenomenon (higher temperature of heat treatment → increase of crystal size → narrower XRD diffraction lines) is usual and it was observed in all the samples prepared by all the methods used.

The particle size from XRD spectra was estimated using Scherrer formula),

$$D_\mathrm{P} = \frac{0.94\lambda}{\beta_{1/2}\cos\theta} \tag{7.1}$$

where D_p represents particle size, λ is the used wavelength of Cu(Kα) radiation (1.542 Å), $\beta_{1/2}$ is full width at half maximum of the line, and θ is the diffraction angle of the given line.

Particle size in case of the most luminescent active powders is larger than 1 μm, which is the limit of using Scherrer equation for particle size determination. In case of the powder CIT B heat-treated at 500°C, we found the particle size from Scherrer equation to be 300 nm.

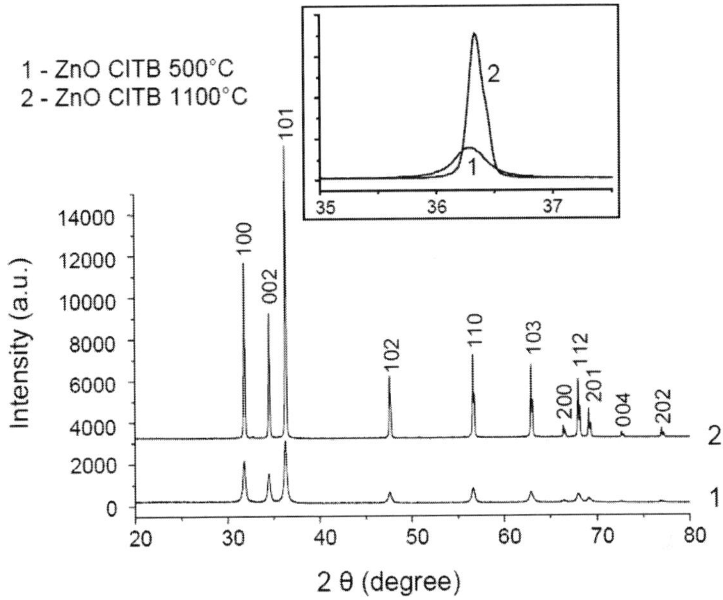

Figure 7.4 XRD spectrum of the ZnO:0.05%Ga sample prepared by citrate method (CIT B) heat-treated at 500°C and at 1100°C. The inset shows the comparison of widths of the most intensive line (101).

The XRD method reveals itself to be very useful in the case of sublimated samples (ZnO sublimates). Figure 7.5 shows the XRD of the sample obtained from sublimated ZnO and the same sample additionally heat-treated with H_2+Ar gas mixture (5% H_2 in argon). Figure 7. 5 shows that the sublimated particles are rather fine (the peaks are broader). When we perform heat treatment of the same sample at the same temperature but in the reduction atmosphere, we observe that diffraction lines become narrower and a little shifted towards the higher value of 2θ diffraction angle (smaller crystal parameters of ZnO unit cell). It means that the reduction atmosphere heat treatment leads to better crystallization of the sample and may as well lead to a better-ordered structure while the sample not treated in this way has smaller crystallites and higher unit cell parameters.

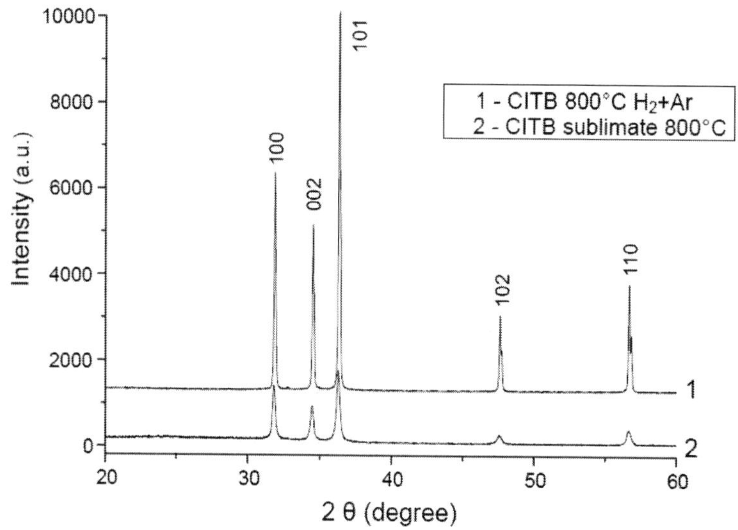

Figure 7.5 XRD of sublimed sample formed during reduction heat treatment at 800°C. Particle size calculated from Scherer equation is 600 nm for the original powder made by citrate method and 300 nm for the sublimed one.

The morphology of the samples was studied using SEM. Figure 7.6 shows the sample prepared using citrate method with a lower amount of citric acid (CIT A – 0.1g/1g ZnO) at 900°C and the sample heat-treated at 1100°C in the evacuated quartz tube. The sample prepared using citrate method shows the average particle size of the order of micrometers. The crystallites are organized into aggregates of the size of tens of micrometers. The sample heat-treated in the vacuum at 1100°C shows fragments with the average particle size of about 50 μm.

Figure 7.7 shows the samples prepared at lower temperature. We observe that the average particle size is lower than 1 μm and the samples are "fluffy." The sublimed sample shows some crystallites in the form of needles.

Figure 7.8 shows the samples prepared using co-precipitation methods. These samples have very regular shape of particles, the sizes of which depend on heat treatment temperature. The sample heat-treated at 900°C has submicrometer particle size while the sample heat-treated at 1100°C shows the particle size of about 3 μm.

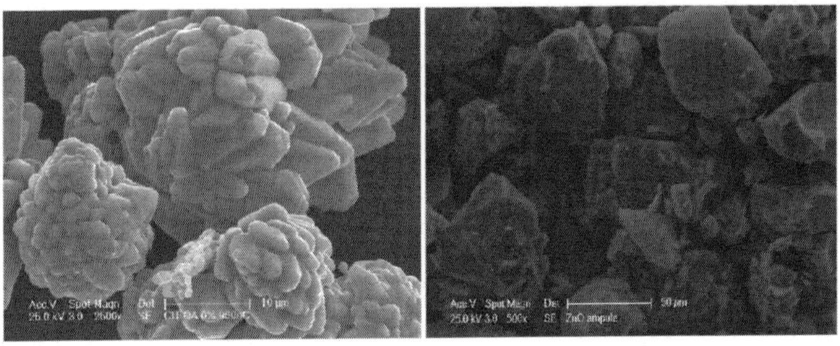

Figure 7.6 SEM of the sample prepared by citrate method (CIT-A) and heat-treated at 900°C (left), and of the sample heat-treated in the evacuated quartz tube at 1100°C (right).

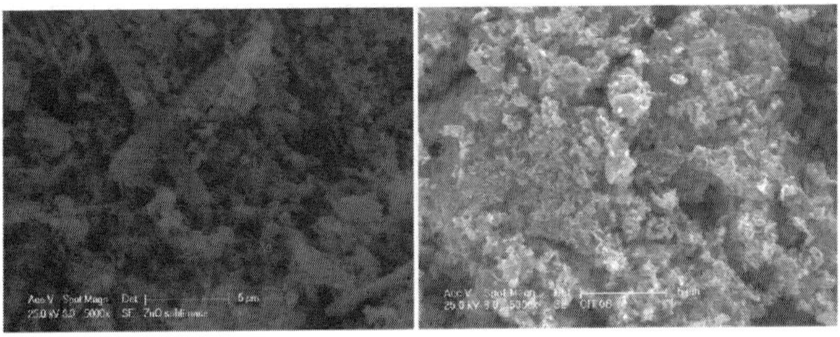

Figure 7.7 SEM of the sample sublimed at 800°C (left), and of the sample prepared by citrate method and heat-treated at 500°C (CIT B) (right).

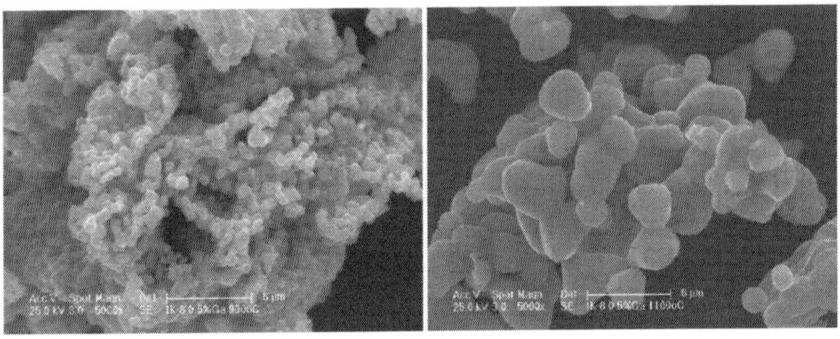

Figure 7.8 SEM of the sample prepared using precipitation method (IK) and heat-treated at 900°C (left), and at 1100°C (right).

The same tendency was observed in the case of the samples prepared by the citrate method (see Fig. 7.9). The crystals formed the aggregates of the submicrometer size in the case of the sample heat-treated at 900°C and of the 2–4 μm size in case of the sample heat-treated at 1100°C.

Figure 7.9 SEM of the sample prepared by citrate methods (CIT B) and heat-treated at 900°C (left), and at 1100°C (right).

The radioluminescence measurement in the reflection geometry at RT was used as the main characterization tool. Powdered bismuth germanate $Bi_4Ge_3O_{12}$ (BGO) was chosen as a scintillation standard for the determination of relative scintillation efficiency of the samples. Two main features were generally observed in the radioluminescence spectra of ZnO samples: the *exciton (band-edge) luminescence* peak centerd at about 390 nm and a broad visible luminescence band centerd at about 500 nm.

Figure 7.10 shows the example of RL spectra of the prepared *undoped ZnO* samples (not treated under the reducing atmosphere). Only the visible luminescence band is present in the spectra with varying intensity. The observed variation in shape and position of the spectra indicates rather different defects of the samples produced by different methods. In Table 7.1 the luminescence maxima and their intensities are summarized for the studied and also several commercial (Lachema, Penta, Aldrich-NP) powder samples and a ZnO ceramic described below. All the samples given in Table 7.1 were heat-treated at 1100°C. This temperature was proven to provide the highest values of luminescence intensity. Table 7.1. shows that the doping by Ga shifts the visible luminescence maximum towards longer wavelengths (red shift). The co-doping with Li leads to further shifts, and the values of wavelength are higher than 600 nm. On the other hand, there is no unambiguous relation between the dopant concentration and the position of visible luminescence maximum, the values of wavelength are therefore given as a range for the respective preparation method (values denoted by asterisk).

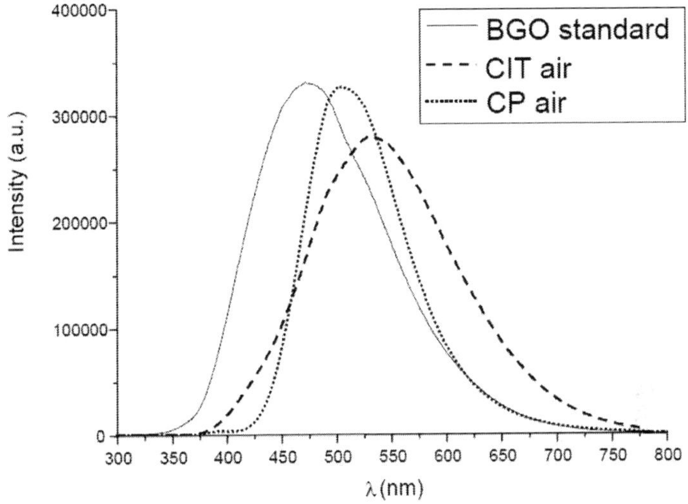

Figure 7.10 Radioluminescence spectra of the "as prepared" undoped ZnO samples. The method is given in the legend.

Because of the existing discussion in the literature concerning the nature of visible luminescence and few data available for its decay kinetics [50, 51], we measured photoluminescence decay at the selected samples from the most intense spectra sample sets IK and CIT B. The decays were always very slow, extending to hundreds of microseconds (see Fig. 7.11a). The fit of the decay by three-exponential function provides "decay time" values of 4–5 μs, 30–40 μs, and 150–170 μs. However, they have no physical meaning and serve merely to evaluate the speed of the decay itself. The decay integral is given in Fig. 7.11b and the evaluation shows that 90% of its intensity is collected within 70 μs. These results are strikingly different from those reported in [50] and [51], where the decays were measured in time scale of units of microseconds or even hundreds of nanoseconds only, respectively, and no decay time values beyond several microseconds were found.

Table 7.1 The survey of peak wavelength values and intensity of the visible luminescence band. The samples were not heat-treated in the reduction atmosphere.

Sample notation	Wavelength of band maximum (nm)	Intensity at maximum (a.u.)
BGO – powder	475	328075
ZnO – Lachema	501	45844
ZnO – Penta	499	85039
ZnO – Aldrich-NP	536	51051
ZnO – CIT B	538	38175
ZnO – CP (IK)	501	96012
ZnO:Ga 0.05-0,2% - Lachema	520–560	5–100 ($\times 10^3$)*
ZnO:Ga 0.05-0,2% - CIT B	540–570	40–500 ($\times 10^3$)*
ZnO:Ga 0.05-0,2% - IK	510–540	50–200 ($\times 10^3$)*
ZnO:GaLi 0.1-2% - CIT B	640–650	10–30 ($\times 10^3$)*
ZnO:GaLi 0.1-2%– IK	620–650	20–70 ($\times 10^3$)*
CERAMIC ZnO:Ga 0,005 and 0.1%	520–540	100–500($\times 10^3$)*

*All samples in these sets have similar values of peak wavelength and intensity without clear tendency. Therefore they are given as a range of values.

Evaluating the decay characteristics over a relatively large set of ZnO:Ga (IK) samples with concentration of Ga within 0.04%–0.4%, we do not observe any reasonable correlation either with radioluminescence intensity of a particular sample or with Ga concentration. Plotting the decay curve in log $I(t)$ – log(t) axes shows in some cases the straight line course through several orders of magnitude (see Fig. 7.12 a,b), which may indicate the tunneling-driven radiative recombination process related to visible luminescence [52]. This possibility is given by the suggested mechanism of visible emission in [50]. The theoretical model available for this kind of decay process [52] concludes that the asymptotic decay form $I(t) \sim t^{-p}$ is physically reasonable for the values of parameter p within 0.95–1.5, while the evaluation of the slope of the curve in Fig. 7.12 a,b provides values within 2–2.5. However, the overestimation of the straight line at early times of the decay may indicate the superposition of an exponential and hyperbolic decay form which would complicate the analysis.

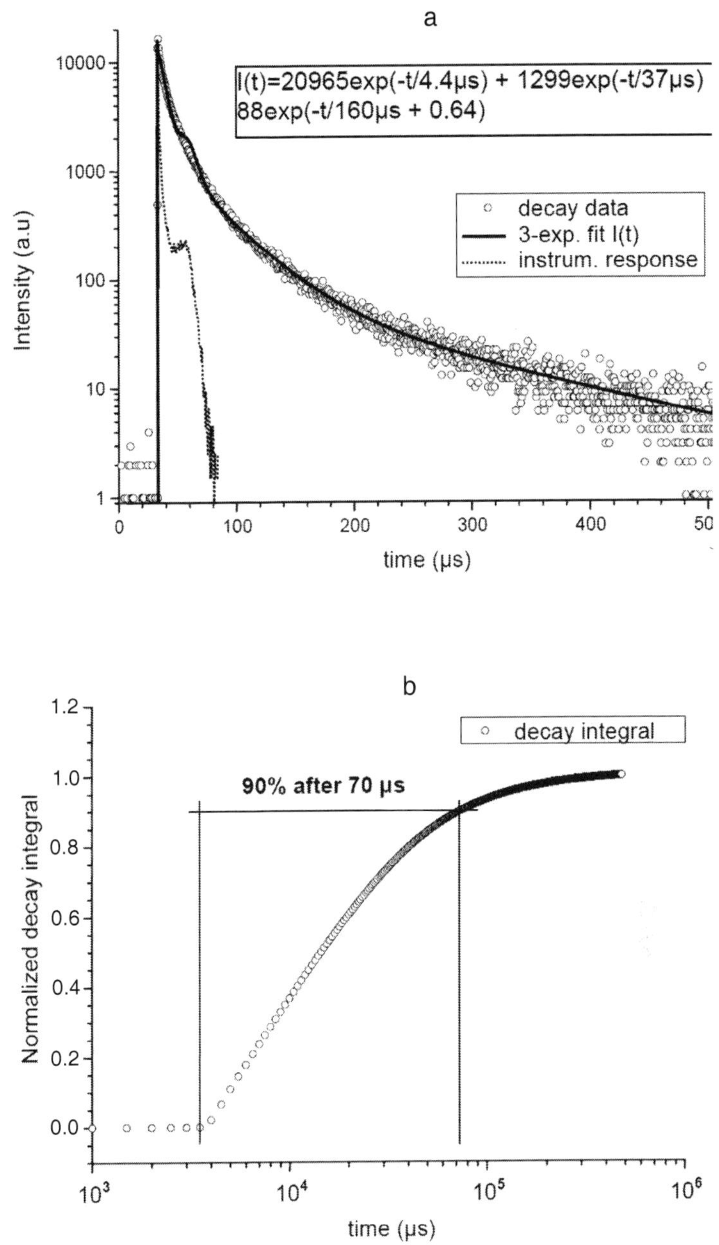

Figure 7.11 (a) Photoluminescence decay of visible luminescence in ZnO, exc = 375 nm, em = 575 nm. Fit $I(t)$ is provided by the solid line. (b) Normalized integral of the decay from (a).

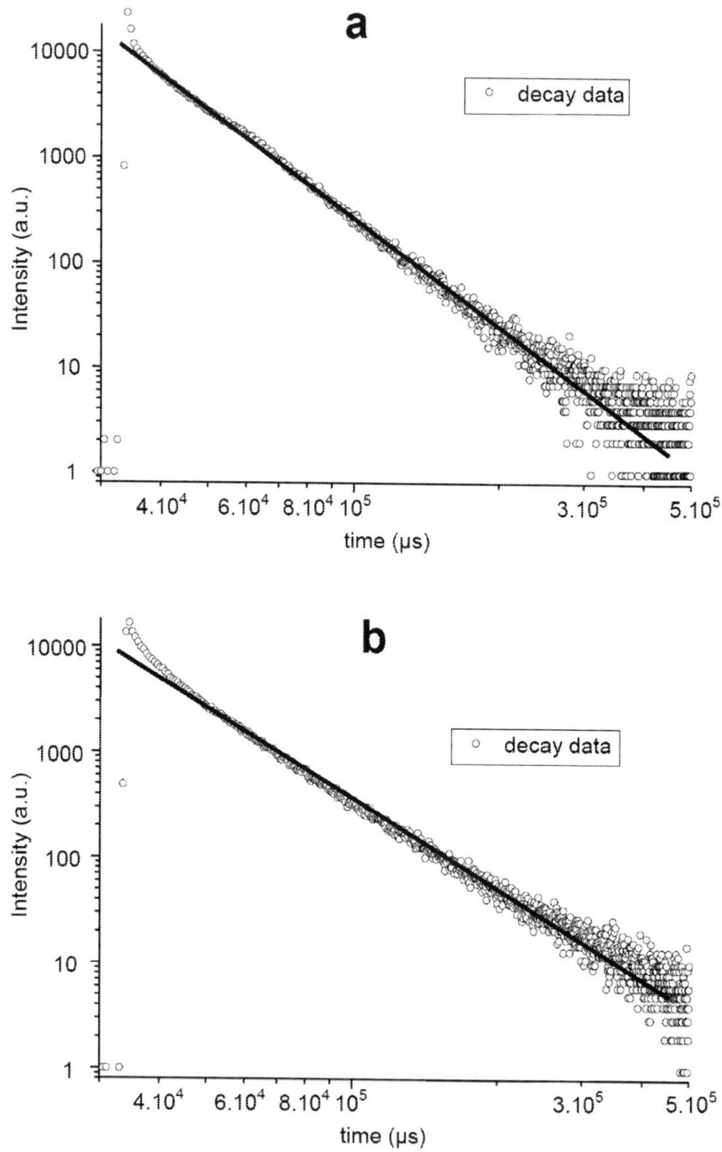

Figure 7.12 Photoluminescence decays of Ga 0.04% (a) and Ga 0.3% (b) plotted in log $I(t)$ – log (t) axes. Exc = 375 nm, em = 575 nm. The solid straight line is just to guide the eye.

In order to improve the ratio of exciton/visible radioluminescence intensity, the heat treatment in the reduction atmosphere was carried out following [22] and [25]. Resulting radioluminescence spectra are shown in the Fig. 7.13. In the most favorable cases the visible luminescence practically disappeared while the exciton one centerd at 395 nm was strongly intensified.

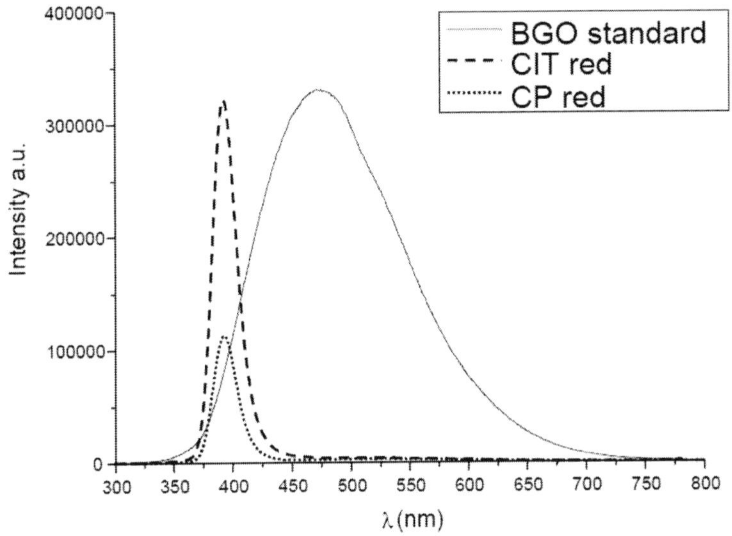

Figure 7.13 RL spectra of the sample heat-treated in reducing conditions (5% H_2 in Ar) compared with BGO standard.

From Fig. 7.14 we obtain the comparison of the exciton luminescence yield of the samples prepared by different methods and depending on Ga concentration. In the most favorable case of CIT B samples the best Ga concentration was between 0.1 and 0.15 mass percent of Ga, which is of the same order as 0.3% stated in [25]. Scintillation decay under ps X-ray pulse excitation [49] was measured at Ga 0.12% CIT B sample (see Fig. 7.15). Obtained single exponential decay with 0.37 ns decay time is in excellent agreement with the photoluminescence one [3]. Within given dynamical resolution, no slower components were noticed.

For the samples prepared by citrate methods, the influence of air thermal pre-treatment (followed by reduction heat treatment at different temperatures) on exciton luminescence yield was studied. The results are shown in Fig. 7.16.

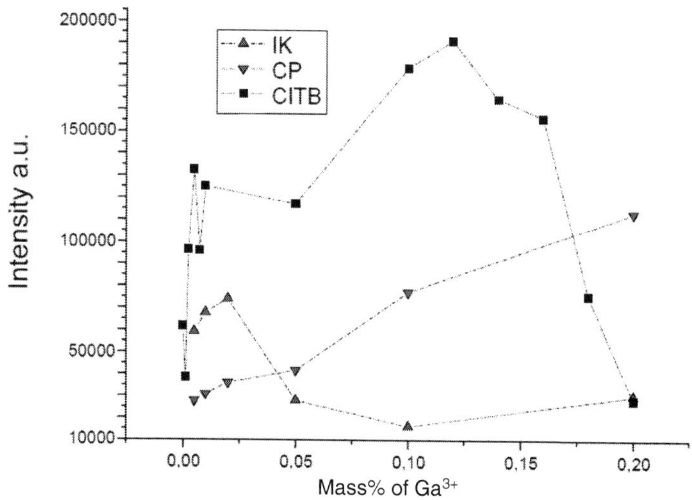

Figure 7.14 Intensity at the maximum of excitonic luminiscence (at 390–395 nm) from RL spectra of the samples prepared by different methods with different Ga-dopant concentrations.

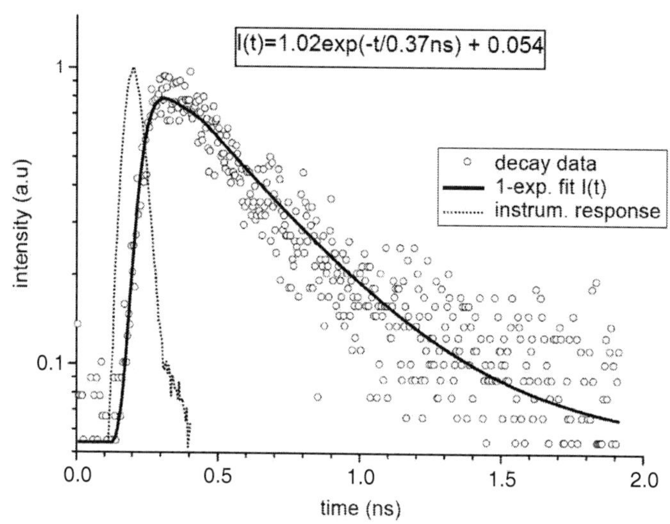

Figure 7.15 Spectrally unresolved scintillation decay under ps X-ray excitation (see [49] for technical details). Function $I(t)$ is convoluted with the instrumental response to excitation pulse. The resulting single exponential fit with 0.37 ns decay time is given by the solid line.

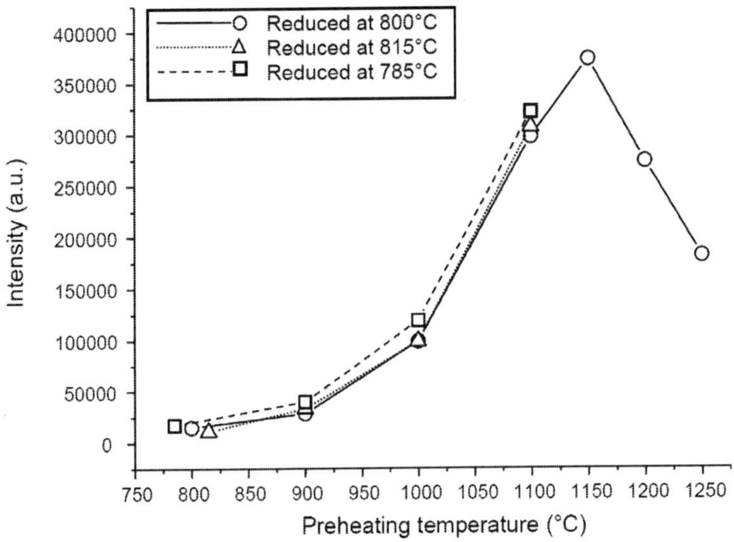

Figure 7.16 Samples prepared by citrate method. Influence of air thermal pre-treatment (followed by reduction heat treatment at different temperatures) on exciton luminescence intensity (excitation X-ray). Samples prepared by the citrate method.

From the experimental data above, it can be concluded that in order to gain the maximum intensity of exciton luminescence, it is advantageous to carry out firstly the thermal pre-treatment at 1100°C accompanied by ZnO crystallites growth. Following reduction heat treatment at 785–815°C leads to good phosphor material with the intense fast exciton luminescence and negligible slow visible one.

Focusing on the powders with the maximum value of exciton luminescence intensity (CIT B and IK at 1100°C), it can be concluded that from the morphological point of view, the best results are obtained in the case of the round crystallites of the size of 1–5 μm. These shapes can be obtained by the heat treatment of amorphous precipitations or gels in the air at the temperatures above 800°C.

The powder prepared by the citrate method (CIT B) was unaxially pressed (200 MPa) into the form of bar, then it was sintered at 1100°C for 2 and then 15 hours. The resulting ceramic material was cut into form of translucent disks (with an approximate thickness of 2 mm) and submitted to reduction heat treatment (5% H_2 in Ar) at 800°C (denoted as RED).

The radioluminescence spectra of ceramic samples are shown in Fig. 7.17 for two different dopant concentrations. Apparently, the higher Ga concentration leads to the decrease in visible luminescence but the relative intensity of the excitonic one is much lower than in the powders themselves. This result needs another study. From a few studies of ZnO ceramics available in the literature [48–50], it can be concluded that it is the visible luminescence which provides the dominant contribution to the light yield. Similar radioluminescence characteristics were obtained in [53] and Fig. 7.17 below. Such ceramics, however, will have very slow scintillation response of the order of tens of microseconds due to visible luminescence decay characteristics (see Fig. 7.11), which is the reason for the time dependence of light yield reported in [50]. Furthermore, the problem of reabsorption of UV fast exciton emission in a bulk ZnO or ZnO:Ga sample naturally remains.

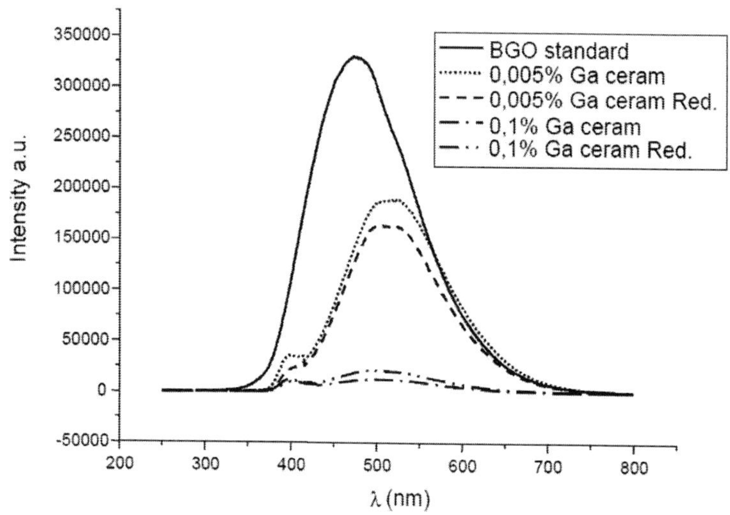

Figure 7.17 Radioluminescence spectra of ZnO ceramics: the influence of Ga concentration and effect of reduction heat treatment (RED).

The hydrogenation in the case of ceramics did not lead to the increase in exciton luminescence and to the suppression of visible luminescence, as can be followed in the powder samples. The luminescence intensities of the reduction heat-treated samples (denoted as RED) are actually slightly lower in comparison with non-reduced ceramics. The reason can be found in worse gas permeability of the bulk ceramic material in comparison with the powder one.

7.4 Summary

An overlook of R&D activities in the field of ZnO-based scintillators was provided. Among the preparation methods of undoped and Ga-doped ZnO powder, the citrate (autocombustion) and precipitate (Ga-nitrate using) methods were found to be the most suitable for the preparation of ZnO:Ga phosphors with high intensity of UV exciton-based luminescence and scintillation response. In addition, slow visible luminescence was diminished almost completely. Annealing in the air at about 1100°C followed by annealing in reduction Ar-H_2 atmosphere at about 800°C appeared to be crucial for the scintillation performance tuning. The best performing morphology of the powders are spherical grains with a diameter of a few micrometers. Measurement of scintillation decay at an optimized ZnO:Ga powder provided 0.4 ns decay time without noticeable presence of slower components. Prepared ZnO:Ga ceramic showed dominant contribution from the slow visible luminescence peaking between 500 and 550 nm in the radioluminescence spectrum. Scintillation response of such a ceramic will be necessarily slow as it follows from the photoluminescence decay characteristics of visible emission that 90% of the intensity is collected within 70 μs. Visible luminescence shows hyperbolic decay, which might be explained by tunneling-driven radiative recombination.

Acknowledgments

The financial support of Czech GA AV KAN300100802 project is gratefully acknowledged. Partial support by the JST Program, Adaptable & Seamless Technology Transfer Program through Target-driven R&D (A-STEP), and Development of Systems and Technology for Advanced Measurement and Analysis is also greatly appreciated.

References

1. Heiland, H., Mollwo, E., and Stöckmann, F. (1959). In *Solid State Physics*, Vol. **8**, F. Seitz, D. Turnbull (eds.), Academic Press: New York, pp. 191–323.
2. Klingshirn, C. (2007). ZnO: from basics towards applications. *Phys. Stat. Sol. B* **244**, pp. 3027–3073.
3. Wilkinson, J., Ucer, K. B., and Williams, R. T. (2004). Picosecond excitonic luminescence in ZnO and other wide-gap semiconductors. *Radiat. Meas.* **38**, pp. 501–505.
4. Nikl, M., Pazzi, G. P., Fabeni, P., Mihokova, E., Pejchal, J., Ehrentraut, D., Yoshikawa, A., and Williams, R. T. (2009). Decay kinetics of the defect-based visible luminescence in ZnO. *J. Lumin.* **129**, pp. 1564–1567.
5. Robin, I. C., Ribeaud, A., Brochen, S., Feuillet, G., Ferret, P., Mariette, H., Ehrentraut, D., and Fukuda, T. (2008). *Appl. Phys. Lett.* **92**, pp. 141101(1)–141101(3).
6. Reshchikov, M. A., Morkoc, H., Nemeth, B., Nause, J., Xie, J., Hertog, B., and Osinsky, A. (2007). Luminescence properties of defects in ZnO. *Physica B* **401–402**, pp. 358–361.
7. Børseth, T. M., Svensson, B. G., Kuznetsov, A. Yu., Klason, P., Zhao, Q. X., and Willander, M. (2006). *Appl. Phys. Lett.* **89**, pp. 262112(1)–262112(3).
8. McCluskey, M. D. and Jokela, S. J. (2009). Defects in ZnO. *J. Appl. Phys.* **106**, pp. 071101(1)–071101(13).
9. Rodnyi, P. A. (2004). Core-valence luminescence in scintillators. *Radiat. Meas.* **38**, pp. 343–352.
10. Derenzo, S. E., Weber, M. J., and Klintenberg, M. K. (2002). Temperature dependence of the fast, near-band-edge scintillation from CuI, HgI_2, PbI_2, ZnO:Ga and CdS:In. *Nucl. Instrum. Methods Phys. Res. A* **486,** pp. 214–219.
11. Nikl, M. (2006). Scintillation detectors for X-rays. *Meas. Sci. Technol.* **17**, pp. R37–R54.
12. Lehmann, W. (1965). Optical absorption edges of ZnO and CdS. *J. Electrochem. Soc.* **112**, pp. 1150–1151.
13. Lehmann, W. (1966). Edge emission of *n*-type conducting ZnO and CdS. *Solid-State Electron.* **9**, pp. 1107–1110.
14. Makino, T., Segawa, Y., Yoshida, S., Tsukazaki, A., Ohtomo, A., Kawasaki, M., and Koinuma, H. (2005). Spectral shape analysis of ultraviolet

luminescence in *n*-type ZnO:Ga. *J. Appl. Phys.* **98**, pp. 093520(1)–093520(4).

15. Wall, R. A., Lipke, K. C., Ucer, K. B., Williams, R. T., Millers, D., Smits, K., and Grigorjeva, L. (2009). Time-resolved absorption and luminescence following electron-hole pair creation in ZnO. *Phys. Stat. Sol. C* **6**, pp. 323–326.

16. Ehrentraut, D., Sato, H., Kagamitani, Y., Yoshikawa, A., Fukuda, T., Pejchal, J., Polak, K., Nikl, M., Odaka, H., Hatanaka, K., and Fukumura, H. (2006). Fabrication and luminescence properties of single-crystalline, homoepitaxial zinc oxide films doped with tri- and tetravalent cations prepared by liquid phase epitaxy. *J. Mater. Chem.* **16**, pp. 3369–3374.

17. Simpson, P. J., Tjossem, R., Hunt, A. W., Lynn, K. G., and Munne, V. (2003). Superfast timing performance from ZnO scintillators. *Nucl. Instrum. Meth. Phys. Res. A* **505**, pp. 82–84.

18. Pejchal, J., Kagamitani, Y., Ehrentraut, D., Sato, H., Odaka, H., Hatanaka, H., Nikl, M. Yoshikawa, A., Fukumura, H., and Fukuda, T. (2007). Luminescence characteristics of the LPE-grown undoped and In-doped ZnO thin films and bulk single crystals. *Phys. Stat. Sol. C* **4**, pp. 942–945.

19. Lorenz, M., Johne, R., Nobis, T., Hochmuth, H., Lenzner, J., Grundmann, M., Schenk, H. P. D., Borenstain, S. I., Schön, A., Bekeny, C., Voss, T., and Gutowski, J. (2006). Fast, high-efficiency, and homogeneous room-temperature cathodoluminescence of ZnO scintillator thin films on sapphire. *Appl. Phys. Lett.* **89**, pp. 243510(1)–243510(3).

20. Schenk, H. P. D, Borenstain, S. I., Berezin, A., Schon, A., Cheifetz, E, Dadgar, A., and Krost, A. (2009). Cathodoluminescence of epitaxial GaN and ZnO thin films for scintillator applications. *J. Cryst. Growth* **311**, pp. 3984–3988.

21. Neal, J. S., Boatner, L. A., Giles, N. C., Halliburton, L. E., Derenzo, S. E., and Bourret-Courchesne, E. D. (2006). Comparative investigation of the performance of ZnO-based scintillators for use as a-particle detectors. *Nucl. Instrum. Meth. Phys. Res. A* **568**, pp. 803–809.

22. Bourret-Courchesne, E. D., Derenzo, S. E., and Weber, M. J. (2009). Development of ZnO:Ga as an ultra-fast scintillator. *Nucl. Instrum. Meth. Phys. Research, A* **601**, 358–363.

23. Yen, W. T., Lin, Y. C., Yao, P. C., Ke, J. H., and Chen, Y. L. (2010). Effect of post-annealing on the optoelectronic properties of ZnO:Ga films prepared by pulsed direct current magnetron sputtering. *Thin Solid Films* **518**, pp. 3882–3885.

24. Choi, Y. S., Hwang, D. K., Oh, M. S., Hong, K. P., Em, V. T., Choi, H. W., and Park, S. J. (2008). Growth and characterization of gallium-doped ZnO films for α-particle scintillators. *J. Electrochem. Soc.,* **155,** pp. H909–H911.

25. Bourret-Courchesne, E. D., Derenzo, S. E., and Weber, M. J. (2007). Semiconductor scintillators ZnO and PbI_2: co-doping studies. *Nucl. Instrum. Meth. Phys. Res. A* **579,** pp. 1–5.

26. Lavrov, E. V., Herklotz, F., and Weber, J. (2009). Identification of two hydrogen donors in ZnO. *Phys. Rev. B* **79**, pp. 165210(1)–165210(13).

27. Wardle, M. G., Goss, J. P., and Briddon, P. R. (2006). First-principles study of the diffusion of hydrogen in ZnO. *Phys. Rev. Lett.* **96,** pp. 205504(1)–205504(4).

28. Neal, J. S., De Vito, D. M., Armstrong, B. L., Hong, M., Kesanli, B., Yang, X. C., Giles, N. C., Howe, J. Y., Ramey, J. O., Wisniewski, D. J., Wisniewska, M., Munir, Z. A., and Boatner, L. A. (2009). Investigation of ZnO-based polycrystalline ceramic scintillators for use as alpha-particle detectors. *IEEE Trans. Nucl. Sci.* **56,** pp. 892–898.

29. Kim, K. H., Bai, J. H., Whang, J. H., and Jun, I. S. (2008). Development of alpha detector module based on large area PIN photodiode. *J. Nucl. Sci. Technol.* Suppl. **5**, pp. 417–420.

30. Lubsandorzhiev, B. K. and Combettes, B. (2008). The Quest for the ideal scintillator for hybrid phototubes. *IEEE Trans. Nucl. Sci,* **55**, pp. 1333–1337.

31. Neal, J. S., Giles, N. C., Yang, X. C., Wall, R. A., Ucer, K. B., Williams, R. T., Wisniewski, D. J., Boatner, L. A., Rengarajan, V., Nause, J., and Nemeth, B. (2008). Evaluation of melt-grown, ZnO single crystals for use as alpha-particle detectors. *IEEE Trans. Nucl. Sci.* **55**, pp. 1397–1403.

32. Tanaka, M., Nishikino, M., Yamatani, H., Nagashima, K., Kimura, T., Furukawa, Y., Murakami, H., Saito, S., Sarukura, N., Nishimura, H., Mima, K., Kagamitani, Y., Ehrentraut, D., and Fukuda, T. (2007). Hydrothermal method grown large-sized zinc oxide single crystal as fast scintillator for future extreme ultraviolet lithography. *Appl. Phys. Lett.* **91**, pp. 231117(1)–231117(3).

33. Shimizu, T., Yamamoi, K., Estacio, E., Nakazato, T., Sakai, K., Sarukura, N., Ehrentraut, D., Fukuda, T., Nagasono, M., Togashi, T., Higashiya, A., Yabashi, M., Ishikawa, T., Ohashi, H., and Kimura, H. (2010). Response-time improved hydrothermal-method-grown ZnO scintillator for soft x-ray free-electron laser timing-observation. *Rev. Sci. Instrum.* **81**, pp. 033102.

34. Kano, M., Wakamiya, A., Sakai, K., Yamanoi, K., Cadatal-Raduban, M., Nakazato, T., Shimizu, T., Sarukura, N., Ehrentraut, D., and Fukuda, T. (2011). Response-time-improved ZnO scintillator by impurity doping. *J. Cryst. Growth* **318**, pp. 788–790.

35. Ohshima, E., Ogino, H., Niikura, I., Maeda, K., Sato, M., Ito, M., and Fukuda, T. (2004). Growth of the 2-in-size bulk ZnO single crystals by the hydrothermal method. *J. Cryst. Growth* **260**, pp. 166–170.

36. Klimm, D., Ganschow, S., Schulz, D., and Fornari, R. (2008). The growth of ZnO crystals from the melt. *J. Cryst. Growth* **310**, pp. 3009–3013.

37. Ehrentraut, D., Sato, H., Kagamitani, Y., Sato, H., Yoshikawa, A., and Fukuda, T. (2006). Solvothermal growth of ZnO. *Prog. Cryst. Growth Charact. Mat.* **52**, pp. 280–335.

38. Yanagida, T., Fujimoto, Y., Yoshikawa, A., Yokota, Y., Miyamoto, M., Sekiwa, H., Kobayashi, J., Tokutake, T., Kamada, K., Maeo, S., (2010). Scintillation properties of In doped ZnO with different In concentrations. *IEEE Trans. Nucl. Sci.* **57**, pp. 1325–1328.

39. Bergman, L., Morrison, J. L., Chen, X.-B., Huso, J., and Hoeck, H. (2006). Ultraviolet photoluminescence and Raman properties of MgZnO nanopowders. *Appl. Phys, Lett.* **88**, pp. 023103(1)–023103(3).

40. Čuba, V., Gbur, T., Múčka, V., Nikl, M., Kučerková, R., Pospíšil, M., and Jakubec, I. (2010). Properties of ZnO nanocrystals prepared by radiation method. *Radiat. Phys. Chem.* **79,** pp. 27–32.

41. Gbur, T., Čuba, V., Múčka, V., Nikl, M., Knížek, K., Pospíšil, M., and Jakubec, I. (2011). Photochemical preparation of ZnO nanoparticles. *J. Nanopart. Res.* **13.** pp. 4529–4537.

42. Anedda, R., Cannas, C., and Musinu, A. (2008). A two-stage citric acid-sol/gel synthesis of ZnO/SiO_2 nanocomposites: study of precursors and final products. *J Nanopart. Res.* **10**, pp. 107–120.

43. Zhang, X. H., Chua, S. J., Yong A. M., Chow, S. Y., Yang, H. Y., Lau, S. P., and Yu, S. F. (2006). Exciton radiative lifetime in ZnO quantum dots embedded in SiO_x matrix. *Appl .Phys. Lett.* **88**, pp. 221903(1)–221903(3).

44. Du, S., Tian, Y., Liu, H., Liu, J., and Chen, Y. (2006). Calcination effects on the properties of gallium-doped zinc oxide powders. *J. Am. Ceram. Soc.* **89**, 2440–2443.

45. Lian, Q., Zhu, J., Guo, Z., Lin, J., Xu, Z., Xu, H., Fang, T., and Juany, X. (2010). Preparation method of zinc oxide varister. Chinese Patent CN101630553 (A).

46. Zhang, H., Cheng, G., Yang, G., Zhang, J., and X. Lu, (2007). Optical properties of amorphous/crystalline ZnO nano-powder prepared by solid state reaction. *J. Mater. Sci: Mater. Electron.* **18**, pp. 381–384.

47. Demidenko, V. A., Gorokhova, E. I., Khodyuk, I. V., Khristich, O.A., Mikhrin, S. B., and Rodnyi, P. A. (2007). Scintillation properties of ceramics based on zinc oxide. *Radiat. Meas.* **42**, pp. 549–552.

48. Gorokhova, E.I., Anan'eva, G.V., Demidenko, V.A., Rodnyi, P.A., Khodyuk, I.V., and Bourret-Courchesne, E.D. (2008). Optical, luminescence, and scintillation properties of ZnO and ZnO:Ga ceramics. *J. Opt. Technol.* **75,** pp. 741–746.

49. Yanagida, T., Fujimoto, Y., Yoshikawa, A., Yokota, Y., Kamada, K., Pejchal, J., Kawaguchi, N., Ishizu, S., Fukuda, K., Suyama, T., Uchiyama, K., Mori, K., Kitano, K., and Nikl, M. (2010). Development and performance test of picosecond pulse X-ray excited streak camera system for scintillator characterization. *Appl. Phys. Exp.* **3**, pp. 056202(1)–056202(3).

50. van Dijken, A., Meulenkamp, E. A., Vanmaekelbergh, D., and Meijerink, A. (2000). *J. Phys. Chem. B* **104**, pp. 1715–1723.

51. Bera, D., Qian, L., Sabui, S., Santra, S., and Holloway, P. H. (2008). Photoluminescence of ZnO quantum dots produced by a sol–gel process. *Opt. Mater.* **30,** pp. 1233–1239.

52. Huntley, D. J. (2006). An explanation of the power-law decay of luminescence. *J. Phys. Cond. Matter* **18**, 1359–1365.

53. Rodnyi, P. A., Gorokhova, E. I., Chernenko, K. A., and Khoduk, I. V. (2012). Scintillating ceramics based on zinc oxide. *Mater. Sci. Eng.* **38**, pp. 012002(1)–012002(7).

Index